Green Giants?

Environmental Policies of the United States and the European Union

edited by
Norman J. Vig and Michael G. Faure

The MIT Press
Cambridge, Massachusetts
London, England

This book was set in Sabon by SNP Best-set Typesetter Ltd., Hong Kong and was printed and bound in the United States of America.

Library of Congress Cataloging-in-Publication Data

Green giants? : environmental policies of the United States and the European Union /
edited by Norman J. Vig and Michael G. Faure.
    p. cm.—(American and comparative environmental policy)
Includes bibliographical references and index.
ISBN 0-262-22068-7 (alk. paper)—ISBN 0-262-72044-2 (pbk. : paper)
1. Environmental policy—United States. 2. Environmental policy—European Union countries. I. Vig, Norman J. II. Faure, Michael (Michael G.)
III. Series.
GE180.G735   2004
363.7′00973—dc22

2003065186

Printed on recycled paper.

10 9 8 7 6 5 4 3 2 1

Contents

Series Foreword   vii

Preface   xi

Introduction   1
Norman J. Vig and Michael G. Faure

I   Comparing Policy Trends: Divergence or Convergence?   15

1   The Precautionary Principle, Risk Assessment, and the Comparative
Role of Science in the European Community and the US Legal
Systems   17
Theofanis Christoforou

2   The Roots of Divergence: A European Perspective   53
Ludwig Krämer

3   Convergence, Divergence, and Complexity in US and European
Risk Regulation   73
Jonathan B. Wiener

II   Regulatory Trends: Institutional and Policy Innovations   111

4   Environmental Federalism in the United States and the European
Union   113
R. Daniel Kelemen

5   Implementation of Environmental Policy and Law in the United
States and the European Union   135
Christoph Demmke

6   Convergence or Divergence in the Use of "Negotiated
    Environmental Agreements" in European and US Environmental
    Policy: An Overview   159
    David J. E. Grimeaud

7   What Future for Environmental Liability? The Use of Liability
    Systems for Environmental Regulation in the Courtrooms of the
    United States and the European Union   183
    Timothy Swanson and Andreas Kontoleon

III Policy Divergence on Global Issues   205

8   The Climate Change Divide: The European Union, the United
    States, and the Future of the Kyoto Protocol   207
    Miranda A. Schreurs

9   Trade and the Environment in the Global Economy: Contrasting
    European and American Perspectives   231
    David Vogel

10 International Development Assistance and Burden Sharing   253
    Paul G. Harris

11 Sustainable Development: Comparative Understandings and
    Responses   277
    Susan Baker and John McCormick

IV Transnational Networks and Dialogue   303

12 Emerging Transnational Policy Networks: The European
    Environmental Advisory Councils   305
    Richard Macrory and Ingeborg Niestroy

13 The Transatlantic Environmental Dialogue   329
    Carl Lankowski

V   Conclusions   345

14 Conclusion: The Necessary Dialogue   347
    Michael G. Faure and Norman J. Vig

About the Editors and Contributors   377
Index   381

Series Foreword

Environmental and other public policy conflicts in recent years have underscored significant differences between the positions adopted by the United States and the member nations of the European Community. Disagreements over the Kyoto Protocol on climate change have been among the most visible in the environmental arena, but they are by no means the only ones. The disparities in outlook among the world's leading industrial nations have important consequences for global environmental governance and achievement of sustainable development. At a time when third-generation environmental challenges such as climate change and loss of biological diversity call for unprecedented cooperation among the world's nations, what are the reasons for these divergent perspectives? What are the implications for concerted international action? For the goal of sustainable development?

In this volume, Norman Vig, Michael Faure, and their collaborators address these questions through original and perceptive assessments of the actions, achievements, and future potential of the United States and the European Union. These two "green giants" of the world together account for at least one-half of the world's GDP and the generation of about 40 percent of global greenhouse gas emissions, as well as much of the world's toxic waste. They offer great potential for resolving global environmental problems because of their wealth, technological expertise, and level of public commitment to environmental protection. Yet their cooperation is by no means assured, nor is their support for global sustainable development. Thus it is imperative to understand why the green giants take the environmental policy paths they do, and the conditions under which the paths are likely to come together or head off in separate directions. A range of variables may be important: differences in political and cultural values, political and institutional processes, legal and regulatory traditions, the role of economic interests, and recent domestic political and ideological shifts.

The sixteen authors represented in this collection offer varying analyses of policy trends and especially of the evidence for divergence or convergence of the United States and the European Union across a range of environmental policy issues. The chapters are particularly attentive to differences and similarities with regard to the regulation of environmental and health risks; trends in reform of environmental regulation, such as decentralization, negotiation, and changes in liability systems; positions taken on global issues such as climate change, international trade, development assistance, and sustainable development; and involvement with transnational networks and environmental dialogue.

In some of these areas, such as the use of risk assessment, there is evidence of convergence between the EU and the United States. But equally striking is the apparently increasing divergence in how best to deal with third-generation global environmental challenges such as climate change. The authors also identify areas in which hybridization, or interaction and mutual learning, in environmental policies is occurring, for example, in a willingness to consider a new generation of "smart regulation" that may work better than early "command-and-control" approaches. In short, convergence, divergence, and hybridization are all taking place, but at different levels of government and in different policy arenas. Taken together, the conclusions in this volume contribute significantly to our understanding of comparative environmental politics and policy, and to environmental policy within the EU.

The analyses represented in this collection illustrate well our purpose in the MIT Press series in American and Comparative Environmental Policy. We encourage work that examines a broad range of environmental policy issues. We are particularly interested in volumes that incorporate interdisciplinary research and focus on the linkages between public policy and environmental problems and issues both within the United States and in cross-national settings. We welcome contributions that analyze the policy dimensions of relationships between humans and the environment from either a theoretical or empirical perspective. At a time when environmental policies are increasingly seen as controversial, and new approaches are being implemented widely, we especially encourage studies that assess policy successes and failures, evaluate new institutional arrangements and policy tools, and clarify new directions for environmental politics and policy. The books in this series are written for a wide audience that includes academics, policymakers, environmental scientists and professionals, business and labor leaders, environmental activists, and students concerned with environmental issues. We hope they contribute to public understanding of the most important environmental problems, issues, and policies that society now faces and with which it must deal.

Sheldon Kamieniecki, *University of Southern California*
Michael E. Kraft, *University of Wisconsin-Green Bay*
American and Comparative Environmental Policy Series Editors

Preface

The Transatlantic Alliance has been the linchpin of relations between the United States and Europe since World War II. It not only ensured western solidarity in containing the communist threat east of the Berlin Wall, but also encouraged the nations of western Europe to build a strong European Community (EC) to foster economic growth and international trade. Indeed, by the mid-1990s the new "single market" of the fifteen-member European Union (EU) had come to rival that of the United States. Although comprising only one-tenth of the world's population, the US and EU accounted for half of global production and an even greater share of consumption. Moreover, the two economic giants had also led the world in enacting new legislation to control pollution and in promoting international agreements to mitigate the impacts of human development on the global environment. They were, relatively speaking, the planet's "green giants."

Yet in the new world of the twenty-first century, relations between the US and Europe have reached their lowest point in more than half a century. Much of this deterioration is the result of the new foreign and security policies of the Bush administration, especially regarding the dangers posed by Iraq. But in many areas, the divergence across the Atlantic precedes the current administration and appears to lie in changing values and priorities within the politics of the United States and the European Union. This book is motivated by a concern that growing differences between the two sides over environmental policy will not only make transatlantic cooperation increasingly difficult, but will seriously weaken the capacity of the international community to deal with a host of global environmental problems. We thus believe there is an urgent need for a new strategic dialogue between the US and EU to clarify and strengthen our common interests in protecting the planet.

Although there is a rich literature on the environmental law and policies of both the United States and the European Union, relatively little comparative analysis has yet been done. This volume attempts to break new ground in comparing the two systems. We have brought together a distinguished group of environmental scholars and practitioners from different backgrounds on both continents to analyze key similarities, differences, and trends in environmental policy and regulation in the EU and US. They have attempted not only to clarify policy differences, but to assess whether developments in different fields of policy are diverging, converging, or (in the apt phrase of Jonathan Wiener) undergoing "hybridization" through joint learning and exchange. They also examine past and ongoing policy dialogues within Europe and across the Atlantic. Overall, the contributors present a mixed picture in which there are numerous common trends in domestic regulatory practices while at the same time there is clearly a growing divergence over a number of international issues and the principles for addressing them. The patterns are thus dynamic and complex, but nevertheless troubling for the future of transatlantic relations and international environmental protection.

We hope this book will help scholars, professionals, and policymakers on both sides of the Atlantic to better understand current tensions between the United States and Europe and that it will stimulate new efforts to rebuild cooperation between the "green giants." The book should thus be of interest to all who care about the global environment, as well as students of comparative politics, comparative federalism, international relations, and environmental policy.

We are indebted to many people who have made this project possible. We especially thank the contributors for their patience, good will, and splendid cooperation throughout the lengthy production of the book. We also owe special gratitude to Mark Pollack and the BP Chair in Transatlantic Relations of the Robert Schuman Center for Advanced Studies at the European University Institute in Florence for hosting a two-day conference in December 2001 at which first drafts of most of the chapters were discussed and refined. We acknowledge the support of our colleagues at Carleton College and Maastricht University. Michael Faure owes special thanks to his secretariat at the Maastricht European Institute for Transnational Legal Research (METRO) for editorial and communications assistance. Norman Vig thanks Carleton College for travel and research support and Tricia Peterson of the political science department for invaluable assistance in preparing the manuscript for publication. Finally, we both express our appreciation to Clay Morgan of the MIT Press for encouragement and advice and to the anonymous referees for their helpful suggestions for revising the manuscript. Any remaining errors are, of course, the authors' own responsibility.

Green Giants?

Introduction

Norman J. Vig and Michael G. Faure

This book is motivated by a growing concern on both sides of the Atlantic Ocean that the United States and the European Union (EU) are following divergent paths in one of the most critical areas of contemporary policy and governance—protection of the natural environment. Particularly in the past decade, the United States has more often than not appeared reluctant to support new national and international initiatives to regulate human impacts on the environment. While the US led the world in establishing environmental policies and institutions for this purpose in the 1970s and 1980s, in the 1990s it appeared to become a laggard in international environmental politics. By contrast, the process of economic and political integration in Europe resulted in a flood of new environmental legislation following the Single European Act of 1987, which called for establishment of a "single market" by 1992. This event coincided with the opportunity for the European Commission to play an active role in representing the European Community (EC) at the UN Conference on Environment and Development (the "Earth Summit") in 1992 and in other international negotiations. The new European Union that resulted from the Maastricht Treaty revisions of 1992 further confirmed the European Community's commitment to a high level of environmental protection within its own borders and signaled a new international leadership role by the EU in promoting global environmental sustainability.[1] The United States, by contrast, was increasingly hesitant to make international commitments, particularly after the Republican sweep of the congressional elections in 1994. Despite attempts by the Clinton administration to promote a transatlantic environmental dialogue and regulatory cooperation, by the end of the decade it appeared that the EU and US were dangerously at odds over a series of high-profile issues ranging from climate change to regulation of genetically modified organisms.

The election of George W. Bush as president brought many of these issues to a head. Bush's sudden withdrawal of the US from the Kyoto Protocol on climate change and his unilateral pronouncement that the treaty was "dead" without consulting European leaders produced an outpouring of criticism from the EU. Despite urgent visits by Chancellor Gerhard Schröder of Germany and EU Environment Commissioner Margot Walström, the Bush administration refused to alter its stance and proceeded to lay out an energy and environmental policy agenda that was radically different from that of the Europeans. In foreign policy generally, the administration made it clear that it would pursue US interests outside of existing multilateral frameworks if necessary and that it would be reluctant to enter into new binding international agreements. The European response—despite great solidarity and support of the US following the terrorist attacks of September 11, 2001—was to proceed with completion and ratification of the Kyoto Protocol and other multilateral initiatives. In short, by the summer of 2002 it appeared that the US and the EU were operating on completely different premises regarding national, regional, and global environmental policy. To some scholars and commentators this was only part of a larger pattern of diverging transatlantic interests and values.[2]

This book stems from our concern that the US and the EU—representing the world's two largest and most developed economic markets—seem increasingly incapable of resolving differences over the priority of environmental problems and methods of addressing them, thus preventing them from taking the kind of joint leadership role that will be necessary to halt environmental degradation on a global scale. The US and EU together account for at least half of the world's gross domestic product and consume a disproportionate share of the world's resources. They also generate about 40 percent of global greenhouse gas emissions and most of the planet's toxic waste. At the same time, they are the source of much of the world's advanced technology needed to reduce pollution and provide alternative sources of energy in the future. Without their support, it is unlikely that the 170 other nations of the world will be willing or able to pursue sustainable development policies in the future.

We are also concerned that environmental policy has been increasingly subordinated to economic and trade issues in the new "transatlantic partnership" established between the United States and the European Union in the late 1990s and in the development of the global trade regime. Most of the transatlantic dialogue thus far has revolved around liberalization of markets, avoidance of trade disputes, and enhancement of business cooperation.[3] Environmental standards and policies have often been reduced to questions of regulatory cooperation to avoid trade disputes and reduce business costs, rather than focusing on improving environmental performance and economic sustainability. On the other hand, because the EU now rivals the US as an international economic actor (its combined economy totals about $9 trillion compared with nearly $11 trillion in the US), it is conceivable that the two "green giants" will engage in a debilitating competition to establish their environmental standards on a global level. Such regulatory competition could have spillover effects both for transatlantic cooperation and for the viability of multilateral environmental agreements.

Finally, the pending enlargement of the European Union in 2004 to include ten new member states—mostly in central and eastern Europe—could redirect much of Europe's attention toward internal economic and environmental problems. Such a turn inward and eastward could reduce the EU's interest in the renewal of the transatlantic partnership and in international environmental leadership outside Europe. The further division of the world into competing regional blocs could halt or even reverse the progress made toward global environmental protection in the past three decades.

Focus of the Book

There is a rich literature on both US and EU environmental policy and law,[4] yet comparative analysis of environmental policy has only recently begun to flourish. Although the initial seminal works comparing US and European environmental law and regulation appeared in the mid-1980s,[5] it was not until the late 1990s that systematic comparative studies of US and EU policies and approaches to environmental issues were under-taken.[6] This renewed interest reflected the enormous development of environmental legislation on both sides of the Atlantic since the 1970s and early 1980s, but also the realization that environmental policy was entering a new phase of development requiring different policy instru-ments and approaches and that many problems initially treated as national or regional were in fact global in scope. Comparative studies have thus begun for the first time to address the interface between domes-tic and international environmental policies in such areas as international trade and sustainable development policies.[7] This has coincided with a revival of interest in transatlantic relations between the US and Europe, particularly in the emergence of a new transatlantic "economic partner-ship" and need for regulatory cooperation between the US and the EU.[8] Recent work on new regulatory instruments, such as environmen-tal taxes, ecolabels, and voluntary agreements, has begun to analyze experimentation with such approaches in different contexts, although there is yet no systematic comparison of policy innovations of this kind in the US and EU.[9] Other recent comparative studies of environmental policy have focused on individual countries rather than on the US and EU per se.[10] Finally, some of the new literature on international envi-ronmental law compares implementation of international agreements and sustainable development policies in different nations and regions, but with only limited comparison of the US and EU.[11]

We believe this volume contributes significantly to the emerging field of comparative US and EU environmental policy. We have attempted to compare environmental policies and related institutional developments in both systems in order to (1) determine what the similarities and dif-ferences are in selected policy areas; (2) gauge whether these policies are on divergent paths or are, on closer inspection, developing new com-monalities; (3) explain some of the sources of these policy trends and differences, e.g., by reference to domestic political pressures and differ-ing institutional structures; (4) examine areas in which mutual learning may be taking place, e.g., in implementation and enforcement strategies, adoption of new policy instruments, and increased intra-European and transatlantic communication; and (5) suggest potentials for greater coop-eration in the future.

One must approach the comparison of large-scale systems and policy trends with caution. It is obviously easy to select issues on which there are clearly differences as representing a general trend toward divergence. However, as Jonathan Wiener argues in chapter 3, there may also be policy areas in which there is growing convergence, and other areas in which neither convergence nor divergence but "hybridization" is occurring as a result of mutual exchange and borrowing of standards and techniques. It is also important to consider the whole policy context for addressing environmental problems. For example, while governmental regulation may appear weaker in certain areas in the US than in Europe, it may also be the case that policies are implemented more stringently in the US because of the constitutional powers of the federal government or because the court system allows private parties to play a more active role in enforcing environmental law. It is also important to distinguish between domestic policies and international or foreign policy positions. It may be that on certain international issues, particularly those involving trade disputes, differences may appear much greater than if one looks at internal regulatory trends.

In fact, it is widely suggested in the literature that both the US and EU are moving toward adoption of new policy instruments to supplement or replace the old "command-and-control" approach to regulation.[12] The first generation of regulations of the 1970s and 1980s relied heavily on central government imposition of emission limits and installation of standard control technologies for large classes of industry in order to clean up the major "point" sources of air and water pollution. Potential violators were to be deterred by threats of fines or legal prosecution and arrest. However, as evidence accumulated that this traditional form of regulation was often economically inefficient and difficult to enforce, as well as inappropriate for many new types of small-scale, dispersed, or "nonpoint" sources of pollution, the search for a second generation of policy instruments that relied more heavily on economic incentives or other noncoercive mechanisms began on both sides of the Atlantic. There has been a great deal of interest in, and limited experimentation with, new approaches such as ecotaxation, emissions trading, green product labeling, environmental auditing and management, and voluntary industry agreements in both the US and Europe, suggesting a potential for

growing policy convergence. However, since the 1980s a third generation of environmental problems demanding international or even global
cooperation has rapidly moved to the fore: issues such as climate change,
loss of biodiversity, collapse of ocean fisheries, deforestation, desertification, international trade, and "sustainable development" generally.
Many of these problems require much more integrated and comprehensive approaches spanning all policy sectors, both within and across countries. It is on these issues that the US and EU are most clearly diverging.

We cannot examine all of these issues in this volume, or discuss all
the measures under consideration for improving policy performance or
addressing global problems. But we will consider both domestic regulatory trends—including the search for new regulatory concepts, policy
instruments, and methods of enforcement—and some of the most controversial international issues. We are also concerned with the underlying principles or policy "frames" that shape environmental policies on
each side. For example, many of the current disputes revolve around
questions of scientific uncertainty, risk assessment, and what role the
"precautionary principle" should play in adopting environmental standards and policies—a question taken up in the first part of the book.

Nor can we consider all of the factors that shape US and EU policies.
There is obviously a rich menu of potential variables, including differences in political and cultural values, economic interests, political-
institutional processes, and legal and regulatory traditions, to mention a
few. We are particularly interested in how differences in the "constitutional" or "federal" structure of the United States and the European
Union may affect policy-making processes and consequently policy
results. For example, how does the fact that the EU is a quasi-federal
system in which member states participate directly in policy formation
as well as in policy implementation affect policy outcomes? What difference does it make that the EU lacks the competence to regulate environmental actors directly, whereas in the US the Environmental
Protection Agency can directly impose standards and sanctions? What
difference does it make that in the US the three coequal branches of government continually compete to shape and enforce environmental law?

A related question is to what extent divergence between US and EU
environmental policies can be explained in terms of domestic political

shifts and electoral and ideological trends. In the 1990s, center-left political coalitions came to power in almost all EU member states and dominated the European Parliament as well, whereas in the US the 1994
congressional and state elections produced a marked shift to the right.
Congress, especially, exerted a strong conservative influence on both
domestic and international environmental policy. In Europe, by contrast,
green parties gained representation in the parliaments of most countries
and entered coalition governments in leading EU states such as Germany
and France.[13] To some extent these differences may also be rooted in
institutional variables. The multiparty, proportional representation electoral systems of most EU countries not only ensure green representation
but also force other political parties to compete for the environmental
vote, magnifying the impact of the green parties. In the US, on the other
hand, the high level of interest aggregation that occurs in a two-party,
winner-take-all system of elections tends to marginalize environmental
interests. It must also be recognized that formal government actors are
not the only source of policy development. Many other actors, including scientific bodies, corporations, and a host of other nongovernmental
organizations (NGOs) participate in lobbying and formation of policy.
In general, environmental organizations have quite good access to policymakers in the EU, whereas in the US different presidential administrations and congressional leaders are more open or closed to green
influences.[14]

The emergence of new transnational policy networks has been especially important in the field of environmental policy in the past decade.[15]
Some scholars have argued that scientists form increasingly influential
"epistemic communities" that influence governments worldwide by
developing scientific consensus on environmental problems and solutions.[16] Others have suggested that a new global "civil society" is emerging in which citizen and stakeholder organizations communicate freely
across borders and shape public opinion on environmental and sustainability issues.[17] This book analyzes two unique experiments in
transnational dialogue and issue advocacy—the European Environmental
Advisory Councils and the Transatlantic Environmental Dialogue.

We reiterate that the chapters that follow do not cover all aspects
of environmental policy. Nor do they reflect a single disciplinary or

methodological perspective. Rather, a diverse group of European and American scholars and practitioners from different intellectual backgrounds have focused on certain issues that have been at the forefront of US–EU relations or that involve critical principles and concepts in international policy debates as well as within the academic community. As such, the book can be read as presenting differing perspectives on a series of fundamental questions rather than as an attempt to provide a definitive set of conclusions.

Organization and Contents

The book is divided into five parts. Part I examines the overall question of convergence and divergence. In chapter 1, Theofanis Christoforou, a legal adviser to the European Commission, discusses the role that the "precautionary principle" has come to play in regulating uncertain health and environmental risks in Europe, compared with the US. He defends the EU against charges that its environmental policies lack a firm scientific basis, pointing out that the EU Treaty and decisions of the European Court of Justice require scientific risk assessment much like that in the US. Nevertheless, he insists that the EU has the right to adopt more protective standards when risks are unknown or uncertain, and when the public regards particular risks as unacceptable. The US, he suggests, has resisted recognition of the precautionary principle in national and international law for economic reasons, and is now falling behind Europe in regulating emerging environmental risks. Chapter 2 by Ludwig Krämer, an expert on European environmental law and a longtime official in the Environment Directorate-General of the European Commission, paints a more sobering picture of growing transatlantic divergence rooted in fundamental political and cultural differences. He expresses a common European view that US environmental policy is increasingly stagnant and dominated by economic interests, whereas the EU is pursuing ever higher levels of environmental protection as part of the integration process.

In chapter 3, Jonathan Wiener, a professor at Duke University Law School, responds to both Christoforou and Krämer from an American perspective. He presents a rich analysis of the complexities and difficulties of comparing US–EU policies, and argues that it is by no means clear that European policies are more protective or "precautionary" than those of the US; rather, the two sides have often chosen to regulate different risks. He also suggests that the concepts of convergence and divergence are too simplistic to capture the continuing evolution of environmental regulation. While there is divergence in some fields of environmental regulation and convergence in others, he argues that it is more useful to look at the cross-fertilization and "hybridization" of legal principles and practices that is taking place across the Atlantic.

Part II then shifts the focus to common regulatory trends and new policy approaches within the US and the EU. In chapter 4, R. Daniel Kelemen compares "environmental federalism" in the two systems and demonstrates how changing patterns of authority allocation among levels of government are leading to more similar regulatory styles on both sides of the Atlantic. Christoph Demmke follows this chapter with an analysis of trends toward more innovative, voluntary approaches to environmental policy compliance and enforcement in both Europe and the US. He also reveals the need for much greater information and analysis of the effectiveness of new policy instruments. In chapter 6, David J. E. Grimeaud examines the use of negotiated voluntary agreements with industry in Europe and the US as a particular case of such regulatory innovation, focusing especially on the example of Project XL. Chapter 7 by Timothy Swanson and Andreas Kontoleon examines the role of US courts in defining how liability for damage to natural resources should be measured and assessed, and suggests how these lessons may need to be adapted to the emerging environmental liability regime in the EU.

Part III contains four chapters analyzing divergence between the two sides on international issues. Chapter 8 by Miranda Schreurs traces differences between the US and EU over strategies for addressing climate change—from negotiations leading to the 1992 United Nations Framework Convention on Climate Change and the 1997 Kyoto Protocol, to US withdrawal from the Kyoto Protocol in 2001 and subsequent EU efforts to implement the treaty without American participation. Chapter 9 by David Vogel explores emerging conflicts between the US and EU over issues of international trade and the role of the World Trade Organization in reconciling conflicts between principles of free trade and environmental protection. In chapter 10, Paul G. Harris then highlights different responses of the EU and US to the need for greater economic assistance to developing countries and more equitable burden sharing to meet the costs of environmental protection and sustainable development. Finally, chapter 11 by Susan Baker and John McCormick takes a critical look at how the EU and US have responded to the broader call for sustainable development since the adoption of sustainable development principles at the 1992 UN Conference on Environment and Development. All of these chapters explain how the US and EU are diverging on critical international environmental issues and trace these divisions to differences of values, ideologies, and/or domestic economic and political pressures.

Part IV turns from policy issues to the emergence of transnational policy networks and civil society dialogues. Chapter 12 by Richard Macrory and Ingeborg Niestroy examines the work of the European Environmental Advisory Councils, an independent federation of more than thirty governmental advisory councils in western and east-central Europe. Though there is no counterpart in the US, the European network of expert and stakeholder councils suggests the role that advisory networks can play in influencing policy formulation across national and international borders—in this case the sustainable development policies of the EU. Chapter 13 by Carl Lankowski then describes and analyzes the Transatlantic Environmental Dialogue (TAED) sponsored by the Clinton administration and the European Commission in 1998–2000 as an experiment in engaging civil society in transnational policy. Though largely unsuccessful as a policy forum, the TAED casts fascinating light on the asymmetrical nature of the US and European environmental policy communities and on the need for more serious and sustained exchanges of this kind.

Finally, in Part V we review the findings of all the chapters of the book and attempt to draw some conclusions about the extent to which convergence, divergence, or hybridization of environmental policies is occurring across the Atlantic. Chapter 14 also discusses the sources of divergence and the implications for the future of transatlantic environmental relations. Overall, we find a mixed pattern of divergence and hybridization. While there is some evidence that the two green giants are moving closer together on basic questions of risk assessment, there are still substantial differences over what role scientific evidence and public opinion should play in determining regulatory policies. The EU has "constitutionalized" a deeper level of environmental commitment by incorporating the precautionary principle and the goal of sustainable development into its governing treaty. At the same time, efforts to move beyond first-generation command-and-control policies to second-generation "smart regulation" on both sides of the Atlantic suggest that a good deal of mutual learning and hybridization is taking place. But it also appears premature to conclude that new environmental policy instruments are likely to do more than supplement traditional forms of regulation in the foreseeable future.

The evidence of policy divergence on third-generation international issues appears overwhelming. The governments of the US and EU differ markedly in their approaches to global climate change, international trade, international aid and burden sharing, and sustainable development. These differences may reflect a larger pattern of divergence rooted in American and European concepts of world order, international cooperation, the role of international law, sovereignty, and national interest. However, close examination of the policies discussed in Part III indicates that rhetorical differences often exceed actual policy results.

Finally, we conclude that while transnational policy networks have been quite successful in building consensus on environmental and sustainable development policy in Europe, transatlantic exchanges have been far less successful in fostering mutual understanding and cooperation. Indeed, the net effect of the two examples examined here, as well as the history of negotiations described by Ludwig Krämer in chapter 2, may have been to intensify policy differences between the US and EU. We thus conclude that a new high-level transatlantic dialogue is urgently needed if actual and perceived policy differences over these issues are not to contribute to a growing sense of alienation between Europe and America.

Notes

1. The European Economic Community (EEC) was established by the 1957 Treaty of Rome, which entered into force in 1958. The Rome treaty was amended by the 1987 Single European Act and then, more importantly, by the 1992 Treaty

on European Union (the "Maastricht Treaty," which entered into force in 1993). The latter treaty created the European Union (EU), replaced the EEC by the European Community (EC), and added two new "pillars" on Common Foreign and Security Policy and Justice and Home Affairs that partly remain separate from the legal jurisdiction of the EC. Further treaty revisions adopted at Amsterdam (1997) and Nice (2000) provided for enlargement of the EU and reform of its institutions as new member states enter (ten new states are scheduled to join in May 2004).

The EU does not technically enact law; rather, the EC legislates in areas of its competence, including environmental law (a domain of shared competence between the EC and the member states). The Commission of the EC (or European Commission) initiates most legislation, which usually requires joint approval by the European Council (representing member states) and the European Parliament (directly elected by voters of the member states). The Commission also oversees member states' implementation of and compliance with EC law, and may represent the EC in diplomatic negotiations. The EU as a unit may become a party to international treaties and conventions if all of the member states are also parties, and EU treaties must be ratified by the parliaments of each of the member states. The terms European Union and European Community are often used interchangeably in this volume, even though they refer to legally distinct entities.

2. See, e.g., Ivo H. Daalder, "Are the United States and Europe Headed for Divorce?" *International Affairs* 77 (2001): 553–557; Robert Kagan, "Power and Weakness," *Policy Review* (June 2002): 3–28; Kagan, *Of Paradise and Power: America vs. Europe in the New World Order* (New York: Knopf, 2003); and Charles A. Kupchan, *The End of the American Era: U.S. Foreign Policy and the Geopolitics of the Twenty-first Century* (New York: Knopf, 2002), especially chap. 4.

3. Mark A. Pollack and Gregory C. Shaffer, eds., *Transatlantic Governance in the Global Economy* (Lanham, M.: Rowman & Littlefield, 2001).

4. On the United States, see e.g., Michael E. Kraft, *Environmental Policy and Politics,* 2nd ed. (New York: Addison Wesley Longman, 2001); Walter A. Rosenbaum, *Environmental Politics and Policy*, 5th ed. (Washington, D.C.: CQ Press, 2002); J. Clarence Davies and Jan Mazurek, *Pollution Control in the United States: Evaluating the System* (Washington, D.C.: Resources for the Future, 1998); Norman J. Vig and Michael E. Kraft, *Environmental Policy: New Directions for the Twenty-first Century*, 5th ed. (Washington, D.C.: CQ Press, 2003); on the European Community, see Albert Weale, Geoffrey Pridham, Michelle Cini, Dimitrios Konstadakopulos, Martin Porter, and Bendan Flynn, *Environmental Governance in Europe* (Oxford: Oxford University Press, 2000); John McCormick, *Environmental Policy in the European Union* (Basingstoke, UK: Palgrave, 2001); Pamela M. Barnes and Ian G. Barnes, *Environmental Policy in the European Union* (Cheltenham, UK: Edward Elgar, 1999); Anthony R. Zito, *Creating Environmental Policy in the European Union* (New York: St. Martin's, 2000); Stanley P. Johnson and Guy Corcelle, *The Environmental Policy*

of the European Communities, 2nd ed. (The Hague: Kluwer Law International, 1995).

5. Eckard Rehbinder and Richard Stewart, *Environmental Protection Policy: Legal Integration in the United States and the European Community* (Berlin and New York: de Gruyter, 1985); David Vogel, *National Styles of Regulation: Environmental Policy in Great Britain and the United States* (Ithaca, N.Y.: Cornell University Press, 1985); Ronald Brickman, Sheila Jasanoff, and Thomas Ilgen, *Controlling Chemicals: The Politics of Regulation in Europe and the United States* (Ithaca, N.Y.: Cornell University Press, 1986).

6. See especially Randall Baker, ed., *Environmental Law and Policy in the European Union and the United States* (Westport, Conn.: Praeger, 1997); and Richard L. Revesz, Philippe Sands, and Richard B. Stewart, eds., *Environmental Law, the Economy, and Sustainable Development: The United States, the European Union and the International Community* (Cambridge: Cambridge University Press, 2000).

7. See Revesz, Sands, and Stewart, *Environmental Law, the Economy, and Sustainable Development*.

8. See especially Pollack and Shaffer, *Transatlantic Governance in the Global Economy*; George A. Berman, Matthias Herdegen, and Peter L. Lindseth, eds., *Transatlantic Regulatory Cooperation: Legal Problems and Political Prospects* (Oxford: Oxford University Press, 2000); and David Vogel, *Barriers or Benefits? Regulation in Transatlantic Trade* (Washington, D.C.: Brookings Institution Press, 1997).

9. See, e.g., Jonathan Golub, ed., *New Instruments for Environmental Policy in the EU* (London: Routledge, 1998); Duncan Liefferik and Mikael Skou Andersen, *The Innovation of European Environmental Policy* (Oslo: Scandinavian University Press, 1997); Neil Gunningham and Peter Grabosky, *Smart Regulation: Designing Environmental Policy* (Oxford: Clarendon Press, 1998); Timothy O'Riordan, ed., *Ecotaxation* (London: Earthscan, 1997); Andrew Jordan, Rüdiger Wurzel, and Anthony R. Zito, eds., *New Instruments of Environmental Governance? National Experiences and Prospects* (London: Frank Cass, 2003).

10. Uday Desai, ed., *Environmental Politics and Policy in Industrialized Countries* (Cambridge, Mass.: MIT Press, 2002); Miranda Schreurs, *Environmental Politics in Japan, Germany and the United States* (Cambridge: Cambridge University Press, 2002); Kenneth Hanf and Alf-Inge Jansen, eds., *Governance and Environment in Western Europe: Politics, Policy and Administration* (Harlow, UK: Addison Wesley Longman, 1998); Martin Jänicke and Hellmut Weidner, eds., *National Environmental Policies: A Comparative Study of Capacity-Building* (Berlin: Springer, 1997); Mikael Skou Andersen and Duncan Liefferink, *European Environmental Policy: The Pioneers* (Manchester, UK: Manchester University Press, 1997); Susan Rose-Ackerman, *Controlling Environmental Policy: The Limits of Public Law in Germany and the United States* (New Haven, Conn.: Yale University Press, 1995).

11. Edith Brown Weiss and Harold K. Jacobsen, *Strengthening Compliance with International Environmental Accords* (Cambridge, Mass.: MIT Press, 1998); David G. Victor, Kal Raustiala, and Eugene B. Skolnikoff, eds., *The Implementation and Effectiveness of International Environmental Commitments: Theory and Practice* (Cambridge, Mass.: MIT Press, 1998); William M. Lafferty and James Meadowcroft, eds., *Implementing Sustainable Development: Strategies and Initiatives in High Consumption Societies* (Oxford: Oxford University Press, 2000).

12. See note 9.

13. Ferdinand Müller-Rommel and Thomas Poguntke, eds., *Green Parties in National Governments* (London: Frank Cass, 2002).

14. Vig and Kraft, *Environmental Policy*, chap. 5; Paul G. Harris, *The Environment, International Relations, and U.S. Foreign Policy* (Washington, D.C.: Georgetown University Press, 2001), chap. 7; Elizabeth R. DeSombre, *Domestic Sources of International Environmental Policy: Industry, Environmentalists, and U.S. Power* (Cambridge, Mass.: MIT Press, 2000). On EU interest group access, see John Peterson and Elizabeth Bomberg, *Decision-Making in the European Union* (New York: St. Martin's, 1999); and Jeremy J. Richardson, *European Union: Power and Policy Making*, 2nd ed., (London: Routledge, 2001), chap. 11.

15. See, e.g., Thomas Princen and Matthias Finger, *Environmental NGOs in World Politics* (London: Routledge, 1994); Paul Wapner, *Environmental Activism and World Politics* (Albany: State University of New York Press, 1996); and Margaret E. Keck and Kathryn Sikkink, *Activists Beyond Borders: Advocacy Networks in International Politics* (Ithaca, N.Y.: Cornell University Press, 1998).

16. Peter M. Haas, "Introduction: Epistemic Communities and International Policy Coordination, " *International Organization* 46 (winter 1992): 1–36.

17. See, e.g., Ronnie D. Lipschutz, with Judith Mayer, *Global Civil Society and Global Environmental Governance* (Albany: State University of New York Press, 1996); and Miranda A. Schrears and Elizabeth C. Economy, eds., *The Internationalization of Environmental Protection* (Cambridge: Cambridge University Press, 1997).

I

Comparing Policy Trends: Divergence or Convergence?

1

The Precautionary Principle, Risk Assessment, and the Comparative Role of Science in the European Community and the US Legal Systems

Theofanis Christoforou[1]

The precautionary principle applies to scientific uncertainty and risk regulation. It permits regulatory authorities to take action or adopt measures in order to avoid, eliminate, or reduce risks to health, the environment, or in the workplace. The precautionary principle may also oblige the regulatory authorities to take action when this is necessary to avoid exceeding the acceptable level of risk.

This chapter provides a short analysis of the rise, development, and growing gap in the application of the precautionary principle in the respective legal systems of the two "green giants." It also briefly examines whether science has an increasing role to play in the regulation of environmental risk and the extent to which sophisticated risk assessment techniques and expert scientific committees are likely to reduce the growing regulatory gaps in the two systems and prevent trade wars. For analytical purposes, the review of the regulatory history on environmental protection in the two green giants is divided into three phases: the early phase (up to 1970s), when the regulation of risk on the basis of precaution in the United States was more rigorously applied; the second phase (up to 1990s), when the European Community accomplished tremendous progress in regulating risk to health and the environment and nearly closed the gap with the United States; and the final phase (from the early 1990s to the present), in which more stringent regulation of risk on the basis of precaution has become greater in the European Community than in the United States. It is important to bear in mind, however, the obvious difficulties of developing in the short space of a chapter more than the bare outlines of an analysis of a subject of such a nature and complexity.

A Historical Perspective on the Application of Precaution in the Regulation of Health and Environmental Risks

First Phase: Parallel Developments in Applying Precaution to Protect Human Health

The basic duty of governments to act cautiously or to err on the side of safety in protecting public health has been a long-standing principle in the legal systems of nearly all major jurisdictions, including those of the United States and the member states of the European Community.[2] Moreover, the basic elements of the precautionary principle—that is, uncertainty, risk, and lack of a direct causal link—have been applied, consciously or unconsciously, in both legal systems since public health was threatened by diverse technological sources.[3] This may also be observed from the long-standing regulatory systems requiring pre-marketing approval for medicinal products, veterinary drugs, pesticides, contaminants, additives, and other substances.[4] During this first phase of environmental regulation, the coverage of the US legal system was more advanced than that of the European Community. Until the mid-1970s, the European Community was primarily concerned with laying down general rules to facilitate the free movement of goods in the internal market, very often leaving it to the competent authorities of its member states to set the desired level of health protection.[5] After the mid-1970s, however, the regulatory system for protecting human health in the European Community began to develop essentially in parallel with that in the United States. Both systems recognized and allowed the application of precaution in public health on broadly the same grounds. Until the 1990s the US system was probably more rigorous in its risk analysis requirements and aimed to achieve a level of health protection, both broadly and in concrete cases, higher than that sought by the European Community and its member states.[6] The precise reasons for this divergence in the two regulatory regimes on the acceptable level of health risk have not, to the best of my knowledge, been the subject of any detailed study and comprehensive review until now.[7]

In the US, many laws and regulations require that action be taken to anticipate, prevent, or reduce risk where there is scientific uncertainty or a lack of clear evidence of risk. Many activities or substances are included in such statutes: the Delaney clauses in the 1959 Food, Drug and Cosmetic Act concerning food additives, colorants, new drugs, and pesticides; the 1966 Endangered Species Act; the 1970 Clean Air Act; the 1972 Clean Water Act; the 1972 Federal Insecticide, Fungicide and Rodenticide Act; the refusal by the US authorities to approve certain medicinal products such as thalidomide in 1962; the automobile emission standards of 1970; a number of pesticides (aldrin, dieldrin) whose use has been severely restricted by the Environmental Protection Agency (EPA) since 1974; and the 1978 ban on certain types of chlorofluorocarbons (CFCs) used in cans.[8]

The developments are roughly the same in the legal systems of the member states of the European Community and in the European Community itself. The statutes that require a manufacturer to demonstrate safety before a product is approved, or that stipulate that certain activities or the use of certain substances are prohibited in the absence of clear evidence of no harm, include the 1965 Council Directive 65/65/EEC relating to medicinal products, which requires approval prior to marketing; the 1967 Council Directive 67/548/EEC relating to the classification, packaging, and labeling of dangerous substances, which aims to warn consumers and prevents dangerous substances from being placed on the market; the 1970 Council Directive 70/524/EEC concerning additives in feeds that required prior approval upon showing of safety by manufacturer; the 1975 Council Directive 75/319/EEC relating to medicinal products; the 1973 Swedish Act on hazardous chemical products, which explicitly mentioned precaution in order to prevent or minimize damage to man or the environment;[9] the 1975 Council Directive 75/442/EEC on waste to reduce or avoid risk; the 1976 Council Directive 76/117/EEC prohibiting certain dangerous pesticides and Directive 76/895/EEC specifying maximum levels of pesticide residues in fruits and vegetables; the 1979 Council Directive 79/112/EEC relating to the labeling, presentation, and advertising of foodstuffs for sale to the ultimate consumer; the 1981 Council Directive 81/851/EEC on veterinary medicinal products, which clearly placed the burden of demonstrating safety on the applicant manufacturer before marketing authorization could be granted; the prohibition of some polychlorinated biphenyls (PCBs) already in 1972 and the early ban on asbestos in some member states, etc.[10]

Court decisions in both legal systems have also supported the practice of precaution on matters of public health even in cases where clear evidence of risk was uncertain or missing. For example, the Delaney clauses imposing a no-risk policy to humans on the basis of tests on laboratory animals alone were upheld in 1987 in the case *Public Citizen v. Young* [831 F. 2d 492 (D.C. Cir. 1987)]; the prohibition on discharging asbestos fibers in Lake Superior was upheld in 1975 in the case *Reserve Mining Co. v. EPA* [514 F. 2d 492 (8th Cir. 1975)] despite the lack of clear evidence that ingested—as opposed to inhaled—asbestos is harmful; the EPA's regulation of PCBs under the 1972 Clean Water Act was upheld in 1978 in *EDF v. EPA* [598 F. 2d 62 (D.C. 1978)], where the court found that where initial, but not conclusive, evidence suggests a danger, preventive action can be taken in advance of obtaining more definitive data. A similar finding was applied by the court in 1979 in *Lead Industries Association v. EPA* [647 F. 2d 1130 (D.C. Cir. 1979)] regarding the EPA's regulation of airborne lead under the Clean Air Act. The court held that Congress directed the agency "to err on the side of caution" in making regulatory decisions because one of the Act's purposes is to "emphasize the preventive or precautionary nature" of the actions taken under it.[11]

In exactly the same way, the European Court of Justice ruled in a number of cases in the late 1970s and early 1980s that "In so far as there are uncertainties in the present state of scientific research with regard to the harmfulness of a certain additive, it is for the Member States, in the absence of full harmonization, to decide what degree of protection of the health and life of humans they intend to assure, in the light of specific eating habits of their own population" [e.g., *Sandoz BV* case, 174/82 (1983) ECR 2445, at paragraph 16]. This case law recognized the right of cautious member states to block imports into their territory on the ground of threats to human health when there was scientific uncertainty about the harmfulness of a product.

In sum, during this first phase, precaution was mainly applied to activities, processes, or substances that appeared to pose a direct risk to public health or that threatened safety in the workplace. Environmental protection was envisaged, initially in the United States, essentially as preventing or avoiding harm to human health indirectly through environmental exposure, not so much as protecting the environment per se.[12]

Second Phase: Convergence in the Application of Precaution to Protect the Environment

During the second phase, essentially after 1972, the European Community began to develop its environmental policy.[13] Despite the fact that the EC Treaty at that time lacked a specific provision on which environmental protection could be based, the Community institutions adopted with astonishing speed quite stringent secondary legislative measures in the areas of water, air, and noise pollution; pesticides; biocides and other dangerous substances; waste treatment; major accident hazards from industrial activities; environmental impact assessment requirements; etc., by having recourse to the general enabling clause of the EC Treaty. When the EC Treaty was amended in 1987 to include specific provisions on environmental protection and again in 1992 to explicitly incorporate the precautionary principle, much of the Community's environmental policy was already in place.[14]

Although it reflected the spirit of laws and practices that were initially meant to protect public health, the term "precautionary principle" was first coined in Europe in the late 1970s and the early 1980s in reference to the need to protect the environment itself. This concept of precautionary action was brought into environmental policy and law for a number of reasons that are common to both regulatory systems. First, increasing environmental damage was observed that could not be clearly attributed to a specific agent or source of contamination or pollution. This created overall scientific uncertainty, which meant that such problems could not be approached on the basis of the old principle that allowed intervention only in situations of full scientific knowledge and established causality.[15] This fact explains the specific reference made to the lack of a clear and direct causal link between the suspected cause and the observed damage that is found in a number of early national laws and international environmental agreements and conventions that contain an explicit reference to precaution. However, the progressive recognition and subsequent wider acceptance of the precautionary principle have made such specific reference to the lack of a direct causal link

increasingly rare in the more recent international agreements and conventions, and this has been practically abandoned in the latest ones (e.g.,
2000 Cartagena Protocol on Biosafety). For reasons explained later, the
interaction of these aspects of uncertainty and the lack of a causal link
continues to play an important role in the interpretations of precaution
in the European Community.

Second, the theory that the limits of the assimilative capacity of the
environment had been reached (at the beginning especially in regard to
chemicals and marine pollution in the North Sea) motivated the regulatory authorities in Europe to tackle scientific uncertainty and to provide
economic incentives for the private sector to take the steps necessary to
reduce or eliminate pollution at the source.[16] Third, since there was no
prior consent and approval procedure in the regulation of a number of
potentially harmful agents, activities, and substances (essentially in the
area of chemicals), the objective of avoiding or reducing environmental
damage is thought to have played a role in the development of the precautionary principle in Europe. Fourth, public pressure was increasing
to protect the environment as such (e.g., through sustainable development), in addition to avoiding harm to public health caused indirectly
by environmental exposure. Fifth, because damages as a general rule
cannot be obtained unless it is established that the harmful emission or
damage was intended or reasonably foreseeable by the polluter, this
favored anticipatory action.

Contrary to conventional wisdom, therefore, the true origin of the precautionary principle does not seem to be in the area of environmental
protection. It appears, instead, that national environmental legislation
and international agreements and conventions have borrowed from the
area of health and transferred into the area of environmental protection
the basic rationale and core value of the precautionary principle—that
is, to err on the side of caution in the case of scientific uncertainty when
regulating risk.

The precautionary principle was brought into environmental protection by the need to prevent environmental degradation, which was
perceived to be growing rapidly. The term precautionary *principle* was
clearly inspired by a desire to create a normative basis for action even
in the absence of clear evidence of harm and causality. It aimed, therefore, to achieve and maintain a high level of health and environmental

protection and facilitate the decision-making process in the complex area of risk regulation. While precaution as a risk assessment and risk management principle arose in Europe, it is consistent with both the letter and spirit of many US laws, regulations, and court decisions.[17] For example, the US has accepted the 1985 Vienna Convention for the Protection of the Ozone Layer, including its 1990 London Amendments, which envisages the adoption of precautionary measures to protect the ozone layer; it signed the 1992 Rio Declaration of the UN Conference on Environment and Development, which is considered to have established the precautionary principle in international environmental law; it signed the 1973 and 1994 Convention on International Trade in Endangered Species (CITES), which explicitly mentions the precautionary principle as a basis for action if there is scientific uncertainty. Action based on Article 5.7 of the Agreement on Sanitary and Phytosanitary Measures (SPS) is also considered to reflect the precautionary principle [e.g., World Trade Organization (WTO) Appellate Body report in *Meat Hormones* case, at paragraph 124]. On the other hand, the European Community and its member states are parties to an even larger number of multilateral agreements and conventions that explicitly allow the adoption of precautionary measures when there is scientific uncertainty, risk to the environment, and lack of a direct causal link.[18]

Equally, the courts in the legal systems of both the European Community and the US have upheld the application of precaution when there is scientific uncertainty, even in the absence of clear evidence and causality of harm to the environment from a given substance, process, or activity. Thus, the US Supreme Court in *Maine v. Taylor* [477 U.S. 131, 148–149 (1986)] held that "[The state] has a legitimate interest in guarding against imperfectly understood environmental risks, despite the possibility that they may ultimately prove to be negligible. The constitutional principles underlying the commerce clause cannot be read as requiring the State . . . to sit idly by and wait until potentially irreversible environmental damage has occurred . . . before it acts to avoid such consequences." In the *Smitch* case [1994 U.S. App. LEXIS 6028 (20 F. 3d 1008 at 1017)], a U.S. Court of Appeal (9th circuit) held that "Particularly when the extent of the risks is in dispute, the Department is clearly permitted to err on the side of excess in taking precautionary measures."

In the European Community, the European Court of Justice held for the first time in 1985 in the *Waste Oils* case [240/83 (1985) ECR 531] that the protection of the environment is one of the essential objectives of the Community, which as such may justify certain limitations on the principle of free movement of goods. It also held in the same case that Council Directive 75/439/EEC on the disposal of waste oils "requires the Member States to prohibit *any* form of waste-oil disposal which has harmful effects on the environment." In the *Danish Bottles* case [302/86 (1988) ECR 4607 at paragraph 20], the European Court of Justice accepted a national system for returning beer and soft drink bottles aiming to ensure "a maximum rate of re-use and therefore a very considerable degree of protection of the environment." In the *Mirepoix* case [54/85 (1986) ECR 1067 at paragraphs 13–14] the European Court of Justice accepted that "pesticides constitute a major risk to human and animal health and to the environment," and in the case of scientific uncertainty, because "the quantities absorbed by the consumer can neither be predicted nor controlled" it justified strict measures intended to reduce the risks faced by the consumer, provided that the member state undertook to review the prohibition when new information from scientific research became available. These cases, although relevant because they indicate the acceptance by the European Court of Justice of strict prohibitions to protect the environment itself when there is scientific uncertainty, do not mention nor do they provide a definition of the precautionary principle. This definition came some years later in the *Bovine Spongiform Encephalopathy (BSE)* case [C-157/96 (1998) ECR I-221 at paragraph 63], in which this court held that "Where there is uncertainty as to the existence or extent of risks to human health, the institutions may take protective measures without having to wait until the reality and seriousness of those risks become fully apparent." This judgment contains all the necessary elements of a general definition of the precautionary principle that can be applied in all areas of Community law, i.e., uncertainty, risk, and lack of proof of a direct causal link.[19]

In sum, during the second phase, the regulation of risk in the European Community and its member states in the areas of health and environmental protection advanced quickly and caught up both in scope and degree of protection with that in the United States.[20] Moreover, as will be seen later, in a number of respects and substances (in particular, the bans on meat hormones and the milk hormone recombinant bovine somatrophin (rBST), antibiotics in farming, the total ban on asbestos, the Kyoto Protocol, and the regulation of many pesticides and genetically modified organisms, GMOs), the European Community had already begun to overtake the United States in both quantitative and qualitative terms of protection, in particular by laying down substantially more stringent restrictions that aimed to achieve a much higher level of health and environmental protection in situations of scientific uncertainty and genuine consumer concerns.

Third Phase: Divergence in the Application of Regulatory Precaution
Before 1990, some European legislation occasionally laid down stricter standards than the corresponding legislation in the United States in the area of health or environmental protection (e.g., asbestos, introduction of new chemicals, marine pollution). However, it was the introduction in 1989 of the total ban on the use of hormones to promote animal growth (Council Directive 88/146/EEC) that clearly marked a conscious departure by the European Community from the US standard of protection despite the visible trade tensions this policy was expected to raise. This departure was followed in 1990 by the moratorium on the use of another recombinant hormone (rBST) to increase milk production (Council Decision 90/218/EEC), and again in the same year with the adoption of legislation on the deliberate release into the environment of GMOs (Council Directives 90/219/EEC and 90/220/EEC), which laid down prior testing and approval requirements before GMOs could be released on the market. Since then it has become almost a constant trend to see more and more legislation being planned or adopted in Europe that sets higher standards to protect health or the environment than those in the United States (e.g., on biodiversity in 1992; ecolabeling in 1992; packaging wastes in 1994; GMO seeds, food, and feed in 1994, 1997, 2000, and 2001; the permanent ban on rBST in 1999; climate change in 1997 and 2001; food irradiation in 1999; pesticides in baby food in 1999; certain antibiotics in animal feed in 1999; maximum acceptable levels for certain aflatoxins in dried fruits in 1998; phthalates in toys in

2000; chemicals in 2001; automobile and electronic recycling in 2000 and 2002; and the new food safety law in 2002). In several of the these instances the United States either has no legislation or has adopted legislation that provides a lower level of health or environmental protection. Conversely, the European Community has been maintaining or even increasing its level of protection nearly every time it reviews or amends its existing legislation (e.g., in meat hormones, on rBST, on GMOs). The regulatory approach of the European Community on meat hormones, rBST, and the GMOs is discussed in some detail later because it provides useful insights into the reasons the European Community and its member states adopted a more risk-averse attitude. There are a few instances, however, where the regulation in the United States has occasionally been more precautionary than that in Europe (e.g., early animal feed regulation to avoid BSE, blood donations), but these relate mainly to areas where the competence in the European Community was essentially still with the member states.

During this period, participation of the United States at the international law level has also been constantly shrinking. The 1987 Montreal Protocol on Substances That Deplete the Ozone Layer, the 1992 Rio Declaration, and the 1995 WTO/SPS Agreement were the last important international agreements accepted by the US that embody or contain specific references to the precautionary principle in health and environmental protection. In fact, since 1992 the United States has not signed or ratified a number of important international agreements and conventions, such as the 1989 Basel Convention on hazardous wastes, the 1992 Convention on Biological Diversity, the 1997 Kyoto Protocol on carbon dioxide emissions, and the 2000 Cartagena Protocol on Biosafety; and it seems unlikely at present that it will approve the 2001 Stockholm Convention on persistent organic pollutants and the 2001 International Undertaking on Plant Genetic Resources.

Also during this same period, the work in the three most prominent international institutions aiming to merge economic (Organization for Economic Cooperation and Development, OECD), trade (WTO), and food standard-setting [World Health Organization (WHO)/Codex Alimentarius] policies on a global scale has been a battleground between the two green giants on the issue of applying precaution to health and environmental protection. It should be clarified that the United States does not deny that precautionary *measures* or a precautionary *approach* may be adopted to regulate risk.[21] What it is contesting is the existence or emergence of a precautionary *principle* that can trump or override provisions in existing agreements. It is therefore the status of precaution under international law that explains US resistance, as well as the wider discretion that a general principle or rule of customary law would provide to cautious states, allowing them to legally apply strict regulation even when there is no positive proof of harm.

Reasons Underlying the Growing Divergence in the Application of Precaution

European Community: Risk Averse and Higher Level of Health and Environmental Protection

The European Community had fixed 1992 as the target date by which to achieve completion of the internal market. Although it initially progressed by small steps, the 1992 target sparked a frenzy in the adoption and implementation of a long list of secondary legislation on human, animal, plant, and environmental protection, including consumer protection and occupational health and safety. This legislation was achieved by a teleological and quite expansive interpretation and application of the available E.C. Treaty provisions as well as by its successive amendments. What was initially quite general, framework legislation became progressively quite detailed and precise, thus constantly shrinking the member states' authority to enact domestic legislation (principle of preemption).[22] On the other hand, the standards of living in Europe rose generally during that period, and the higher the per capita gross national product (GNP), the greater the demand to adopt and implement legislation to protect health and the environment. Moreover, during the 1970s and 1980s, environmental protection took center stage internationally.[23] This is the general backdrop against which one has to add some other factors: tradition, education, and culture;[24] a deepening commitment to human rights; and attachment to moral and equity considerations and principles (e.g., sustainable development, animal welfare).[25] For example, the first European Community measure restricting the domestic use of hormones to promote growth (Directive 81/602/EEC) had a strong ethical and moral background because it sprang from scandals (use of illegal substances and high concentrations of residues in baby food) in Germany and Italy (where very young children exhibited serious symptoms of early puberty).[26] In addition, the first moratorium on the use of rBST (Council Decision 90/218/EEC) and in particular its subsequent permanent prohibition (Council Decision 1999/879/EC), were based on uncertainty but also on animal health and animal welfare considerations.

These factors all interact and play an important role in the definition of the acceptable level of risk in the European Community. Yet they alone cannot possibly explain the much lower level of risk from technological processes, activities, or product that Europeans find acceptable,[27] compared with the attitude of people in the United States (see chapter 9). What is it then that makes people in the European Community so risk averse?[28] We will review this question from four angles: institutional, scientific, economic, and legal. They are not exclusive and not necessarily the best that one may employ to approach the complexity of the issues involved.

Institutional Structures, Societal Values, and Democratic Rules

Traditionally, the regulation of the level of acceptable risk in the member states of the European Community has varied, sometimes considerably. As a result, the early attempts to harmonize Community legislation on health and environmental protection inevitably progressed on the basis of the generally accepted average, which sometimes led to agreeing to lower standards. In the 1987 amendment of the EC Treaty (Single European Act), however, qualified majority voting was established and it was provided that first, the regulation of risk in the Community will aim to attain a "high level of protection" [Article 100a(3)]; and second, the member states wishing to apply an even higher level of protection would be permitted to do so [Article 100a(4)].[29] As a result, in order to discourage the member states from continuing to apply disparate national standards that might undermine free circulation in the internal market, the European Community legislation on health and environmental protection tended to preempt national action by choosing very high levels of protection [see *Danish Bottles* case, 302/86 and the *PCP* case, C-41/93 (1994) ECR I-1829]. By 1992, the abolition of internal controls on free movement in the Community made an upward surge in the adoption of even higher levels of health and environmental protection inevitable in order to maintain the cohesion and competitiveness of the internal market. Thus, the 1992 Maastricht and the 1997 Amsterdam amendments to the EC Treaty specified in several places that one of the objectives of the European Community is to aim for a "high level of health protection," e.g., in Articles 3(1)(p) and 152(1) and (4) on regulation of public health and agriculture; in Article 153 on consumer protection; in Article 174(2) on environmental regulation; and in Article 95(3) on internal harmonization measures.

It should be noted, however, that the European Community's success in establishing a "single" market by 1992 also came at a price. The European institutions (in particular the European Commission) were blamed for excessive regulatory zeal and lack of democratic control and legitimacy (the "democratic deficit"). Consequently, by the early 1990s there was a growing concern that elaborate and detailed EC provisions imposed excessive burdens on the member states and their industries and people. This concern gave rise to the principle of subsidiarity and its inclusion, together with the principle of proportionality, in Article 3b of the Treaty on Union in 1992. During this same period, the political and civic landscape has also been changing rapidly with the appearance of environmental nongovernmental organizations (NGOs), with ecology appealing to even wider sections of the population, and the need to achieve sustainable development constantly gaining ground (which was also written into Article 3c of the Treaty).[30] Moreover, the rise of green parties in some powerful member states changed their politics, with political parties in the center and left of the political spectrum in many member states also espousing broad environmental concerns. "Greening" the laws and the trade rules were claims and slogans that caught the attention of many people.[31] These profound societal changes were reflected also in the composition of the European Parliament, whose powers have been constantly increasing in all successive Treaty amendments.[32] In this context, the principle of precaution was then written into Article 130r of the Treaty in 1992 to anticipate regulatory action and halt environmental degradation.

There is no doubt, therefore, that all the above structural, societal, and institutional changes have created the dynamics for a regulatory policy in Europe that continuously aims for stricter standards in health and environmental protection in order to complete the internal market and maintain its cohesion, avoid regulatory failures in the member states, and regain democratic legitimacy in the representation and defense of the basic interests of ordinary people in Europe (see also chapter 9). In contrast, the US regulatory system has not gone through such profound structural or institutional changes. It appears to be exhibiting instead the symptoms of a mature regulatory system in decline and is increasingly criticized on a number of counts, most notably for excessive regulation of "small" risks.[33]

Regulatory Failures and the Interface of Science, Law, and Risk Regulation

Science has always been the basis of risk regulation in the European Community. As explained, this reliance on science was necessary in the early regulatory measures in order to establish the internal market and to resist national protectionism. When the aim to achieve a high level of health and environmental protection was given the status of a general objective in the EC Treaty, Article 100a, which laid down a general harmonization clause, was amended in 1992 to dispel any doubt that this should be based "on scientific facts" [see Article 100a(3)]. Consequently, the member states were allowed to adopt stricter standards for health and environmental protection (Article 130t) only on condition that these measures were "based on new scientific evidence relating to the protection of the environment or the working environment" [see Article 100a(5)]. This mandatory requirement to base regulation on science has underlain all secondary legislation on risk regulation. The EC Treaty provision also explicitly prohibits the adoption of stricter national measures that are "a means of arbitrary discrimination or a disguised restriction on trade between Member States" and those that "constitute an obstacle to the functioning of the internal market" [see Article 100a(6)].

In a series of seminal judgments, the European Court of Justice has clearly examined the scientific basis of many national or Community measures purporting to regulate risk to human health or the environment. Thus, the *Angelopharm* case [C-212/91 (1994) ECR I-171] made consultation of the relevant scientific committees mandatory. The *Cassis de Dijon* case [120/78 (1979) ECR 649], the *German Beer* case [178/84 (1987) ECR 1227], and the *Danish Bottles* case (302/86) have all made it clear that any measure purporting to regulate risk should be based on scientific evidence and should respect the principle of proportionality.

If this is so, how then can one explain the almost constant claim by the United States that many European regulatory measures lack scientific basis and constitute disguised protectionism (e.g., meat hormones, rBST, GMOs)? First, it should be noted that none of the regulatory measures on these substances was adopted by the European Community for international trade protection. On the contrary, they were all clearly motivated by the desire to achieve a very high level of health and environmental protection in the face of scientific uncertainty. It should be noted that since 1989 the European Community has applied the prohibition on the use of hormones on a national treatment basis, that is, without distinguishing between animals and meat treated with hormones in the Community and imports from other countries. Thus, in 1990 the European Court of Justice in the *Fedesa* case [C-331/88 (1990) ECR I-4023] upheld the total ban on the use of hormones because of uncertainty and genuine consumer concerns.

The WTO Appellate Body in the *Hormones* case in 1998 reversed one of the previous panel's findings and accepted that the Community legislation on meat hormones was *not* a disguised restriction on trade because the record showed that it was not motivated by protectionism, as was claimed by the United States and Canada, but by the "depth and extent of the anxieties experienced within the European Communities concerning the results of the general scientific studies showing the carcinogenicity of hormones, the dangers of abuse (highlighted by scandals relating to black-marketing and smuggling of prohibited veterinary drugs in the EC) of hormones and other substances used for growth promotion and the intense concern of consumers" (WT/DS26/AB/R and WT/DS48/AB/R, at paragraph 245). The initial moratorium and subsequent permanent ban on the use of the milk-enhancing hormone rBST were also motivated by similar concerns of European consumers and farmers.[34]

When the moratorium on rBST was initially imposed in 1990, few scientific data were available other than the data produced by the industry wishing to commercialize the rBST. The resistance of European consumers and farmers to rBST and meat hormones also reflected a certain antipathy to artificial stimulation of agricultural production by new technologies, the harmless nature and long-term effects of which were not clearly established scientifically. For European regulators and consumers, this was an instance of scientific uncertainty that, in the cases indicated earlier, included real situations of lack of evidence and even of ignorance. Indeed, it took 9 years before more evidence finally became available in both cases. In 1999, the European Community introduced a permanent ban on the use of rBST within its territory (but not on the tiny quantities of imports from countries outside the Community) when clear evidence showing detrimental effects on animal health and welfare became available (Council Decision 1999/879/EC of December 17, 1999, OJ No L 331, December 23, 1999, p. 71). In September 2000, the European Council also decided *not* to fix a maximum residue limit for rBST under another Community regulation [Council Regulation 2377/90/EEC, and case C-248/99P, *France v. Monsanto (BST)*, judgment of January 8, 2002], on the grounds of scientific uncertainty in the risk assessment and the precautionary principle (EC Council doc. 11307/00 of September 21, 2000). It should be pointed out that Canada, in a similar reevaluation of the risk, in 1998 also decided to withdraw the authorization of rBST on grounds of animal health and welfare. As of today, the United States is one of the very few countries that continue to allow the use of rBST in its territory.

In regard to meat hormones, after the findings of the WTO Appellate Body in 1998,[35] the relevant scientific committees of the EC found, in three risk assessments carried out in 1999, 2000, and 2002 that on the basis of the latest scientific evidence available and the results of specific research, meat hormones have a number of potential adverse effects, in particular cancer, and that they appear to be more dangerous to children than to adults.[36] Not all of this evidence was available before, but it is known that when the US authorities evaluated these hormones in the 1960s and 1970s, they did not find that these hormones are risk free, but that they did not pose a "significant" risk.[37] On the basis of those risk assessments, the United States allowed the use of six hormones for growth promotion and continues to do so today. Conversely, on the basis of its latest risk assessments, the European Commission decided again to propose maintaining the total ban on the use of one hormone and, on the basis of the precautionary principle, the provisional ban on the other five hormones both within the EC and on imports from other countries (EC OJ No C 337, November 28, 2000, p. 163, and EC OJ No C 180, June 26, 2001, p. 190).[38]

It is clear that the rBST ban carries economic costs for Community producers. And in the case of meat hormones, the maintenance of the total ban, despite the continuous application of trade sanctions by the United States following the WTO dispute settlement, is also imposing additional economic costs on the Community and its farmers. The same is true, for instance, in regard to the restrictions on GMO authorizations. Given these costs, there must be some reason that EC authorities decided to maintain their regulatory choices. Food scares and scandals have taken place in both the US and the European Community, although there have been more such regulatory failures recently in Europe (e.g., the BSE crisis, dioxins in Belgium, hoof-and-mouth disease in the United Kingdom, blood contamination by AIDS, and asbestos regulation in France). Similar examples in the United States, but on a smaller scale, include *Escherichia coli* outbreaks in 1993, 1996, and 1997; hepititis A from Mexican strawberries; and cyclospora in Guatemalan raspberries, which resulted in President Clinton's Food Safety Initiative in 1997.[39] There is little doubt that such failures and scandals fuel consumer demand for more restrictive measures.[40]

Food scandals and regulatory failures are therefore both the causes and expressions of a deeply rooted desire by consumers to reduce further the level of risk considered acceptable by the regulatory authorities. But in my view, strict regulatory standards are imposed or maintained in Europe solely to achieve a high level of health protection—higher it would seem than that pursued by the United States at present (see chapters 2 and 9).[41] As the WTO Appellate Body put it in 1998 in the *Meat Hormones* case, "We are unable to share the inference . . . that the import ban . . . was not really designed to protect its population from the risk of cancer" (WT/DS26/AB/R and WT/DS48/AB/R, at paragraph 245).[42]

It follows that these cases, as well as the regulation of GMOs in the two systems, underscore the fundamental differences in the understanding by each party and its consumers of what science is and its role in risk assessment and risk regulation.[43]

Possible Solutions on the Horizon

There are different ways of assessing risk to health, to the environment, or in the workplace, and international practice in this regard is far from being coherent (e.g., the Codex Alimentarius Commission 2001). As explained, in the Community legal order, regulatory action is nearly always based on a risk assessment of the highest possible quality [e.g., Directive 2001/18/EC, OJ No L106, April 17, 2001, p. 1; and Regulation (EC) 178/2002, OJ No L 31, February 1, 2002, p. 1]. Past experience has shown that lack of evidence establishing a direct causal link between an activity, process, or substance and an identified risk has always been at the root of applying precaution. But there are obviously limits to scientific knowledge at any given moment.[44] Moreover, there are risks that can be caused by multiple, confounding factors that sometimes take time to materialize. This poses serious problems for regulatory authorities because it makes causality difficult to establish. Allowing fears that arise from pure ignorance and indeterminacy to guide any risk regulation is, however, likely to halt technological progress and impose heavy regulatory and financial burdens, if regulations are enforced inflexibly. On the other hand, in the past the mistake has been made (in probably too many cases) of requiring scientific certainty before deciding to take restrictive or protective action.[45]

Normally two reasons appear to have led to such a regulatory attitude in the past in both Europe and the United States. First, there is the positivist view of science, considering it to be a powerful and neutral tool capable of predicting risk and causality.[46] This view has been demonstrated to be wrong in several cases, because the experts' judgments appear to be prone to many of the same mistakes and biases as those of the general public, particularly when experts are forced to go beyond the limits of available information and data and rely on assumptions and intuition.[47] Second, existing risk assessment methodologies are inherently biased in favor of avoiding overinclusive regulatory measures (i.e., the inclination is to avoid false positives) for fear of imposing undue costs on technological progress and society.[48]

Because uncertainty and lack of causality normally undercut the ability to prove negligence in litigation, it would be legally inappropriate and wrong to require scientific certainty before allowing action to be taken to protect health or the environment.[49] Research has demonstrated that risk means more to people than the expected number of fatalities based on probabilistic quantitative assessments, which is the usual way experts assess risk.[50] Indeed, the perception people have of risk is wider than that of experts and reflects a number of legitimate concerns (e.g., familiarity with the risk, catastrophic potential, irreversibility of harm, threat to future generations, risk control possibilities, and voluntariness of exposure), which are frequently omitted from an expert risk assessment.[51] It is also well established in rational choice theory that there can in principle, in purely logical terms, be no effective analytical means to definitively compare the intensities of subjective preference displayed by different social agents in a pluralistic and multirisk society.[52]

Regardless of whether objective or subjective methods are used to evaluate uncertainties in a risk assessment, the specific parameters used in the assessment remain pivotal and have important normative implications for implementing improved risk assessments that acknowledge uncertainty.[53] When science does not provide a definitive answer as to which data, models, or assumptions should be used in a risk assessment, it is normally the task of the risk managers to provide the assessors with guidance on the science policy that should apply in the assessment.[54] Therefore, although they are distinguishable, scientific uncertainty and ignorance may coexist in a risk assessment and can further increase the potential for error in the degree of confidence regarding the harm to health, the environment, or in the workplace.

It follows from the preceding analysis that Community risk management measures, instead of trying to scientifically patronize consumers, increasingly take into account their genuine and legitimate concerns (or the public's perception of risk), as opposed to the mere consumer (commercial) preference or choice that is also addressed, but by other less trade-restrictive measures, such as providing consumer information and labeling (see the EC Commission 2001 proposals on GMO traceability and labeling, OJ No C 304E, October 30, 2001, p. 327). It has been argued that attempts to "educate" the public in order to bring their perceptions of risk in line with those of experts are in most cases unlikely to succeed, especially for risks that are genuinely perceived to be unknown and potentially catastrophic.[55] A recent survey on consumer acceptance of GMOs in Europe appears to confirm this proposition, since it has shown that there was no "knowledge/education effect," although it is generally observed that the more knowledge people have, the more favorable they are to scientific and technological progress. This was not true with GMOs—those persons ranked as having the greatest knowledge of science based on other evaluations still tended to say they did not want this type of food (65.4 percent).[56] Moreover, detailed studies of expressed consumer preferences indicate that people tend to view current levels of risk as unacceptably high for most activities and substances. Studies have also shown that the gap between perceived and desired risk levels suggests that people are not satisfied with the ways in which the market and regulatory authorities have balanced risks and benefits.[57] Therefore, being able to accurately define the acceptable level of risk (or chosen level of health or environmental protection) is fundamental in *risk management* and the application of the precautionary principle. In simple terms, therefore, the objective in both US and EU regulatory systems should be to discover *how safe is really safe enough* for people.

It is generally agreed that defining the level of acceptable risk is a normative decision that belongs to the democratically elected and accountable institutions of a state.[58] The regulation of risk entails making important decisions about how much health and safety people wish and can afford.[59] Since this touches upon the basic functions and mission of a democratic system of government, that is, to protect inter alia the life and health of its people and the environment, decisions about the level of acceptable risk cannot be made only by scientific or other kinds of experts who are not accountable. It follows that in any democratic system of government the electorate must have an opportunity for the final say about which risks it will bear and which benefits it will seek to obtain.[60] This is essentially the reason that in the Community legal system, as in many other systems, the opinions of technical and scientific committees are of an advisory nature only,[61] which means that their opinion is *a necessary but not sufficient condition* for risk regulation.[62] This also explains the fact that the work of all international standard-setting bodies on substances, agents, activities, or processes is voluntary and nonbinding, unless the parties to an international agreement or convention have clearly and explicitly renounced their autonomous right to set the level of protection or the level of risk considered acceptable by their people.[63]

As a general rule, people and regulatory authorities normally pursue policies that seek to avoid risk to health or the environment unless this avoidance becomes a burden that is clearly too great for them or their society to bear.[64] Pursuing zero-risk policies, therefore, is not uncommon in any legal system, and the right to choose a zero level of risk from a particular substance, process, or activity has been upheld explicitly by both national and international courts and tribunals.[65] The fact that in our technologically complex societies there are multiple sources of risk, including risks to which people voluntarily expose themselves, does not cancel out the legitimate objective to aim, whenever possible, for a zero-risk level of health or environmental protection.[66] In addition, the fact that subsequent implementation and enforcement measures cannot always eliminate risk is not itself a reason to refrain from aiming for a zero-risk policy. Pursuing a zero-risk level of protection, therefore, is not always synonymous with effectively achieving no risk, but with minimizing the identified risk as much as possible.

Arguments have also been made in favor of requiring a detailed cost-benefit analysis in nearly all risk management decisions, based inter alia on the multirisk nature of our world and on reasons of efficient allocation of resources.[67] Although they are understandable, these arguments are not only misconceived and flawed, but may also be dangerous.[68] First, voluntary exposure to risk by some must not enter into any type of balancing exercise against unintended, involuntary exposure to the same or other type of risk by other people. Contrary to what some authors have suggested,[69] the fact that people face multiple sources of risk in our society is not an argument in favor of an averaging or a balancing exercise.[70] Second, the right to life and health is the most fundamental of all human rights, which implies that no restriction should in principle be placed on this right without proper consideration.[71] Indeed, as a matter of principle, reasons of justice, fairness, and morality militate against a balancing exercise based on broad considerations of cost and efficient allocation of resources.[72] It is important to note that European consumers normally expect more positive and active intervention by state authorities in the regulation of risk than is probably the situation in the United States (compare, e.g., the regulatory approaches to GMOs in the two systems). Third, at least in Europe, the Court of Justice has held several times that in a risk management and balancing exercise, considerations of health should take precedence over economic or commercial considerations [e.g., case C-183/95, *Affish* (1997) ECR I-4315, at paragraph 43].

Unlike the situation in United States law, there is no general guideline in Community law that obliges the regulatory authorities to systematically analyze the economic impact or cost of risk management measures. However, risk management does play an important role in improving the overall well-being of the member states and their citizens in the Community and for that reason there is no barrier to having the regulatory authorities, whenever feasible, measure and report upon the economic impact of their decisions so as to inform themselves and the public. Indeed, the regulatory authorities in the European Community sometimes make, consciously or unconsciously, gross estimates of first-level, direct cost and benefit analyses of their decisions, despite the difficulties inherent in such an exercise because of the scientific uncertainty involved.[73] For those reasons, considerations of the level of economic impact or the cost of adopting a future precautionary action do not play a decisive role in the determination of whether to adopt a measure, only in the actual choice or design of the measure to be taken and the acceptable level of risk. In the European Community legal order, as explained, it is the principle of proportionality that is used to check the balance between the health or environmental objective pursued and the restrictive effects of the precautionary measure. It follows that the principle of proportionality in risk management decisions in the Community requires tailoring the measures to the chosen level of health or environmental protection.[74]

It is also important to note that European risk-averse societies are likely to be reluctant to trade a chosen high level of health protection for unpredictable uncertainty of possible harm.[75] The problem of understanding and defining uncertainty in the context of a risk assessment can be large, complex, and nearly intractable, unless the analysis is structured into small and simpler concepts for each stage and component of the risk analysis. It follows that it is of paramount importance for risk assessors in both systems to explain in detail any kind of scientific uncertainty they encounter in every step of their analysis and the techniques, assumptions, and values they employ to eliminate or reduce it. This has not been done with the required degree of detail and consistency in either the US or EC system. Residual uncertainties, however, are most likely to remain when there is a lack of pertinent scientific knowledge or there is ignorance of the nature and extent of risk, despite the efforts employed by scientists to reduce the potential for error.[76] Precaution can be applied therefore and has actually been applied both by the scientists completing the risk assessment on the basis of science policy guidelines that can be issued to them only by the risk management authorities, as well as by the regulatory authorities themselves, who have to draw the necessary regulatory implications. It is known that both risk assessors and risk managers attribute at any given moment different subjective values to available scientific data, the risks, and the nature of possible adverse effects.[77] Precaution applied by scientists in a risk assessment does not therefore eliminate the need to also allow risk managers to apply precaution to the same agent, activity, or process when taking regulatory action (as does Directive 2001/18/EC, Annex II, B, on GMOs). This is a proposition that is forcefully denied by the US internationally, basically for economic and trade policy considerations, and general litigation and negotiation tactics.[78] In Europe, risk assessors' technical precaution (when modeling and interpreting evidence and data) is therefore distinguishable from the risk managers' regulatory precaution (when taking normative regulatory action).

Dealing with scientific uncertainty becomes an issue when it is institutionalized in a democratic decision-making process, because regulators and judges are obliged to make decisions, sometimes within short time limits, even when scientific evidence in a risk assessment is inconclusive.[79] Contrary to conventional wisdom in Europe, the stringency of control by judges is not much different than that conducted in the US especially after the establishment in 1992 of the Court of First Instance in Europe, because in reality in both systems, in solving a specific legal dispute the courts are required only to decide whether the authorities have used their regulatory discretion in an arbitrary and unjustifiable manner.[80] But the courts are not required nor are they epistemically capable of resolving the underlying basis of scientific uncertainty.[81] On the other hand, the regulatory authorities' main cause of concern consists of the potential effects of uncertainty and risk on health, the environment, or in the workplace. The difficult decision, therefore, should rest ultimately with the regulatory authorities that are accountable to the people. As explained, in the Community legal system the objective of any risk management measure is to achieve a "high level" of health or environmental protection. One of the means of achieving this is the mandatory requirement to base the measure or action on the precautionary principle. This requires that appropriate consideration be given to the interaction between the level of acceptable risk and the lack of conclusive evidence on risk and causality. It is in this interaction that the precautionary principle functions as a catalyst by obliging the regulatory authorities to err on the side of safety in order to achieve the desired level of health or environmental protection.[82] Therefore, the precautionary principle in the European Community legal system plays an important role in that it gives the regulated or potentially affected natural or legal persons the means to control, if necessary by court action, the way risk management agencies make their normative decisions when they evaluate scientific uncertainty and risk, as well as in the way they balance costs and benefits. This entails both ex ante and ex post control of the measures taken to regulate risk.[83]

Conclusion

In the legal system of the European Community, the precautionary principle has constitutional status because it is explicitly mentioned in Article 174(2) of the EC Treaty, and it is firmly enshrined in implementing legislation and in the case law. It is a principle binding on the Community institutions that can be used to ensure that the societal values and policy choices pertaining to the desired level of health and environmental protection are fulfilled. The same applies in most of its member states under their domestic systems of law. At present, the precautionary principle does not seem to enjoy the same status in the legal system of the United States. This does not mean, of course, that the United States has not been applying precaution in its regulation of risk to the environment, health, or in the workplace. To the contrary, precaution has long been well embedded in the regulation of risk in the United States. There are many reasons and factors that explain the current divergence in the regulatory approach of the two green giants. Some of these factors include social, economic, legal, scientific, cultural, ethical, political, and regulatory policy choices. They all interact and play an important role, although the relevance of one or the other of these factors may be different, depending on the circumstances of each case. There are two factors, however, that appear to play a dominant role: the Europeans' desire to achieve and maintain a high level of health and environmental protection, on the one hand, and the Americans' greater reliance on economic cost-benefit and market-oriented values on the other. Despite the efforts that are being undertaken by both sides to reach consensus or reduce the gap in their regulatory approach to risk, at present the prospects do not seem promising because of the powerful economic and trade interests in the United States and the potential health and environmental effects that are at stake. This does not of course mean that these efforts should be abandoned, but rather that they should be reinforced in order to attempt to reach a better understanding of the underlying causes and differences in the approach to regulation of risk by the United States and the European Community.

Every society is free to choose the level of risk to health or the environment that it considers acceptable. In the European Community, the precautionary principle provides a way for both the regulatory authorities and the regulated natural or legal persons to ensure that this democratic, societal choice is achieved. First, it *enables* and sometimes *obliges* the regulatory authorities to take action when there is scientific uncertainty and risk but a direct causal link cannot be established. This is the most important normative function of the principle. Second, in a pre-marketing authorization procedure, the precautionary principle some-times entails placing the burden of proof on the applicant manufacturer, who has to demonstrate that a product is safe or that the level of accept-able risk will not be exceeded. Third, the precautionary principle also makes it possible for the affected persons to control, if necessary by court action, the exercise of regulatory discretion in risk management. These are the three basic normative functions the precautionary principle per-forms in European Community law. This principle is firmly based on science because its application is normally warranted only when uncer-tainty and the lack of a causal link between the risk and harm is scien-tifically established. Since it also reflects a principle of common sense, that is, to err on the side of caution in case of uncertainty, its normative force in the legal systems of the two green giants and in international law cannot be denied.

Notes

1. The author is expressing his personal views only.

2. See D. P. Fidler, *International Law and Public Health* (Ardsley, N.Y.: Trans-national Publishers, 2000); J. S. Applegate, "The Precautionary Preference: an American Perspective on the Precautionary Principle," *Human and Ecological Risk Assessment* 6 (2000): 413; D. Bodansky, "Scientific Uncertainty and the Precautionary Principle," *Environment* (September 4, 1991); N. de Sadeleer, *Les Principes du Polluer-Payer, de Prévention et de Précaution* (Brussels: Bruyland, 1999); and United Nations Environment Programme, "The Legal Implications of the Precautionary Principle for Multilateral Trade Rules," draft paper from the Center for International Environmental Law, March 24, 2000, available from the author.

3. European Commission, *Communication on the Precautionary Principle*, COM (2000) 1 final (Brussels: European Commission, 2000); European Environment Agency, *Late Lessons from Early Warnings: The Precautionary Principle 1896–2000* (Brussels: European Commission, 2001); C. Larrere, "Le Contexte Philosophique du Principe de Précaution," in C. Leben and J. Verhoeven, eds., *Le Principe de Précaution—Aspects de Droit International et Communautaire* (Paris: Editions Panthéon Assas, 2002), pp. 15–28.

4. European Commission, *Communication on the Precautionary Principle*; N. de Sadeleer, "Le Statut Juridique du Principe de Précaution en Droit Commu-nautaire: de Slogan a la Regle," *Cahiers de Droit Europeen* 91 (2001): 90–132; Applegate, "Precautionary Preference," p. 427.

5. E. Vos, *The Institutional Framework of Community Health and Safety Regulation—Committees, Agencies and Private Bodies* (Oxford: Hart, 1999).

6. S. Breyer and V. Heyvaert, "Institutions for Regulating Risk," in R. L. Revesz, P. Sands, and R. B. Stewart, eds., *Environmental Law, the Economy, and Sustainable Development: The United States, the European Union and the International Community* (Cambridge: Cambridge University Press, 2000), pp. 283–352.

7. David Vogel, "Risk Regulation in Europe and the United States," prepared for the *Yearbook of European Environmental Law*, Vol. 3 (New York: Oxford University Press, forthcoming 2003), provides the most accurate description and thoughtful analysis of the factual, legal, and economic developments this author has seen on this subject.

8. Ibid.

9. C. Raffensperger and J. Tickner, eds., *Protecting Public Health and the Environment: Implementing the Precautionary Principle* (Washington, D.C.: Island Press, 1999).

10. T. Christoforou, "The Origins, Content and Role of the Precautionary Principle in European Community Law," in Leben and Verhoeven, *Le Principe de Précaution*, pp. 205–230.

11. Applegate, "Precautionary Preference."

12. M. MacGarvin, "Precaution, Science and the Sin of Hubris," in T. O'Riordan and J. Cameron, eds., *Interpreting the Precautionary Principle* (London: Cameron and May, 1994), pp. 69, 74; Bodansky, "Scientific Uncertainty and the Precautionary Principle," pp. 204–208.

13. L. Krämer, *EC Environmental Law*, 4th ed. (London: Sweet and Maxwell, 2000); Krämer, *Focus on European Environmental Law*, 2nd ed. (London: Sweet and Maxwell, 1997).

14. Breyer and Heyvaert, "Institutions for Regulating Risk," p. 328.

15. Bodansky, "Scientific Uncertainty and the Precautionary Principle;" O'Riordan and Cameron, *Interpreting the Precautionary Principle*; P. Sands, *Principles of International Environmental Law*, Vol. 1 (Manchester, UK: Manchester University Press, 1995); D. Freestone and E. Hey, eds., *The Precautionary Principle and International Law: The Challenge of Implemention* (The Hague: Kluwer Law International, 1996).

16. MacGarvin, "Precaution, Science and the Sin of Hubris," p. 69.

17. Vogel, "Risk Regulation in Europe and the United States."

18. For example, P. Sand, "The Precautionary Principle: A European Perspective," *Human and Ecological Risk Assessment* 6 (2000): 445–458.

19. The Court of First Instance of the European Communities (CFI) handed down two seminal judgments on September 11, 2002, upholding the European Community's ban on the use of certain antibiotics in animal farming (i.e., virginiamycin, bacitracin zinc, spiramycin, and tylosin phosphate). The cases are important because for the first time the court discussed in great detail the scope

and conditions of application of the precautionary principle in Community law and the power of the relevant Community institutions in risk regulation. In the decisions—which are closely in line with the analysis in this chapter and reaffirm the European Commission's Communication on the Precautionary Principle of February 2000—the court held that it is possible to take preventive measures without having to wait until the reality and seriousness of the risks perceived become fully apparent. In the court's view, the concept of risk entails some probability that the negative effects, which the precautionary measure is specifically designed to prevent, will occur. The court concluded that despite uncertainty as to whether there is a link between the use of those antibiotics as additives and the development of resistance to them in humans, the ban on the products is not a disproportionate measure in comparison with the objective pursued; namely, the protection of human health. See judgments of September 11, 2002 in Case T-13/99 *Pfizer* (2002) ECR II-3305, and in Case T-70/99, *Alpharma* (2002) ECR, II-3495.

20. D. Vogel and T. Kessler, "How Compliance Happens and Doesn't Happen Domestically," in E. Brown Weiss and H. Jacobson, eds., *Engaging Countries: Strengthening Compliance with International Environmental Accords* (Cambridge, Mass.: MIT Press, 1998), pp. 19–37.

21. S. Shaw and R. Schwartz, "Trade and Environment in the WTO: State of Play," *Journal of World Trade* 36 (2002): 129–154.

22. Breyer and Heyvaert, "Institutions for Regulating Risk," p. 315.

23. Vogel and Kessler, "How Compliance Happens and Doesn't Happen," pp. 30–32.

24. M. A. Echols, *Food Safety and the WTO: The Interplay of Culture, Science and Technology* (London: Kluwer, 2001); M. A. Pollack and G. C. Shaffer, "The Challenge of Reconciling Regulatory Differences: Food Safety and GMOs in the Transatlantic Relationship," in M. A. Pollack and G. C. Shaffer, eds., *Transatlantic Governance in the Global Economy* (Lanham, Md.: Rowman & Littlefield, 2001), pp. 153–175.

25. Sand, *The Precautionary Principle*; Christoforou, "The Origins, Content and Role of the Precautionary Principle in European Community Law."

26. G. M. Fara, G. Del Corvo, S. Bernuzzi, A. Bigatello, C. DiPictro, S. Scaglioni, and G. Chiumello, "Epidemic of Breast Enlargement in an Italian School," *Lancet* 2 (1979), 295; G. Chiumello, M. P. Guarneri, G. Russo, L. Stroppa, and P. Squaramella, "Accidental Gynecomastia in Children," in A. M. Anderson, M. Grigar, E. Rajpert-de Meyts, H. Leffers, and N. E. Skakkebaek, *Hormone and Endocrine Disrupters in Food and Water: Possible Impact on Human Health* (Copenhagen: Munksgaard, 2001), p. 203; A. Perez Comas, "Precocious Sexual Development: Clinical Study in the Western Region of Puerto Rico," *Bolletin—Associacion Medica de Puerto Rico* 74 (1982): 245–251; C. A. Saenz, M. Toro-Sola, L. Conde, and N. P. Bayonet Rivera, "Premature Thelarche and Ovarian Cyst Probably Secondary to Estrogen Contamination," *Bolletin—Associacion Medica de Puerto Rico* 74 (1982): 16–19.

27. Larrere, "Le Contexte Philosophique du Principe de Précaution."

28. J.-P. Dupuy, *Pour Un Catastrophisme Eclairé: Quand l'Impossible est Certain* (Paris: Seuil, 2002).

29. See, e.g., Commission Decision 1999/832/EC of October 26, 1999 concerning the national provisions notified by the Netherlands concerning the limitations of the marketing and use of creosote (OJ No L 329, December 22, 1999, p. 25), where the European Commission approved the stricter measures of the Dutch explicitly on the basis of scientific uncertainty and the precautionary principle.

30. A. Boyle and D. Freestone, eds., *International Law and Sustainable Development: Past Achievements and Future Challenges* (Oxford: Oxford University Press, 1999).

31. De Sadeleer, "Le Statut Juridique du Principe de Précaution en Droit Communautaire."

32. Vos, *Institutional Frameworks of Community Health and Safety Regulation.*

33. For example, S. Breyer, *Breaking the Vicious Circle: Toward Effective Risk Regulation* (Cambridge, Mass.: Harvard University Press, 1993).

34. E. Millstone and P. van Zwanenberg, "The Scientific Basis of Applying the Precautionary Principle in Biotechnology-Related Potential Trade Conflicts: A Draft Report on the UK," June 2002 (draft available from the author).

35. V. R. Walker, "Keeping the WTO from Becoming the 'World Trans-science Organization': Scientific Uncertainty, Science Policy, and Fact-finding in the Growth Hormone Dispute," *Cornell International Law Journal* 31 (1998): 251–320.

36. A. -M. Anderson et al., *Hormone and Endocrine Disrupters in Food and Water*; A. -M. Anderson and N. E. Skakkebaek, "Exposure to Exogenous Estrogens in Food: Possible Impact on Human Development and Health," *European Journal of Endocrinology* 140 (1999): 477–485; D. T. Zhu and A. H. Conney, "Functional Role of Estrogen Metabolism in Target Cells: Review and Perspective," *Carcinogenesis* 19 (1998): 1–27; J. Liehr, "Genotoxicity of the Steroidal Oestrogens Oestrone and Oestradiol: Possible Mechanism of Uterine and Mammary Cancer Development," *Human Reproduction Update* 7 (2001): 273–281.

37. R. Hertz, "The Estrogen-Cancer Hypothesis with Special Emphasis on DES," in H. H. Hiatt, J. D. Watson, and J. A. Winston, eds., *Origins of Human Cancer*, Vol. 4, Cold Spring Harbor Conference on Cell Proliferation (Cold Spring Harbor, N.Y.: Cold Spring Harbor Laboratory, 1977); S. S. Epstein, *The Politics of Cancer Revisited* (New York: East Ridge Press, 1998), appendix XI, p. 585; and O. Schell, *Modern Meat* (New York: Vintage Books, 1985).

38. The latest risk assessment on the six meat hormones by the Scientific Committee on Veterinary and Public Health Measures of April 10, 2002 is available at http://europa.eu.int/comm/food/fs/sc/scv/out50_en.pdf.

39. Cited by T. Josling, *EU-US Trade Conflicts over Food Safety Legislation: An Economist's Viewpoint on Legal Stress Points that Will Concern the Industry* (Helsinki: The Mentor Group, House of Estates, 1998), p. 11; Vogel, "Risk Regulation in Europe and the United States."

40. Pollack and Shaffer, "Challenge of Reconciling Regulatory Differences," p. 158; Echols, *Food Safety and the WTO*.

41. See also Josling, *Trade Conflicts over Food Safety Legislation*.

42. It is interesting to note that according to recent studies by the World Health Organization's International Agency for Research on Cancer (IARC), the United States has a higher incidence of breast and prostate cancer (i.e., predominantly in hormonally dependent tissue of the human body) than the rest of the world, while the rate in Europe is on average about 20 percent lower than that in the US. See D. M. Parkin, S. L. Whelan, J. Ferlay, L. Ruymond, and J. Young, eds., *Cancer Incidence in Five Continents*, Vol. VII, Scientific Publication No. 143 (Lyon, France: IARC, 1997).

43. See, e.g., European Environment Agency, *Late Lessons from Early Warnings*; Walker, "Keeping the WTO from Becoming the 'World Trans-science Organization';" and T. Christoforou, "Science, Law and Precaution in Dispute Resolution on Health and Environmental Protection: What Role for Scientific Experts?" in J. Bourrinet and S. Maljeay, eds., *Le Commerce International des Organismes Génétiquement Modifiés* (Paris: La Documentation Française, 2002), pp. 213–283.

44. D. Shelton, "The Impact of Scientific Uncertainty on Environmental Law and Policy in the United States," in Freestone and Hey, *The Precautionary Principle and International Law*, p. 228; R. A. Bohrer, "Fear and Trembling in the Twentieth Century: Technological Risk, Uncertainty and Emotional Distress," *Wisconsin Law Review* (1994): 83–128.

45. European Environment Agency, *Late Lessons from Early Warnings*; Raffensperger and Tickner, *Protecting Public Health and the Environment*.

46. P. Bordieu, *Science de la Science et Réflexivité* (Paris: Seuil, 2001); B. Wynne, "Establishing the Rules of Law: Constructing Expert Testimony," in R. Smith and B. Wynne, eds., *Expert Evidence: Interpreting Science in the Law* (London: Routledge, 1989), pp. 23–55; B. Wynne, "Scientific Uncertainty and Environmental Learning," *Global Environmental Change* 3 (1992): 111; Dupuy, *Pour un Catastrophisme Eclairé*.

47. B. Fischhoff, S. Lichtenstein, P. Slovic, S. L. Derby, and R. L. Keaney, *Acceptable Risk* (Cambridge: Cambridge University Press, 1981); P. Slovic, "Perceptions of Risk," *Science* 236 (1987): 280–285; S. Jasanoff, "The Problem of Rationality in American Health and Safety Regulation," in Smith and Wynne, *Expert Evidence*.

48. Breyer, *Breaking the Vicious Circle*; C. F. Cranor, *Regulating Toxic Substances: A Philosophy of Science and Law* (New York: Oxford University Press, 1993); J. D. Graham and J. Wiener, *Risk vs. Risk: Tradeoffs in Protecting Health*

and the Environment (Cambridge, Mass.: Harvard University Press, 1995); J. D. Graham, opening remarks at a conference on "U.S., Europe, Precaution and Risk Management: A Comparative Case Study Analysis of the Management of Risk in a Complex World," Bruges, November 1, 2002 (found at http://www.uspolicy.be/Issues/Biotech/precprin.011502.htm); European Environment Agency, *Late Lessons from Early Warnings*; N. Ashford, "Implementing a Precautionary Approach in Decisions Affecting Health, Safety, and the Environment: Risk, Technology Alternatives, and Tradeoff Analysis," in E. Freytag, T. Jake, G. Loibl, and M. Wittmann, eds., *The Role of Precaution in Chemicals Policy*, Favorita Papers 01/2002 (Vienna: Diplomatische Akademie Wien, 2002), pp. 128–140.

49. European Environment Agency, *Late Lessons from Early Warnings*.

50. Fischhoff et al., *Acceptable Risk*; Slovic, "Perception of Risk."

51. This has been elegantly described by the appellate body in the *Hormone Beef* case as follows: "It is essential to bear in mind that the risk that is to be evaluated in a risk assessment under Article 5.1 is not only risk ascertainable in a science laboratory operating under strictly controlled conditions, but also risk in human societies as they actually exist, in other words, the actual potential for adverse effects on human health in the real world where people live and work and die." See the WTO Appellate Body report in *EC Measures Concerning Meat and Meat Products (Hormones) (European Communities—Hormones)*, WT/DS26/AB/R, WT/DS48/AB/R, adopted February 13, 1998, at paragraphs 187 and 194.

52. K. Arrow, "Behavior Under Uncertainty and its Implications for Policy," in D. E. Bell, H. Raiff and A. Tversky, eds., *Decision Making* (Cambridge: Cambridge University Press, 1988); S. Funtowicz and J. Ravetz, *Uncertainty and Quality in Science for Policy* (Amsterdam: Kluwer, 1990); S. Funtowicz and J. Ravetz, "Three Types of Risk Assessment and the Emergence of Post-Normal Science," in S. Krimsky and D. Golding, eds., *Social Theories of Risk* (Westport, Coun.: Praeger, 1992), pp. 251–273; A. Stirling, "Sciences et Risques: Aspects Theoretiques et Pratiques d'une Approach de Précaution," in E. Zaccai and J. N. Missa, *Le Principe de Précaution: Signification et Consequences* (Brussels: Éditions de l'Université de Bruxelles, 2000).

53. National Research Council, National Academy of Sciences, *Science and Judgment in Risk Assessment* (Washington, D.C.: National Academy Press, 1994).

54. U.S. Environmental Protection Agency, *Science Policy Council: Guidance for Risk Characterization* (Washington, D.C.: EPA, 1995); Walker, "Keeping the WTO from Becoming the 'World Trans-science Organization.'"

55. Slovic, "Perception of Risk," pp. 284–285.

56. This survey on the acceptability of GMOs is part of the report: "Europeans, Science and Technology," DG Research, *Eurobarometer* No. 55.2, (Brussels: European Commission, 2001).

57. For instance, this appears to be the situation in Europe in regard to the use of biotechnology in agriculture. See, e.g., B. Wynne, P. Simmons, and S. Weldon, *Public Attitudes to Agricultural Biotechnologies in Europe*, Final Report of Project PABE, 1997–2000, DG Research (Brussels: European Commission, 2000) See also Slovic, "Perception of Risk" and G. Gaskell and M. W. Bauer, *Biotechnology 1996–2000: The Years of Controversy* (London: Science Museum, 2001).

58. For instance, in the context of international trade it is accepted that defining the acceptable level of risk is the sovereign or autonomous right or prerogative of each state. See the WTO Appellate Body report in *EC Measures Concerning Meat and Meat Products (Hormones)*, paragraph 172; and the appellate body report in *Australia—Measures Affecting Importation of Salmon (Australia—Salmon)*, WT/DS18/AB/R, adopted November 6, 1998, at paragraph 199.

59. Shelton, "The Impact of Scientific Uncertainty on Environmental Law and Policy in the United States," p. 210.

60. A. O. Sykes, *Product Standards for Internationally Integrated Goods Markets* (Washington, D.C.: Brookings Institution Press, 1995).

61. See, e.g., Case C-120/97, *Upjohn* (1999) ECR I-223, at paragraph 47. See also Case C-405/92, *Armand Mondiet* (1993) ECR I-6133, at paragraph 31.

62. In its judgments of September 11, 2002 on the use of antibiotics in farming, the Court of First Instance stressed the conditions with which the public authority must comply in its risk assessment. It placed particular emphasis on the essential role of scientists in this context and concluded that the view of the competent scientific committees must be obtained, even if their opinion is only advisory or even if this is not specifically provided for by legislation, unless the public authority can ensure that it is acting on an equivalent scientific basis. However, the court pointed out that the decision to ban a product is not a matter for the scientists to decide, but rather is one for the public authority to whom political responsibility has been entrusted and who can claim democratic legitimacy—as opposed to scientific legitimacy—in risk regulation.

63. In the US legal system, this has been accurately explained in the Statement for Administrative Action for the WTO Agreements as follows: "The SPS Agreement thus explicitly affirms the right of each government to choose its level of protection, including a 'zero risk' level if it so chooses. A government may establish its levels of protection by any means available under its law, including by referendum. In the end, the choice of the appropriate level of protection is a societal value judgment. The Agreement imposes no requirement to establish a scientific basis for the chosen level of protection because the choice is not a scientific judgment." See *US Statement of Administrative Action for WTO/SPS Agreements (1994)*: 103d Congress, 2d Session, H.D. 103–316, 745 (September 27, 1994). See also R. H. Steinberg, "Trade-Environment Negotiations in the EU, NAFTA, and the WTO: Regional Trajectories of Rule Development," *American Journal of International Law* 91 (1997): 231.

64. Slovic, "Perception of Risk."

65. See, e.g., the WTO Appellate Body report in *Australia—Measures Affecting Importation of Salmon (Australia—Salmon)*, at paragraph 125.

66. M. Geistfeld, "Implementing the Precautionary Principle," *Environmental Law Reports News & Analysis* 31(2001): 11326; V. R. Walker, "Some Dangers of Taking Precautions Without Adopting the Precautionary Principle: A Critique of Food Safety Regulation in the United States," *Environmental Law Reports* 31(2001): 10040.

67. J. Wiener, "Precaution in a Multi-Risk World," 2001 (paper available from the author); C. Sunstein, "Beyond the Precautionary Principle," John M. Olin Law and Economics Working Paper No. 149, University of Chicago, April 2002 (preliminary draft).

68. Indeed, it has been demonstrated that cost-benefit analysis, combined with capture of an agency and manipulation and litigation by an interest group, can underestimate scientific uncertainty and unduly delay precautionary action, leading to further loss in health or environmental protection. See, e.g., European Environment Agency, *Late Lessons From Early Warnings: The Precautionary Principle 1896–2000*, chap. 4 (on benzene regulation), chap. 5 (on asbestos regulation), chap. 9 (on antimicrobials regulation), and chap. 15 (on BSE regulation). For another concrete example in the US, see *Corrosion Proof Fittings v. EPA*, 947 F. 2nd 1201 (5th Cir. 1991) (invalidating the ban on asbestos). See also R. Glicksman and C. H. Schroeder, "EPA and the Courts: Twenty Years of Law and Politics," *Law and Contemporary Problems* 54 (1991): 249.

69. Wiener, "Precaution in a Multi-Risk World"; Sunstein, "Beyond the Precautionary Principle"; G. Majone, "The Precautionary Principle and Regulatory Impact Analysis," 2001 (paper available from the author).

70. Geistfeld, "Implementing the Precautionary Principle."

71. See B. Toebes, "The Right to Health," in A. Eide, C. Krause, and A. Rosas, *Economic, Social and Cultural Rights,* 2nd ed. (Dordrecht the Netherlands: Nijhoff, 2001) pp. 169–190; and Committee on Economic, Social and Cultural Rights, General Comment No. 14: *Right to the highest attainable standard of health (Article 12 of the International Covenant on Economic, Social and Cultural Rights)*, adopted on May 11, 2000, 22nd Session (2000), UN Doc. E/C.12/2000/4, 8 IHRR 1 (2001). In regard in particular to the relationship between human rights and the protection of the environment, see G. Handl, "Human Rights and the Protection of the Environment," in Eide, Krause, and Rosas, *Economic, Social, and Cultural Rights,* pp. 303–328.

72. R. Dworkin, *Taking Rights Seriously* (London: Duckworth, 1987); A. Sen, *On Ethics and Economics* (Oxford: Blackwell, 1986); Geistfeld, "Implementing the Precautionary Principle."

73. D. L. Bazelon, "Science and Uncertainty: A Jurist's View," *Harvard Environmental Law Review* 5 (1981): 209.

74. The European Court of Justice has defined the principle of proportionality in Community law as follows: "It must be recalled that the principle of proportionality, which is one of the general principles of Community law, requires that measures adopted by Community institutions do not exceed the limits of what is appropriate and necessary in order to attain the objectives legitimately pursued by the legislation in question; when there is a choice between several appropriate measures recourse must be had to the least onerous, and the disadvantages caused must not be disproportionate to the aims pursued." See Case C-157/96, *The Queen v. Ministry of Agriculture* (BSE) (1998) ECR I-2211, at paragraph 60. See also Case C-331/88, *Fedesa and others* (1990) ECR I-4023, at paragraph 13; and Joined Cases C-133/93, C-300/93, and C-362/93, *Crispoltoni* (1994) ECR I-4863, at paragraph 41.

75. Geistfeld, "Implementing the Precautionary Principle."

76. A. Stirling, O. Renn, A. Klinke, A. Rip, and A. Salo, "On Science and Precaution in the Management of Technological Risk," Institute for Prospective Technological Studies (Brussels: European Commission, 1999).

77. D. Shelton, "The Impact of Scientific Uncertainty on Environmental Law and Policy in the United States," p. 225.

78. See, e.g., US Food and Drug Administration and Department of Agriculture, *United States Food Safety System: Precaution in U.S. Food Safety Decision-making: Annex II to the United States' National Food Safety System Paper* (March 3, 2000), in response to the OECD Ad Hoc Group on Food Safety, to be found at http://www.foodsafety.gov/~fsg/fssyst4.html (visited on December 1, 2001). For a critical review of the US paper, see Walker, "Some Dangers of Taking Precautions Without Adopting the Precautionary Principle." The official US position given in this document and previous US positions expressed in the context of discussions and negotiations in a number of other international forums, such as those held by the WTO, the Food and Agricultural Organization, the Codex Alimentarius Commission, the OECD, the Biodiversity Convention and in the European Commission's *Communication on the Precautionary Principle* of February 2, 2000 have sparked rather voluminous written exchanges and a direct dialogue between the administrations of the US and Canada, on the one hand, and the European Commission on the other. See US Department of State, *Questions from the US on the Commission's Communication on the Precautionary Principle*, March 2000; European Commission, *Comments, Responses and Further Questions to the US on the Precautionary Principle*, June 27, 2000; and US Department of Agriculture, *United States Response to Questions on the Use of Precaution in Risk Analysis Raised by the European Commission in its 27 June, 2000 Comments to the US on the Matter of Precaution* (January 19, 2001); see also Canada, *Comments on the Communication from the Commission of the European Communities on the Precautionary Principle* (letter from the Canadian Ambassador in Brussels of July 25, 2000); and European Commission, *Replies from the European Commission to the Questions from the Canadian Ambassador on the Commission's Communication on the*

Precautionary Principle and Questions from the Commission to Canada (May 10, 2001) (copies of all documents are on file with the author). As already explained, the dialogue between the US administration and the Commission continued in the summer and early September of 2001 in the context of launching the new WTO Doha round of trade negotiations (copies of the minutes are on file with the author).

79. Bazelon, "Science and Uncertainty," p. 209.

80. See, e.g., Orders of the Court of First Instance in Case T-70/99 R, *Alpharma* (1999) ECR II-2027, and in Case T-13/99 R, *Pfizer* (1999) ECR II-1961. See also Sykes, *Product Standards for Internationally Integrated Goods Markets.*

81. S. Brewer, "Scientific Expert Testimony and Intellectual Due Process," *Yale Law Journal* 107 (1998): 1535; T. Christoforou, "Settlement of Science-based Trade Disputes in the WTO: A Critical Review of the Developing Case Law in the Face of Scientific Uncertainty," *N.Y.U. Environmental Law Journal* 8 (2000):622–648.

82. Shelton, "The Impact of Scientific Uncertainty on Environmental Law and Policy in the United States,"p. 210.

83. J. Scott and E. Vos, "The Juridification of Uncertainty: Observations on the Ambivalence of the Precautionary Principle within the EU and the WTO," in C. Joerges and R. Dehouse, eds., *Good Governance in Europe's Integrated Market* (Oxford: Oxford University Press, 2002), pp. 253–286.

2

The Roots of Divergence: A European Perspective

Ludwig Krämer[1]

Different Points of Departure

Active protection of the environment began in both the United States and Europe in the 1960s, although many measures in the areas of water management, nature protection, town and regional planning, and waste management were adopted earlier. The political, legislative, and administrative actions in the years following the publication of Rachel Carson's famous book *Silent Spring* led, on both sides of the Atlantic, to more organized, deliberate, and planned measures which, since that time, have come to be grouped under the term "environmental policy."

Yet this coincidence in time clouds the fact that the points of departure for the United States and Europe were completely different. Indeed, in the 1960s the European Union[2] did not even exist by its present name, and the underlying argument in this chapter is that a comparison between the United States and "Europe" neither does justice to the European integration process nor does it help much to facilitate an understanding of present or future developments.

When the United States started to develop an active environmental policy, it was a sovereign nation-state that possessed all the constitutional, institutional, economic, and political requirements to conceive and implement a coherent and consistent environmental policy at home and abroad. However, until the 1960s, water and air issues were mainly dealt with at the level of the individual states within the United States. Growing public concern about environmental pollution caused Congress to adopt federal air pollution legislation in 1965 and 1967 that was considerably reinforced by the Clean Air Act Amendments of 1970, which were, in later years, extended and fine tuned. A similar development occurred in the water sector. Relatively soft federal provisions of 1965 were considerably sharpened and "nationalized" by the Federal Water Pollution Control Act Amendments of 1972. President Richard Nixon established the Environmental Protection Agency (EPA), which received powerful regulatory functions from Congress. Responsibility for other parts of environmental policy was largely in the hands of Congress; product and process legislation was traditionally dealt with by Congress under the interstate commerce clause of the Constitution. The fact that the federal government owned about one third of the land in the United States facilitated nature conservation measures without serious interference with private property or the prerogatives of the states. Furthermore, Congress had the power to levy taxes and to provide subsidies, which it used in particular to encourage state environmental measures. Overall, since the end of the 1960s, a number of strong, extremely detailed, and prescriptive legislative instruments have been adopted which, together with federal executive institutions, have formed the backbone for US environmental policy ever since.

The European Union was in a quite different situation. It was not a nation, but a supranational joint venture of nation-states (fifteen at present) that could act only where the European Community (EC) Treaty expressly so provided. Its member states had very different perceptions of and objectives for the European integration process; this in turn influenced their attitude in day-to-day Community decisions.

Environmental concerns in Europe developed at the level of EC member states; they concerned different subjects with variable intensities, consequences, and reactions from the national legislatures and national policymakers. The European situation should be compared, not with that of the United States, but with that of all the states of North and Central America in order to understand the importance of the "sovereignty" of the nation-state. Sovereignty was at the core of all sorts of difficulties that slowed down European integration and consequently the adoption of common European environmental standards.

The EC Treaty of 1958 did not contain any explicit reference to the environment or to environmental policy; explicit provisions on environmental policy were not introduced in it until the Single European Act of 1987.[3] Also, the Treaty was not—and is not—a constitution for the European Union. Some of the key institutional differences between the US and the EU include the following:

• The Treaty allows the Community institutions—the "federal level" in US terminology—to act only when they are entitled to do so under the Treaty provisions. The basic competence for dealing with (environmental) matters is vested in the EU member states. While in theory this might not seem very different from US law, Congress can, in practice, deal with almost all matters of environmental law and policy, particularly in regard to pollution control, environmental subsidies, product and production standards, and land use.

• There is no European "Congress." European Union environmental legislation is adopted jointly by the council of ministers, which is composed of representatives of the governments of member states, and by the European Parliament, the members of which are directly elected. The European Parliament cannot overturn decisions of the council; and the European Commission, composed of appointed persons who act in the general interest of the European Community, has only the right to propose legislation, not to adopt it. This means that member states have a decisive influence on the question of which environmental matters they want regulated at "federal level" and which they prefer to keep for themselves.

• The European Union does not own land and, for the most part, EU member states do not either.

• The European Union has practically no income of its own; it receives a fixed percentage—1.27 percent—of the national income of member states, a fact that makes it practically impossible to influence environmental changes within the European Union by economic or fiscal incentives or subsidies.

• The European Union has no power to levy environmental taxes unless all member states unanimously agree in council—something they have not done so far.

To these "constitutional" differences have to be added the different political, economic, social, cultural, and environmental differences among the constituent members of the European Union; the absence of a European media (television, press, radio); of a European public opinion; and consequently of a European-wide common interest on many issues.

In the following sections I compare the development of the internal and external environmental policies of the United States and the European Union during two periods: (1) up to the mid-1980s, when both sides enacted their initial environmental legislation and became active in international environmental negotiations and (2) since the mid-1980s, when the US and EU have moved in different directions.

The Period to the Mid-1980s

Environmental policy in the United States was marked at the beginning of the 1970s by a strong degree of centralization that can be seen in the adoption of federal legislation concerning air and water pollution, industrial permitting, nature protection, and soil cleanup policies. It also had powerful enforcement mechanisms, in particular via the EPA. During the 1970s, the EPA and other federal agencies pursued a vigorous and robust policy of standard-setting and enforcement of environmental standards.

However, this centralized policy approach, although it might not have been all-embracing and comprehensive, came progressively under attack from sources that favored setting environmental policy at the state level and, more important, from economists and the regulated businesses. The EPA's activity came to be seen as excessively interfering with the market and not taking sufficient account of the economic costs of regulations. In the early 1980s, deregulation was started by the Reagan administration, and while the basic environmental legislation adopted by Congress was not abolished, the regulatory responsibilities of the EPA were limited and measures were taken to give the states greater responsibility for regulating the environment. President Reagan's Executive Order 12291 required the EPA and other federal regulatory agencies to conduct cost-benefit analyses of all regulatory proposals and adopt the most economically efficient or cost-effective alternative. Compliance with those requirements was policed, not by the courts, but by the Office of Management and Budget.[4] Also, economic impact assessment requirements and other economic barriers to environmental regulations were established.

The United States was represented in international environmental negotiations by the State Department and the Department of Commerce. Little consideration was given to creation of a cabinet-level Department of the Environment or to EPA participation in international negotiations. This demonstrates that environmental concerns remained secondary to trade and economic considerations in US external policy.

In the European Community, environmental policy developed only slowly, with the adoption of specific measures aimed at addressing specific problems. The first EC environmental directives (as opposed to free-trade directives) date from 1975 and dealt with waste oils, the quality of surface waters, wastes generally, and the quality of bathing (swimming) waters.[5] They were followed by product-related provisions and subsequently, after the end of the 1970s, by provisions on protection of nature and air quality. Industrial accidents and the problem of *Waldsterben* (dying forests), which was attributed to environmental changes caused by human activities, increased public and political concern in Western Europe. This allowed the adoption at EC level of new environmental directives that showed strong concern for health issues, frequently took a preventive approach, and progressively encompassed all areas of environmental policy. Thus, when the EC Treaty was amended in the mid-1980s, there was a general consensus among the member states that provisions for a comprehensive European Community environmental policy should be added. The new Treaty provisions in the Single European Act that entered into effect in 1987 laid down inter alia objectives and principles of environmental policy based on the goals and principles the EC and its member states had already agreed upon in 1973, thus ensuring the continuity and consistency of this policy. Cost-benefit considerations were mentioned, but only in the sense that actors should take account of the advantages and costs of environmental action or the lack of it.[6]

Environmental legislation was negotiated, not by the member states' foreign affairs ministries or trade departments, but by the environmental departments that had progressively been established within the member states since the early 1970s. Because the European Commission, which has a monopoly on initiating legislative proposals under the EC Treaty, also had an environmental department, and since environmental legislation was enacted by the council of ministers meeting as separate groups of ministers for each policy sector, environmental matters were from the very beginning of European environmental policy kept outside the direct influence of the member states' foreign and trade policies, and EC environmental policy was accepted as being independent of commercial and foreign policy. This was a major difference from the situation in the United States.

At the international level, the European Community had no overall general competence to act. It had responsibilities for commercial matters, but the exact extent and nature of this competence was constantly disputed by member states who, in the name of national sovereignty, preferred to be represented separately on the international scene rather than as part of the EC. These differences of view on commercial policy issues also favored the development of a foreign environmental policy that was independent of commercial policy and general foreign policy.

As a consequence, when the European Community appeared at international meetings for discussions on environmental matters, it was mainly represented by the environmental directorate-general of the European Commission and by (some or all) environmental departments of the EC member states. In order to find a common European position, long consultations prior to and during international negotiations were necessary; and where a consensus was not reached, the Commission of the European Communities defended what it considered to be the EU interest, while individual member states often promoted their own interests. This inability to speak with one voice often irritated representatives of other nations, who did not fully understand these consequences of the European efforts to progressively integrate sovereign states into one European Union. In general, prior to 1987 the European Community was almost never viewed as being a single autonomous body in international environmental negotiations; rather, the larger EC member states such as France, the United Kingdom, or Germany dominated the scene. This is the reason the publications of this period hardly ever mention the European Union's foreign environmental policy and law.[7]

Prior to 1985, the documents produced by global environmental conventions generally could only be signed by states, not by regional bodies

such as the European Community. However, European regional environmental conventions increasingly provided for signature by the EC
from the mid-1970s on. At a global level, the first important convention
to provide for the European Union's accession was the Convention on
Long-range Transboundary Air Pollution (LRTAP) of 1979.[8] That
convention was generated by efforts after 1975 to improve East–West
political relations. The European Community had asked to have a clause
inserted into the convention according to which "regional economic integration organisations" could also accede to it. The Soviet Union, which
was very interested in establishing the convention, opposed such a clause;
thus, the United States could not, as a member of the Western camp,
oppose it too vehemently. Finally, the Soviet Union accepted the clause
and the United States was satisfied to bring the European EC into the
East–West dialogue.

However, after 1981, the United States, led by the State Department,
changed its policy and opposed European Community accession to global
environmental conventions. The US tried for several years to allow such
accession only under two conditions. The first was that the European
Community make a precise statement on the Community's competence
in the subject matter dealt with by the convention in question (a declaration of competence). This was difficult for the EC because its founding
treaty is not a constitution and therefore the allocation of competencies
between the EC and its member states is not static, but evolving. The
second condition was that a majority of European Community member
states had individually ratified the convention in question.

The European Union invoked the precedent of the LRTAP Convention
and slowly obtained inclusion in other agreements of the same clause
used in that convention. From time to time it made a declaration concerning competence.[9] However, these declarations did not really clarify
anything, and the disagreement with the United States on the European
Community's accession to conventions did not disappear. For instance,
in 1983 the European Community achieved an amendment to the Convention on International Trade in Endangered Species (CITES) to allow
its accession. The United States was not in favor of this accession and
has not, to this day, ratified this amendment. Moreover, it seems to have
encouraged other contracting states not to ratify it either. As a consequence, the amendment has not yet been ratified by the necessary number

of contracting parties, so that the European Union cannot adhere to the CITES Convention. While the European Community had completely incorporated the requirements of the CITES Convention into European law, it was nevertheless formally barred from speaking with one voice at the CITES conferences, and neither the United States nor other contracting parties have made particular efforts to improve this unpleasant situation.

Bilaterally, as early as 1974 the United States and the Commission of the European Communities exchanged letters to promote cooperation in environmental matters.[10] It is rather typical that these letters were signed for the European Commission by the commissioner responsible for environmental affairs, and for the United States by the assistant secretary of state responsible for environmental affairs (among other things). The cooperation was to concentrate on the exchange of information on environmental issues. Since the United States did not have an environmental department and might have been unwilling to let the Environmental Protection Agency initiate this cooperation, the Department of Commerce and the State Department saw the bilateral meetings from their very beginnings as an exchange of information under trade and commercial auspices. However, such discussions had less interest for the European Commission, for which the environmental directorate was the leading representative. Thus the bilateral meetings that were organized more or less every 2 years focused on matters that concerned potential trade conflicts. Intensive technical cooperation took place in matters such as chemical and air pollution, and useful results were reached. In contrast, hardly any time was devoted to questions of how environmental degradation could be prevented or repaired at the national or international level, what lessons were to be learned from legislation adopted so far, and what new concepts or measures might be developed to combat environmental damage in the future.

The Period after the Mid-1980s

In the United States, environmental protection measures have mainly focused on the administration of federal statutes and attempts to establish cost-benefit analyses and risk assessment as conditions for federal action.[11] A divergence of views between the executive branch and Congress on basic questions has frequently paralyzed legislative measures and prevented innovative new protection measures.

In Europe, after the Single European Act, the evolution of environmental policy was marked by the reevaluation of objectives; continued attempts to integrate environmental requirements into other policy areas such as transport, energy, regional policy, agriculture, and industry; the achievement of greater coherence and the covering of new areas of environmental legislation to progressively align national environmental policies; and increasing attention to climate change issues, which gradually became a top political priority. Also, Europe imported tools such as environmental impact assessments and gained access to information and environmental management systems from the United States. Other tools, though, were rejected, such as a Superfund system for repair of environmental damage, an environmental liability system, and an enforcement agency modeled after the EPA.

On the international scene, when the Vienna Convention on the Protection of the Ozone Layer was negotiated in 1985 under the auspices of the United Nations Environment Program (UNEP), the European Community achieved, over considerable objections, more from the United States than from the Soviet Union, the insertion of a provision that allowed regional economic organizations accession to the convention.[12] As a consequence, the Montreal Protocol negotiations that restricted the production, use, and consumption of ozone-depleting substances were, for the European side, to a large extent led by the European Community, which successfully managed to find common language for all its member states and to speak with one voice. This joint European position produced a protocol in which the United States did not fully impose its position, but had to accept considerable concessions. The EC even obtained a clause that allowed for joint implementation of the obligations under the protocol.[13]

The negotiations on the Montreal Protocol were the first at the international level in which the European Community and the United States confronted each other on environmental matters. The member states of the Community realized that their negotiating position was greatly improved by acting under the umbrella of the EC and that the collective gains achieved by this approach outweighed the political advantages of each state negotiating for itself. They also discovered that the fact that the negotiations were being led by the environmental department of the European Commission, together with the EC Council presidency, did not mean that their national or joint European economic interests would be neglected.

This outcome encouraged the European Community to appear more frequently in international environmental negotiations with an agreed upon negotiating mandate, and to try to speak with one voice. Despite many setbacks, this policy was, overall, successful owing in particular to the following factors:

• The Single European Act of 1987 gave the European Community a mandate to contribute to the search for solutions to global environmental problems and specified that the EC had the competence to act internationally, both aside from and jointly with its member states. The new obligation under the Treaty to promote a high level of environmental protection within the EC and worldwide favored efforts to reach environmentally sound solutions in international negotiations. Hence the European Community did not try to subordinate environmental interests to commercial or economic interests, and it did not enter international negotiations with the explicit or implicit concept of agreeing only to solutions that were profitable to the European economy.

• Environmental legislation within the European Community progressively covered more areas, became more coherent, and provided a political and legal framework for environmental measures in all member states. Former national policies in the area of the environment were thus increasingly brought into alignment. The solutions found at the EC level then served as the basis for positions and compromise proposals that were put forward during international negotiations.

• Europe was normally represented at international environmental conferences, meetings, and negotiations by the environmental departments of member states and the European Commission's Directorate-General for environmental affairs; the member state holding the presidency of the EC Council and the commission acted as spokesperson. This was in marked contrast to the United States, whose delegations were normally led by the State Department or the Department of Commerce, but practically never by the EPA, and in which the state-level environmental offices (of California, Texas, etc.) were never represented.

When the Berlin Wall came down in 1989, the Soviet Union collapsed, and the countries in central and eastern Europe emerged as fully sovereign nations, the United States remained the only truly global actor. Some cooperation was established between the US and the European Community in central and eastern Europe, particularly in setting up the Regional Environmental Center for Central and Eastern Europe in Hungary in 1990. However, this cooperation remained marginal because each side tried to promote its own way of life as a model for Eastern Europe, even in environmental matters. While the United States acted much more speedily and efficiently in the beginning, the European Community took a progressively stronger position as countries in central and eastern Europe began to seek membership in the European Union and thus started to adapt their environmental legislation and institutional systems to those of the European Union. At the global level, many in the European Union got the general impression that worldwide environmental problems were seen by the United States mainly in terms of economic globalization. This impression was based on:

• Discussions in the Uruguay Round of the General Agreement on Tariffs and Trade (GATT), in which the United States opposed consideration of environmental aspects;

• Negotiations at the Rio Conference of 1992, where the United States rejected precise targets and timetables for greenhouse gas reductions and, more generally, refused to accept broad environmental texts on which to base global environmental measures for the next decade; e.g., the thorough scrutiny of the Declaration on Environment and Development by the State Department, which led to the rejection by the United States of the words "precautionary principle" in favor of the words "precautionary approach";

• Discussions on the North American Free Trade Agreement (NAFTA), where only strong internal pressure from environmental groups got some environmental considerations incorporated into the agreement and accompanying side agreement;

• The negotiations of various global conventions, in particular the Basel Convention on the international shipment of hazardous waste, the protocols to the Geneva Convention on long-range transboundary air pollution, the Framework Convention on Climate Change and the Kyoto Protocol, as well as others. In all these international discussions, the United States was seen as trying to subordinate environmental questions to economic and trade issues and to avoid, if possible, any substantive environmental provisions at all;

• Continued attempts by the United States to have points removed from the agenda of the United Nations Environment Program and to reduce the funds made available to UNEP.

The fact that Al Gore, the author of *Earth in the Balance: Ecology and the Human Spirit*, in which he pleaded for a global Marshall Plan for sustainable development, became vice-president of the United States but was unable to politically advance the environmental ideas he had expressed in his book, convinced many Europeans that US economic interests had excessive influence on environmental policy. Several further specific examples of this preponderant influence can be cited.

One illustration was the controversy on the noise level of airplanes. Since the International Civil Aviation Organization (ICAO) had not revised the international noise standards for airplanes since 1977, the EU finally adopted stricter European standards in 1999 that the United States considered protectionist and discriminating.[14] The European Union offered to consider delaying their application if the US showed a willingness to push ICAO to adopt more stringent global standards. The United States, however, filed an official complaint and asked the European Union to adopt a more economically attractive solution.

In waste management, the United States favored the so-called prior informed consent (PIC) approach for exports of hazardous waste to developing countries. Under this PIC approach, the importing country obtains the relevant information and then decides whether it will accept the material. The European Union, however, accepted the argument put forward by nongovernmental organizations (NGOs) that, in principle, hazardous waste should not be exported to developing countries at all. The EU thus negotiated and agreed to the introduction of such an export ban under the Basel Convention on the shipment of hazardous waste.[15] In contrast, the United States did not ratify the Basel Convention or its amendment on the export ban.

In regard to the export of chemicals, the European Union progressively sharpened its position, first accepting the PIC approach and then moving toward the elaboration of an international convention under which the most dangerous chemicals would be banned altogether. These efforts led to the signature of the Stockholm Convention on Persistent Organic Pollutants (POPs) in 2001, which the United States signed but has not yet ratified.

The same pattern can be seen earlier on the issue of leghold traps, which the European Union had banned from use. To protect the welfare of wild animals, the EU had added an import ban on furs from specific wild animals that came from countries that had not banned leghold traps. While the European Union hoped for worldwide standards on humane trapping of animals, it was forced by several countries led by the United States to withdraw its import ban. No serious effort was subsequently made to enact worldwide humane trapping standards.

Other examples that cannot be discussed here for lack of space concern the negotiation of the Cartagena Protocol on Biosafety and general discussions on biotechnology, on growth hormones in meat, on heavy metals in specific products (batteries, cars, electrical and electronic goods), and on ecolabels and standards for environmental management systems.

The increasing differences of view on global environmental issues culminated in the discussions on climate change and the conclusion of the Kyoto Protocol. The European Union saw the Kyoto Protocol as an extension of the commitments accepted under the Climate Change Convention.[16] The United States considered the Kyoto Protocol flawed for two reasons. The first was that it contained obligations for industrialized countries to reduce greenhouse gas emissions, but not for developing countries. The US considered climate change a long-term problem that could and should be thoroughly researched before action was taken, and in this long-term perspective (as far out as 2100) the US argued that developing countries, too, should contribute to the reductions. Second, the Kyoto Protocol did not expressly enable industrialized countries to comply with their reduction commitments by investing in reduction technologies in developing countries or otherwise allow industrialized countries to meet their obligations in ways that would not require emission reductions at home (see chapter 8). For Europeans, it was remarkable that the United States did not offer an alternative solution to reducing greenhouse gas emissions and that it did not pursue any consistent policy at home.

Finally, bilateral environmental meetings between the US government and the European Commission took place at almost annual intervals during the 1980s and the first half of the 1990s. These meetings covered a broad range of subjects, such as what attitude to adopt in the different international forums on product-related issues and on biotechnology and biodiversity. Again, the US focus was largely on the prevention of barriers to trade rather than on optimum protection of the environment. This emphasis, together with formal procedures within the context of the World Trade Organization (WTO) and other forums such as ICAO, gradually reduced the importance of the environmental aspects of the meetings. In contrast, bilateral technical discussions on specific questions such as analysis standards or test methods for specific products continued and often produced satisfactory results.

Divergences and Their Causes

From my perspective, the main differences between the United States and the European Union in approaching environmental problems can be summarized as follows:

As it does in its internal situation, the European Union sees globalization more and more clearly as including—with the same degree of importance—trade issues, environmental concerns, and social questions. A correct balance among these competing interests has to be found on a case-by-case basis. In contrast, the United States works toward global institutions and instruments that give greater importance to the economic aspects of free trade than to environmental protection. In this view, environmental considerations should interfere with the global market as little as possible. Globalization is thus as far as possible economic globalization.

Since the European Union does not see itself as a global player—perhaps apart from agricultural matters not discussed here—its foreign environmental policy looks for multilateral solutions that are globally acceptable. These solutions might even appear not to be to the best advantage of European economic interests. The United States, as already mentioned, tends to perceive international environmental negotiations as international trade negotiations. This leads it to defend interests that sometimes appear to be those of US industry, not those of the global environment.

The nation-states forming the European Union accept that their sovereignty is affected by the Treaty on European Union and that the European Court of Justice controls their legislative, regulatory, and administrative activity to ensure its compatibility with the EU Treaty and its principles, as well as with legislation adopted by the European Union. Therefore, not only do they have few fundamental problems in accepting global solutions that do not entirely conform to their economic interests and preferences, but they are also prepared to accept compliance mechanisms and control procedures that further encroach on their sovereignty. By contrast, the United States appears to accept binding commitments and obligations by the international community that influence its policy at home only when this brings economic advantage. While internal enforcement mechanisms and control procedures by administrative agencies and the courts are quite strong, the United States does not seem to accept that international environmental law also requires strong compliance mechanisms that might even impinge on national sovereignty.

These differences have many causes, among which the following appear to be the most important:

1. Traditionally, Europe has had a stronger commitment to social and more recently to environmental concerns than the United States. The idea of Adam Smith in *Wealth of Nations* that an individual who acts in his own self-interest and intends only his own gain "is led by an invisible hand to promote . . . the public interest" has had strong support in US economic theory, legislation, and regulatory practice, but it has never gained the same importance in Europe. Governments were seen as charged not only to promote individual life, liberty, and the pursuit of

happiness, but also to reduce inequalities in society. This attitude has led to far-reaching interventions in the social and more recently the environmental area. There is—with many nuances from one member state to the other—a sort of consensus in Europe that public intervention must also ensure a decent state of the environment, and that environmental protection cannot be left to market forces. Thus, while many businesses in the United States might be philosophically opposed to the current regime of environmental regulation and consider it illegitimate,[17] this attitude does not exist in this form in Europe, where the environmental departments are less likely to act as spokespersons for vested economic interests.

2. In the United States, environmental protection policy was perceived as a centralizing policy and attracted criticism from conservative circles that opposed state intervention in the market, and from those favoring states' rights. This opposition has gained considerable influence in policy circles as well as in academic and public opinion. In Europe, the majority of EU member states are convinced of the necessity to pursue a vigorous and active environmental policy that includes market interference, and since the European Commission and all EU member states have established environmental departments of their own, EU measures are seen less as centralizing than as integrating or harmonizing measures. There are certainly conservative and business objections to aspects of European environmental policy as well, and under their influence, policy has sometimes undergone considerable changes. However, the objections to European environmental policy have not taken on a fundamental character, and EU member states would probably prefer to pursue their national policies again rather than accept a European policy that gives too much weight to business interests.

3. The United States considers discussions within the European Union and internationally as often not "scientifically sound," since cost-benefit considerations and risk assessments do not play a preponderant role in them. It takes the view that the US approach to cost-benefit and risk analyses constitutes such sound science. In Europe, approaches based on economic theory have not gained the same influence over environmental policy, particularly since opinions other than those of economists— such as those from natural science (biology, geology, geography),

philosophy, religion, social science (history, political science, law)—are voiced in public and contribute to forming public and political opinion. Furthermore, the concepts of cost-benefit analysis, risk assessment, and life-cycle analysis are not regarded as scientifically sound because economists have not managed to develop generally acceptable, reliable standards for measuring the benefits of an unimpaired environment or for expressing in monetary terms such things as the loss of biodiversity (see chapter 7). Market instruments such as environmental taxes and charges and emissions trading are also used within the European Union, but more cautiously and without the belief that the market is a remedy for all or most environmental problems.

Expressed in simple terms, the general feeling in Europe was and is that there are environmental assets that money cannot buy, and that the United States considers "cost" to be the cost of a measure to business, but does not include in its cost-benefit considerations the advantages and disadvantages of a measure for society as a whole, including future generations.

In this context, it should be noted that the US Congress does not apply the principles of cost-benefit analysis, life-cycle analysis, or risk assessment to its own legislative decisions; rather, these principles are applied by the EPA and other agencies. Since Europe has no regulatory agency comparable to the EPA (either at the EU or member state level), most environmental regulatory measures are adopted by legislative bodies. Europeans thus frequently consider American demands for more consideration of the above-mentioned economic principles to be misplaced.

4. Overall, in Europe, protection of the environment—like social rights, gender equality, or human rights—is perceived as part of the foundations of any society. All opinion polls show that there is a consensus on the need to protect the environment, to reduce pollution, to protect biodiversity, and to promote changes that go beyond the consumption society; and that people are gradually becoming accustomed to the idea that changes in lifestyle are necessary. It is true that the green political parties that have appeared since the early 1980s in several member states seldom represent more than 10 percent of the electorate. However, the influence of their political thinking goes far beyond that percentage and has brought considerable changes in traditional political parties and

general policy thinking. And EU member states that promote a strong, consistent, and progressive environmental policy have not fared less well economically than those with weaker environmental policies. They believe that investing in clean technologies, alternative energy sources, and new environmental techniques pays off, and that the environmental challenge is a powerful stimulus to innovation and modernization. This consensus has also often been influenced by environmental accidents or setbacks that demonstrated that public authorities cannot be allowed to neglect environmental concerns.[18]

In contrast, discussion of environmental issues within the United States and by the US in international forums often gives Europeans the impression that environmental policy is considered a fad, without much consequence for things that really matter in society. Notions of "prevention" and "precaution," the principle of "polluter pays," and the need to integrate environmental requirements into energy, transport, agriculture, industrial, and foreign policy do not seem to play an important role in current American political debate.

Conclusion

In conclusion, it is submitted that at the global level, conceptual differences between Europe and the United States have led to different approaches to environmental issues. These differences began to manifest themselves during the 1980s, but the end of the East–West conflict did not contribute significantly to the divergences. Rather, by that time, on the one hand the European Union was more systematically represented on the international scene, backed by a generally accepted set of principles and objectives in the EU Treaty and by fairly strong internal legislation that facilitated a consensus among Europeans in global discussions; while on the other hand, the economy-oriented approach to environmental policy that had prevailed in US policy since the early 1980s has come to dominate its external as well as its internal environmental policies. These trends have contributed significantly to the present state of affairs, which is marked by divergence on several important global environmental issues and by a relatively cool and distant bilateral diplomatic relationship.

Notes

1. The author expresses only his personal opinion. He attaches importance to the fact that he was never directly involved in bilateral discussions with the United States.

2. The term "European Union" is used throughout this chapter, although the original EC Treaty of 1958 established the European Economic Community. The term "European Union" has existed only since 1993; it was introduced by a treaty amendment to underline the political objective of the European Community and to take account of and further promote progressive economic and political integration of the member states. The European Union—unlike the European Community, which continues to exist as one of the "pillars" of the European Union—does not have a legal personality; therefore environmental and other legislation is adopted by the European Community, not by the European Union.

3. See Articles 174 to 176 of the EC Treaty (environmental policy); also Articles 2 (objectives of EC), 6 (integration of environmental requirements into other policies), and 95 (environment and EC-wide free trade) of the EC Treaty.

4. See T. Smith, "Regulatory Reform in the USA and Europe," *Journal of Environmental Law* 8:2 (1996): 263; Ekhard Rehbinder and Richard Stewart, *Environmental Protection Policy: Legal Integration in the United States and the European Community* (Berlin and New York: de Gruyter, 1985), p. 303.

5. See Directive 75/439 (waste oils), OJEC 1975, L 194, p. 23; Directive 75/440 (surface water) OJEC 1975, L 194, p. 26; Directive 75/442 (waste) OJEC 1975, L 194, p. 39; Directive 76/160 (bathing water) OJEC 1976, L 31, p. 1.

6. This is the wording of ten of the eleven official languages of the EC Treaty. The English version alone states "the potential benefits and costs of action or lack of action," since it was considered by the drafters that "benefits and costs" in the English language also included societal advantages and costs.

7. See Rehbinder and Stewart, *Environmental Protection Policy*; Mostafa Tolba and Osama El-Kholy, *The World Environment 1972–1992* (London: Chapman and Hall, 1992) Stanley Johnson and Guy Corcelle, *L'autre Europe "Verte": La Politique Communautaire de l'Environnement* (Paris-Brussels: Nathan, 1987); Commission of the European Communities, COM (80) 222 of May 7, 1980.

8. Geneva Convention on Long-range Transboundary Air Pollution of November 13, 1979, UN Doc. ECE/HLM.1/R.1.

9. See, for instance, Declaration on the Montreal Protocol on Substances that Deplete the Ozone Layer, OJEC 1988, L 297, p. 8: "the Community has competence to take action relating . . . to the environment. The Community has exercised its competence . . . in adopting . . . The Community may well exercise its competence in the future by adopting further legislation in this area."

10. Commission: Method for cooperation between the Commission and the Government of the United States in environmental matters—exchange of letters, SEC (74) 2518 of July 1, 1974.

11. For more details, see T. Smith, "Regulatory Reform in the USA and Europe," p. 257; Jerry Anderson, "US Environmental Law: the Challenge of the Next Generation," *Environmental Law Review* 9 (2000): 61.

12. Vienna Convention for the Protection of the Ozone Layer of March 22, 1985, OJEC 1988, L 297, p. 10.

13. Montreal Protocol on Substances that Deplete the Ozone Layer of September 16, 1987, OJEC 1988, L 297, p. 21, Article 2(8).

14. Regulation 925/1999, OJ 1999, L 115, p. 1.

15. Basel Convention on the Control of Transboundary Movements of Hazardous Wastes and their Disposal of March 22, 1989, UN Doc. UNEP/IG.80/3; amendment of 1995 on the export ban (not yet entered into force).

16. See UN Framework Convention on Climate Change of May 9, 1992, OJEC 1994, L 33, p. 13, Article 4(2): "The developed country parties . . . commit themselves specifically as provided for in the following: (a) Each of these Parties shall adopt national policies and take corresponding measures on the mitigation of climate change, by limiting its anthropogenic emissions of greenhouse gases and protecting and enhancing its greenhouse gas sinks and reservoirs. These policies and measures will demonstrate that developed countries are taking the lead in modifying longer-term trends in anthropogenic emissions."

17. See Jim Lofton, "Environmental Enforcement. The Impact of Cultural Values and Attitudes on Social Regulation," *Environmental Law Reporter* 31 (2001): 10906; Richard Stewart, "Antidotes for the 'American Disease,'" *Ecology Law Quarterly* 20 (1993): 85.

18. *Waldsterben* [dying forests] (Germany, early 1980s); Rainbow Warrior incident (France, 1985); Chernobyl accident (1986, Italy, Austria, Sweden); Braer accident (United Kingdom, 1993); mad cow disease (United Kingdom, 1990s); Donana accident (Spain, 1998), etc.

3

Convergence, Divergence, and Complexity in US and European Risk Regulation

Jonathan B. Wiener

National comparisons, whether of cuisine, driving habits, romance, or law, are apt to invoke archaic stereotypes and provoke hurt feelings. The renewed interest of late in comparing US and European health and environmental regulatory policies has been spurred in part by a series of transatlantic conflicts—often rather acrimonious—over trade restrictions and international treaties. Recently this discord has been compounded and largely eclipsed by the post-cold war rift regarding terrorism, security, war, and Iraq; there has been a corresponding rise in mutual nationalist antipathy to those on the other side of the Atlantic.[1] The comparative studies of environmental regulation themselves sometimes succumb to unkind stereotyping. But there is much to be gained from comparative analyses, if they can be serious, respectful, and open-minded. Differences in regulatory policies can be the source of insight rather than discord. Our goal should be constructive dialogue and mutual learning.

A prominent viewpoint these days is that US and European health and environmental policies have been diverging since roughly the 1980s, with Europe adopting more stringent regulations under the banner of the "precautionary principle," while the US resists precaution and focuses on regulatory reform (see chapters 1 and 2).[2] Evidence for this proposition includes the adoption of the precautionary principle in EU law, the more stringent European restrictions on hormones in beef and genetically modified (GM) foods, the US withdrawal from the Kyoto Protocol on climate change, the emphasis by US presidents on cost-benefit analysis of new regulations, the increasing influence of environmental organizations and parties in European regulatory politics, and the growing role of European institutions born of European integration. These are clearly important developments; both US and European environmental policies are clearly evolving, and each can learn much from the other's emerging policy experience.

I argue, however, that this picture is incomplete, and that the reality is much more dynamic and complex. The stereotype of risk-averse European precaution confronting blithe American technological optimism is hardly new; Oscar Wilde cheerfully lampooned that notion in 1887.[3] Serious students of regulatory policy can see that caricature for what it is. Scholars have long argued, for example, that US environmental law was substantially more precautionary than European environmental law in the 1970s,[4] contradicting the crude stereotype. But I dispute the claim that the situation since the 1980s has now reverted to the stereotype (though my point is not to claim that the US remains more precautionary than Europe). Today, both the US and Europe have quite active risk regulatory systems. The US has hardly ceased regulating. Both the US and Europe are often highly precautionary—and on several prominent contemporary examples, including air pollution by particulates, terrorism, and mad cow disease in blood, it is the US that is now regulating in the more precautionary manner. The reality is that the US and Europe do not diverge much—or as much as is claimed—on the general embrace of precaution in regulation. But they often do diverge on the particular question of which risks they select to worry about and regulate most. This particularized divergence in risk selection can give rise to visible conflicts.

Moreover, convergence and divergence are both concepts too simple to capture the interactive reality of transatlantic regulatory relations. The US and European regulatory systems are not large unified blocs that are racing to be more or less precautionary across the board; rather, they are multinodal webs, complex networks of multiple components that are evolving simultaneously in different ways and sharing elements with each other. Although there is divergence on some issues, there is much convergence on others, including the basic criteria for regulation (with Europe also moving to adopt cost-benefit analysis), the choice of policy instruments, and the hierarchical level of governmental authority. The reality is a process of "hybridization," in which both systems are borrowing legal concepts from each other in a complex and continuous mutual evolution.

Hazards of Hasty Comparisons

Quick and broad comparisons of national regulatory policies are fraught with peril. Recent efforts to compare US and European environmental policies illustrate these pitfalls. First, these comparisons frequently leap to macroscale conclusions from just one or a few highly visible examples of conflict, such as the recent controversies over genetically modified foods and climate change, thereby succumbing to the availability heuristic (exaggerated attention to recent crises) while failing to undertake the more serious study of a broad array of comparative data.

Second, comparisons written from one side or the other frequently commit the sins of ignorance and even disrespect of foreign law,[5] claiming that so much has happened over here while so little has happened over there, when the reality is hardly so one-sided. For example, it is not accurate to assert that Europe has enacted many important environmental measures since the 1980s while the US has done little or has retrenched (see chapter 2). The reality is that in the past two decades, while Europe was indeed adopting many important measures, the US (across the governments of both Democrat and Republican political parties) was enacting the 1984 Hazardous and Solid Waste Amendments, the 1986 Superfund Amendments (including tough cleanup standards and the path-breaking Toxics Release Inventory), the 1990 Oil Pollution Act, the 1990 Clean Air Act Amendments (including tight technology controls on air toxics, and the hugely successful national sulfur dioxide allowance trading system to combat acid rain), the 1996 Safe Drinking Water Act amendments, the 1996 Food Quality Protection Act, new laws on youth violence and public smoking, and numerous stringent agency regulations [including the 1987 Top-Down Best Available Control Technology (BACT) policy, the 1989 ban on British beef, the 1997 Ozone and PM2.5 national ambient air quality standards,[6] the 1999 and 2001 bans on European blood, the 2001 standard on arsenic in drinking water, and the 2002 standard on diesel engine emissions]. This is not to say that all of these policies have been desirable, or that countries should compete to enact more laws, or to ignore differences among presidents; it is just to say that American inactivity is not the reality. Likewise, there may have been more policy action in Europe in the 1970s than is typically recognized today. That is, after all, when the notion of precaution blossomed in German, Swedish, and Swiss environmental law.

Third, comparisons along one dimension, such as whether a particular principle (say, precaution) has been adopted in each legal system, frequently neglect the surrounding context of other principles, rules, institutions, and equivalent doctrines under other names, as well as the distinction between the law on the books and the law in action,[7] so that the comparison falsely finds divergence when the reality in toto is functional similarity. For example, the claim that American regulation is governed by cost-benefit analysis, while European regulation is not, neglects several contextual facts: that despite requirements for such analysis issued by every president since Jimmy Carter, including both Ronald Reagan and William Clinton, important areas of American regulation (such as the ambient air quality provisions of the Clean Air Act) remain statutorily immune to cost considerations;[8] that European regulatory policy often also officially espouses cost-benefit or economic analysis, as it does in the European Commission's Communication on the Precautionary Principle,[9] in the commission's 2002 action plan on improving regulation, and often in member state law;[10] and that the principle of proportionality applied in European law[11] amounts to a weighing of benefits and costs that limits the reach of the precautionary principle. Or, to take another example, countries may adopt a degree of precaution that reflects their combination of both ex ante regulation and ex post tort law remedies. Thus, criticism of US regulatory law as inadequately precautionary may neglect the active role of tort liability as a deterrent and as a backstop if preventive regulation misses new risks. Meanwhile, criticism of European regulatory law as excessively precautionary may neglect the relative absence (until recently) of strong tort law in Europe, so that ex ante regulation was the only real option.

Fourth, broad comparisons often neglect great variation within each legal system, such as among the EU member states and among the states of the US, or across different agencies and statutes within each system;

that internal variation can exceed the differences claimed across the two aggregated systems.

Fifth, broad comparisons sometimes take a snapshot of current events but overlook dynamic changes through time, not only in the past, but also into the future. Current events may seem to represent a climax or ending when in fact they are part of an ongoing transition that is difficult to perceive from within.

Sixth, compounding these factors may be the tendency, observed by social psychologists, of group members to assert judgmental distinctions between one's own group and other groups, even when the members are sorted into the groups on a wholly arbitrary basis.[12] The US and Europe may be citing contrasts that would be nearly indistinguishable to outside observers, or far less apparent than the similarities and intermingling between US and European regulatory policies. This is particularly likely with regard to relative precaution, where (if such broad depictions have any validity) both the US and Europe undoubtedly lie at the highly precautionary end of the global spectrum. Debates between the US and Europe over who is "more precautionary than thou" may look baffling and hairsplitting to the billions of people who live in countries that (compared with either the US or Europe) have less stringent environmental standards, less institutional capacity to enforce those standards, less scientific capacity to detect and warn of future risks, and much more pressing immediate crises in hunger, health, and environmental quality.

To be sure, all of these shortcomings in comparative legal analysis may be unintended. But they may also be consciously or unconsciously committed, so that the comparative description becomes less an exercise in dispassionate social science than a vehicle for the author's normative argument about what kind of law is desirable.[13] Advocates of precaution may be using the descriptive claim that Europe is now more precautionary than the US in order to pressure both systems to increase the stringency of their regulatory postures. Critics of precaution may be using the same descriptive claim, that Europe is now more precautionary, in order to warn against such a trend in the US.

Even utterly disinterested observers will find it methodologically vexing to buttress the descriptive claim that one legal system is more precautionary than the other (or not). We cannot "prove" such broad empirical claims unless we can select and compare a representative sample of policies from the population of relevant regulatory actions.[14] Citing a few cases is insufficient to support a broad systemwide claim. A rebuttal based on several contrary cases casts doubt on the initial claim, but is not necessarily sufficient to support a contrary systemwide claim. Both sets may be subject to the criticism that they are a skewed sample of the larger reality.

In short, the fundamental fact of comparative legal analysis is that things are "more complicated than you thought."[15] Broad and catchy depictions miss the true complexity and dynamism of vast and interactive social and legal systems. The same is true of regulatory policy itself; seductively simple prescriptions tend to fail when tested against the complexity of real-world systems.[16] We need caution about precaution, and about comparisons of national precaution. That does not mean, however, that we should look only at the details and never step back to see the bigger picture; on the contrary, we must look at both details and whole systems. A main problem with the recent claimed distinctions between US and European environmental policies is that they focus narrowly on one issue (such as the precautionary principle, or genetically modified foods, or climate change) and neglect the broader systems (such as the proportionality principle, tort law, and a broader sample of risks).

Convergence, Divergence, and Hybridization

Thus, on the question of whether US and EU environmental policies are converging or diverging, my answer is both and neither. US and EU environmental policies are both converging and diverging, because the reality differs in different strata of policy development and implementation. And US and EU environmental policies are neither converging nor diverging, because a better model is one of hybridization: iterative exchange of legal ideas, tools, and approaches through a process not dissimilar to interbreeding among populations in nature. Hybridization involves "legal borrowing" or "legal transplantation"[17] or cross-fertilization,[18] earlier called mimesis,[19] and more generally the diffusion of social concepts.[20] The social, cultural, or legal concepts exchanged are sometimes called memes,[21] as an analogy to the genes or traits exchanged in hybridization among populations. Hybridization in nature was long thought to be of minor evolutionary significance, but careful empirical investigations in the past few decades have revealed its widespread and often crucial role in survival, reproduction, and the emergence of new species.[22] In comparative regulatory policy, we are both observing and participating in the exchange of legal traits; we can both document and shape the process.

Hybridization can contribute to more efficient evolution than purely within-system selection pressures would. Exchange across species and across legal systems can foster success and efficiency by offering a wider array of choices; it helps diversify the portfolio of available tools and thereby helps equip the borrower to survive future challenges. Whereas within-system selection pressures leave as survivors those who have bested past environments (potentially yielding local but not overall optima), intersystem exchange creates hybrid offspring that may be better suited to surviving in the environment yet to come. Most of the hybrid offspring do not prosper while the environment is stable, but when the environment changes (as it always does), the hybrids can become the basis for successful new species and new legal approaches. Indeed, hybridization is an especially appropriate model for the evolution of environmental law, because the essence of environmental problems is interconnectedness.

As a model for contemporary legal evolution, hybridization seems considerably more realistic than convergence or divergence. Whereas convergence and divergence can both occur with no interaction between the systems, hybridization necessarily involves exchange across systems, which seems obvious in an age of globalization and international trade. Whereas convergence and divergence imply curves heading toward or away from a single point (or line) on a plane, as though legal systems had some determinate and common starting or ending points and moved in large unified blocs, hybridization implies an interactive interface between two particle clouds or webs that are continuously exchanging components across one or many planes, thereby reaching and even creating new points on an unfolding multidimensional frontier. Rather than two lines converging or diverging, one can envision two fractals interacting at many junctures as they both evolve. Whereas models of convergence or divergence depict each legal system as a discrete aggregate entity moving in one direction, a model of hybridization corresponds better to a view of legal systems as complex, disaggregated, multinodal webs or networks, with multiple actors pursuing multiple directions at once and interacting across system boundaries in many places at once.[23]

Hybridization of law (or a species) might look like convergence—the generation of a new approach shared by both systems—but it need not. Hybridization can imply a complex web of borrowings of particular features applied to different problems, institutions, and levels of government—a hodgepodge of *bricolage*[24]—that yields a diffuse and cloudy pattern rather than a tight convergence to a new line. One might observe divergence in one example, convergence in another, many aspects heading in different directions all at once. Or hybridization might give rise to a new version that is quite different from both parental approaches, and that appears during the transitional process to be divergent from both original systems.

In order to understand US and European environmental policies in this context of complexity, the Duke Center for Environmental Solutions and the European Commission's Group of Policy Advisers have undertaken a project on "The Reality of Precaution."[25] The project includes participants from both the US and Europe in order to overcome the problems of ignorance of foreign legal systems. The initial products of this effort include a series of transatlantic dialogue meetings[26] and a jointly authored research paper.[27] A central finding from this work is that the US and Europe are not diverging or flip-flopping, with Europe becoming more precautionary than the US across the board. Rather, both the US and Europe are taking a precautionary approach to the regulation of many risks, but they differ on which risks they choose to worry about and regulate most. Examples are discussed in the next section.

Comparisons at Several Strata

The complexity of both convergence and divergence between US and European environmental policies is apparent from a disaggregated analysis of several strata of the regulatory system. By dividing the analysis into component parts of the regulatory process—issue framing, risk assessment methods, risk management standards, choice of risks to regulate, choice of policy instruments, degree of integration across hazards and media, enforcement mechanisms, and hierarchical level of government—one can appreciate the more multifaceted relations between US and European environmental policies. There is both convergence and divergence, depending on the component being examined.

Issue Framing

The EU has advocated the precautionary principle in international fora, while the US (under both Bill Clinton and George W. Bush) has expressed reservations about this principle. This divergence at the level of issue framing or high rhetoric has led to frequent claims that Europe has become more precautionary than the US. The notion of precautionary regulation is not new; prominent endorsements have appeared in both Europe and the US since at least the 1970s.[28] But while US law continues to express an informal precautionary preference,[29] European law has formally adopted precaution as an overarching principle to govern risk regulation,[30] and the European Environment Agency (EEA) has published a book on the advantages of precaution.[31] The EU has championed, and the US has resisted, statements of the precautionary principle in several international treaties, such as the Cartagena Protocol on Biosafety. At the same time, the US has agreed to statements endorsing precaution in the 1992 Rio Declaration, the 1992 Framework Convention on Climate Change, and the 2001 Stockholm Convention on Persistent Organic Pollutants (all signed by Republican presidents—the two Bushes).

Today, the prominent view is that Europe endorses the precautionary principle and seeks proactively to regulate risks, while the US opposes the precautionary principle and waits more circumspectly for evidence of actual harm before regulating.[32] In 1999 the trade commissioner of the European Commission, Pascal Lamy, was quoted asserting that "in the US they believe that if no risks have been proven about a product, it should be allowed. In the EU we believe something should not be authorized if there is a chance of risk."[33] As early as 1992, a senior environmental official of the European Commission had said that the US "was definitely leading European policy back in the 1970s and early

1980s" but now "Europe has certainly managed to catch up" and on some issues "has taken over the role as world leader."[34]

Fifteen years ago, comparisons of US and European regulation found different procedural approaches but similar degrees of regulatory stringency.[35] Today leading scholars of comparative regulation are describing a "flip-flop"; in this view, the US used to be more precautionary than Europe in the 1970s, but Europe has become more precautionary than the US since the 1990s.[36] David Vogel writes: "From the 1960s through the mid 1980s, the regulation of health, safety and environmental risks was generally stricter in the United States than Europe. Since the mid 1980s, the obverse has often been the case."[37] He emphasizes that these trends "have not produced policy convergence. On the contrary, European and American regulatory policies are now as divergent as they were three decades ago. What has changed is the direction of this divergence. In a number of areas, Europe has become more risk-averse, America less so."[38] Normative evaluations of this situation vary. Some observers see a civilized, careful Europe confronting a risky, reckless, and violent America.[39] Others see a statist, technophobic, protectionist Europe challenging a market-based, scientific, entrepreneurial America.[40] But clearly there is a divergence in the rhetorical objectives of environmental regulation.

This divergence may reflect real differences in regulatory policy. Or it may reflect conclusions drawn from a few visible cases (such as GM foods), but not full characterization of the broad array of regulatory policies.[41] It may also reflect a new terrain of international rivalry after the end of the cold war.[42] Given that the US and Europe are both at the highly precautionary end of the global spectrum, and given the finding of simultaneous actual precaution when viewed across a broader set of risks (see following section), the stark claimed divergence between European precaution and US policy seems overdrawn, and the hypothesis of international rivalry seems worth taking seriously.

Risk Assessment
It has long been observed that the US takes a more formal scientific and quantitative approach to risk assessment, while the European approach is more qualitative. The US Supreme Court's *Benzene* decision requiring

the Occupational Safety and Health Administration (OSHA) to conduct a risk assessment before regulating,[43] and a 1983 guidebook from the National Academy of Sciences, spurred widespread adoption of scientific risk assessment as the basis for American risk regulation over the past two decades, while European regulation has remained more qualitative and informal.[44] Yet there are signs of convergence. In its February 2000 Communication on the Precautionary Principle, the European Commission espoused scientific risk assessment as a predicate to any invocation of the precautionary principle.[45] And the European Court of Justice has held, in a case on mad cow disease (bovine spongiform encephalopathy, BSE) quite reminiscent of *Benzene*, that member state governments may not invoke precaution to regulate risks that the commission has deemed insignificant.[46]

On the other hand, in September 2002 the European Court of First Instance issued decisions in two cases that seem to lean against the need for a risk assessment prior to adopting a regulation, *Pfizer Animal Health SA v. Council of the EU*[47] and *Alpharma Inc. v. Council of the EU*.[48] In these cases, the court held that certain antibiotics in animal feed could be banned without a full risk assessment, in the *Pfizer* case notwithstanding a recommendation against the ban by the official scientific committee, and in the *Alpharma* case without even consulting the scientific committee. The two decisions might be read as overriding the requirement of a risk assessment on the narrow ground that the bans were adopted before the European Commission published its Communication on the Precautionary Principle. If so, then for regulations adopted after February 2000, the Communication's requirement of a risk assessment may still be binding. But the court in *Pfizer* also said:

[paragraph 139] where there is scientific uncertainty as to the existence or extent of risks to human health, the Community institutions may, by reason of the precautionary principle, take protective measures without having to wait until the reality and seriousness of those risks become fully apparent.

[paragraph 142] Thus, in a situation in which the precautionary principle is applied, which by definition coincides with a situation in which there is scientific uncertainty, a risk assessment cannot be required to provide the Community institutions with conclusive scientific evidence of the reality of the risk and the seriousness of the potential adverse effects were that risk to become a reality.

[paragraph 143] [But] a preventive measure cannot properly be based on a purely hypothetical approach to the risk, founded on mere conjecture which has not been scientifically verified.

[paragraph 144] Rather, it follows from the Community Courts' interpretation of the precautionary principle that a preventive measure may be taken only if the risk, although the reality and extent thereof have not been "fully demonstrated by conclusive scientific evidence," appears nevertheless to be adequately backed up by the scientific data available at the time when the measure was taken.

These statements are confusing. To be sure, precaution must involve action under uncertainty. But since all decisions involve "situations in which there is scientific uncertainty," the court seems to be saying in paragraph 142 that a risk assessment is *never* required. The court also seems to misunderstand what a risk assessment would do, presuming that it would provide "conclusive scientific evidence," which of course is never available. Then the court holds in paragraph 144 that without such conclusive scientific evidence, the finding of risk must be "adequately backed up" by the "available" scientific data. This new standard (if it can be called that) is highly ambiguous and may generate additional litigation over the colloquial terms "adequately," "backed up," and "available," and perhaps the question of whether preliminary indications of risk qualify as "scientific data."

The court is plainly urging deference to the regulatory body's choice of the level of acceptable risk, and to the regulator's evaluation of the tradeoff between better information and delay—a deference that is familiar in US law. The European Court of First Instance also pointed out that the recommendations of the scientific committee are purely advisory and may be rejected by the commission and the council; that would usually be true under US law as well. But in the US, the courts would likely hold the agency accountable to provide a better explanation of why it set the standard where it did, and why it rejected (or ignored) the scientific committee's advice—a more reasoned and fact-based explanation than just the recitation of the generic goal of protecting public health.

Near the end of its opinion in *Pfizer*, the court mentioned that such precautionary regulations adopted before "full" scientific evidence is available are to be "provisional . . . pending the availability of additional scientific evidence" (paragraph 387). Provisionality is also required by the European Commission's Communication on the Precautionary Principle. However, it remains unclear whose burden it will be to gather such additional information, and when the regulatory body could be required to revise the regulation in light of the additional information.

Appeals will not likely be taken from the decisions in *Pfizer* and *Alpharma* to the European Court of Justice. Yet on appeal, the court could reverse the ruling, along the lines of its decision in the BSE case noted earlier, and prior decisions holding that risk assessment and consultation of scientific committees are required.[49] Or it could limit the decisions to regulations adopted before the February 2000 Communication. Or it could hold that the degree of evidence required before regulating depends, in European law as in American law, on the specific wording of the statute or directive that provides the legal basis for the regulation; in the antibiotics cases, the directive broadly authorized regulation of any antibiotic posing a "danger." Or the court could give more teeth to the provisional character of the regulations, requiring research and reconsideration by some point in time.

In chapter 1 of this volume, Theofanis Christoforou argues at some length that precaution is warranted because governmental risk assessment tends to understate risks (at least compared with public perceptions of risk). Even if this argument were correct, it would not imply that European risk policy is more precautionary. Both US and European policies respond strongly to public perceptions of risk (notwithstanding greater use of formal risk assessment in the US).[50] In any event, the bias Christoforou sees in risk assessment is only one piece of the full picture. The reasons that risk assessments may understate risks include inattention to unforeseen risks, inattention to multiple simultaneous exposures, failure to identify thresholds above which critical damage occurs, difficulty forecasting strategic risk actors such as terrorists and pathogens, and agency capture by the regulated industry. Reasons that risk assessments may overstate risks include linear extrapolation of harm at low doses (whereas low doses may actually be harmless or beneficial), conservative extrapolation from animals to humans, selection of most sensitive animal test species, assumption that harm to one organ can predict harm to other organs, conservative assumption of maximum individual exposure, excessive attention to new risks as opposed to older and more widespread risks, and the regulator's asymmetric incentive to avoid being blamed for allowing harm while not incurring blame for preventing what would not have been harmful. Taking these concerns together, it is not at all clear that risk assessments typically understate risks. It is more likely that risk assessments understate some risks and overstate others, leading to simultaneous paranoia about some risks and neglect of others. Moreover, in contrast to Christoforou's advocacy of deference to public perceptions, public perception of risk does not necessarily weight all risks more heavily than experts do. The public views some risks as more worrisome and others as less worrisome than experts do.[51] Hence deference to public perceptions may affect the distribution of risk priorities but would not obviously increase overall risk protection. Meanwhile, public perceptions of risk may not always be deserving of deference, because they may also be driven by prejudice (such as fear of unfamiliar technologies and races) that is not worthy of respect in a progressive society.[52]

Risk Management: Standard Setting

When actual regulatory policy decisions are made, the trend is toward convergence. As noted, both the US and the European Commission have now adopted risk assessment and cost-benefit analysis as basic criteria for new regulations,[53] and European law adds the closely related principle of proportionality.[54] (Oddly, in chapter 1 in this volume Theofanis Christoforou harshly criticises cost-benefit analysis, yet observes that it is frequently employed by European regulators to good ends; and the European Commission has expressly required cost-benefit analysis in its Communication on the Precautionary Principle and its Action Plan on Improving Regulation.) To be sure, these criteria are not universally applied; for example, as noted earlier, some areas of US environmental law are exempt from cost considerations, and the European Commission has invested far less in the institutional capacity needed to review regulations on cost-benefit criteria than has the US executive branch. But the trend is toward convergence. Both systems also now involve substantial public participation in standard-setting.[55] Both have adopted major environmental legislation over the past two decades, as detailed earlier; the claim that Europe has done so while the US has retrenched since the 1980s is not accurate. David Vogel, who described the transatlantic posture as a reversal of divergent approaches,[56] has more recently written of convergence in US and European regulatory approaches.[57] Similarly,

Robert A. Kagan argues that broadly speaking, the substantive environmental standards in the US and Europe are convergent.[58]

To the extent that standard-setting does differ across the Atlantic, the US may more often employ formal cost-benefit analysis, but sometimes the cost-benefit shoe is on the other foot (or shore). For example, one recent study finds that the US legal regime for air pollution control is more strict and precautionary than the German regime, in part because US law requires standards to be set without considering cost, whereas it is the German approach that applies consideration of benefits and costs under the principle of proportionality.[59] Another study finds that European regulation is less susceptible to the problems of tunnel vision (excessive regulation of minor risks) and random agenda selection that have plagued US regulation.[60]

Moreover, it is not the case (as is often assumed) that cost-benefit analysis always produces weaker regulation. Several of the examples of greater US precaution, including the phaseout of chlorofluorocarbons (CFCs) and the phaseout of lead in gasoline (both in the 1980s), were substantially motivated by cost-benefit analyses. Recently the Office of Management and Budget has initiated a series of "prompt letters" that use cost-benefit analysis to identify and recommend promising new regulations that the agencies ought to consider adopting but have not yet—using economics to spur smart regulation, not just to retard bad regulation.

Further, more precautionary regulation is not always a triumph over industry influence (agency capture). Nor is economic analysis a capitulation to industry. Sometimes industry itself seeks greater regulation for parochial gain, such as to impose costs on its trade rivals.[61]

And, if the contention were true that the use of cost-benefit analysis had led to moderating (or strengthening) some regulations, whether in the US or in Europe, that would not necessarily be unwise—indeed it might be quite sensible. (That is why the European Commission's Communication on the Precautionary Principle itself requires cost-benefit analysis as a predicate to precaution.) More precautionary policies are not always superior to policies chosen by cost-benefit balancing. Precaution may avoid the harms of inaction on false negatives (risks thought to be minor that turn out to be serious), but incur the harms of overreaction to false positives (risks thought to be serious that turn out to be minor). Both types of errors are harmful to society. The harms of ignoring false negatives include the health and environmental damage from the unrestricted risk. The harms of regulating false positives include high costs to consumers and workers, unemployment, the loss of helpful new products, restrictions on personal choices, and public cynicism about exaggerated risks (crying "wolf"). An extreme policy of zero risk would bring valuable activities to a halt; applied broadly it would be impossible. The goal is not zero false negatives, but the best balance of the two types of errors that we can achieve.

The argument that neglecting false negatives yields health damage, but that regulating false positives costs only money and therefore is worth tolerating because health matters more than money,[62] is attractive but flawed. It is flawed because the premise that regulating false positives costs only money is incorrect. Even assuming no costs and inhibitions to innovation from precautionary policies, more precautionary policies can also yield increases rather than decreases in health and environmental risks. Precaution against a target risk can induce increases in other countervailing risks.[63] Hence even ignoring cost-benefit analysis, risk tradeoff analysis is important. To mention just a few of these examples of "risk-risk tradeoffs": Airbags in cars may save adults but kill children. Banning asbestos may reduce cancers but increase highway fatalities because of inferior brake linings. Reducing ozone in smog may protect our lungs but put our skin at risk from increased ultraviolet radiation. The US Food and Drug Administration's (FDA) precautionary measures to safeguard the blood supply against mad cow disease by banning blood from Europe may reduce the availability of blood in hospital emergency rooms. Banning one pesticide (e.g., to protect food consumers from residues) may invite the use of a substitute pesticide (e.g., one that leaves less residue but that is more toxic to uninformed migrant workers). Banning all use of DDT (as opposed to banning just its use in agriculture) may increase the spread of malaria, killing millions. Banning chlorination of drinking water may foster deadly outbreaks of cholera and other microbial pathogens. Promoting fuel-efficient diesel engines to reduce greenhouse gas emissions may increase local air pollution by particulates. The war on drugs may increase inner-city violence. Police chases of fleeing suspects may kill bystanders. Suppressing forest fires may worsen these fires when they occur. And American or European precautionary policies may regressively burden poor countries. For example, wealthy country bans on genetically modified foods may perpetuate hunger in poor countries (a dilemma now facing Zambia and other famine-stricken African nations that are rejecting US offers of donated corn, apparently motivated in part by the fear that US corn might cross-pollinate Zambian corn, rendering future Zambian corn in violation of European restrictions on imports of genetically modified crops).

In short, the phenomenon of risk-risk tradeoffs is ubiquitous. Countervailing risks do not always warrant curtailing precautionary regulations, but ignoring countervailing risks in the pursuit of precaution would perversely lead to systematic increases in overall risk. Hence, even assuming zero financial costs of regulation, the ideal is not maximum precaution, but an optimal precaution that takes into account the tradeoffs among multiple risks.[64] A "race to the top" in precautionary regulation would not be wise even if all one cared about were minimizing risks. The better goal is to minimize the sum of risks and to seek "risk-superior" options that reduce multiple risks in concert.

Citing my work among others, Christoforou argues in chapter 1 that weighing the countervailing risks of a risk reduction policy is misconceived and dangerous. He gives three reasons for this view: that "voluntary exposure to risk by some must not enter into any type of balancing exercise against unintended, involuntary exposure to the same or other type of risk by other people. . . . The fact that people face multiple sources of risk in our society is not an argument in favor of an averaging or a balancing exercise"; that "the right to life and health is the most fundamental of all human rights, which implies that no restriction should in principle be placed on this right without proper consideration"; and, quoting the European Court of Justice, that "considerations of health should take precedence over economic or commercial considerations." The second two points are inapposite to risk-risk tradeoffs. Christoforou appears to conflate cost-benefit and risk-risk analyses, although they are distinct; risk-risk analysis, as noted earlier, does not weigh cost or money or economic considerations against health. Rather it weighs health against health. Regulations causing risk-risk tradeoffs do not pit the "right to life and health" against "restrictions" for other reasons; they pit some life and health interests against other life and health interests. Such a right to life and health could not be inviolate precisely because efforts to reduce target risks often incur countervailing risks. Christoforou's first point is also confusing. Risk-risk analysis does not necessarily compare voluntary and involuntary risks, nor does it seek to favor one over the other. It considers all aspects of risk-risk tradeoffs, including qualitative attributes such as voluntariness.[65] Such a tradeoff might happen to be incurred by some policies, but other policies might have different effects. Some precautionary policies would themselves violate Christoforou's rule by protecting some people from voluntary risks (e.g., food consumption choices they could avoid) while imposing involuntary risks on others (e.g., hunger in poor countries, or exposure of uninformed migrant workers to toxics). None of Christoforou's three points is actually an argument against risk-risk tradeoff analysis. Without such analysis, precautionary policies could often increase overall risk, contradicting Christoforou's interest in safeguarding life and health.

Choice of Risks

The conceptual rhetoric of greater precaution in Europe, based largely on the visible examples of food safety and climate change, does not capture the full reality of actual regulatory policies. Nor does the convergence in standard-setting approaches. Disaggregating the overall convergence in regulatory criteria, one can see differences as to particular risks, but no simple divergence in whether Europe or the US is more precautionary than the other across the board. Nor has the EU in some broad sense moved ahead of the US in relative precaution in the 1990s. The picture is more complex.

Europe appears to be more precautionary than the US on some risks, such as genetically modified foods, hormones in beef, climate change, toxic substances, phthalates, marine pollution, and guns. The US appears to be more precautionary than Europe on other risks, such as mad cow disease (especially in blood donations), air pollution by fine particulate matter (from electric power plants and motor vehicles), nuclear power, teenage drinking, cigarette smoking, hazardous waste disposal, "right to know" information disclosure requirements, youth violence, and terrorism. In the past the US had also been more precautionary regarding new drug approval (e.g., forbidding drugs such as thalidomide, which were licensed in Europe), the 1978 ban on CFCs in aerosol spray cans and the 1970s ban on supersonic transport to protect the stratospheric ozone layer (both adopted years before Europe acted to phase out CFCs), and the phaseout of lead in gasoline (petrol) (adopted years earlier than in Europe), but Europe has now converged on most of those policies.[66]

The picture that emerges is of precaution on both sides of the Atlantic, but regarding different risks. The length of these lists is not important; as discussed earlier, neither set of examples is a representative sample of the full arena and thus neither set "proves" a general characterization. Moreover, the point is not a contest to see who is "more precautionary than thou." This broader set of examples merely indicates that neither the US nor the EU can easily claim to be the more precautionary actor across the board, today or in the past. Simple contrasts, such as that Americans are risk takers while Europeans are risk averse (then how to explain tighter US restrictions on particulate matter, smoking, and BSE in blood?), or that Americans are individualistic and antiregulation while Europeans are collectivist and proregulation (then how to explain tighter US restrictions on smoking and teenage drinking?), are unsupported by the evidence of actual regulatory policies. The better view is that both legal systems are precautionary, but against different risks.

In one example from my list of divergent risk regulations, the US and Europe are simultaneously precautionary about the same technology, but in opposite directions. The US tightly regulates diesel engines to reduce human exposure to fine particulate matter,[67] while Europe promotes diesel engines to reduce carbon dioxide emissions and global warming.[68] Both policies are precautionary, but against different (and countervailing) risks.

To note another example, the US has been highly precautionary about mad cow disease.[69] It banned the import of British beef in 1989, several years before the EU adopted such a ban. The EU has since lifted its ban and sued France in the European Court of Justice to force it to lift its ban,[70] while the US ban remains in place. (Meanwhile, Europe has adopted somewhat more stringent policies than the US on the kinds of protein matter that can be fed to cattle and sheep.) In addition, in 1999 the USFDA adopted a precautionary measure that prohibits blood banks from collecting blood from donors who have spent 6 months or more in the UK, which it since has tightened to exclude donors who have spent time anywhere in Europe. This regulation is especially precautionary given that there is no evidence of transmission of the disease via blood donations, and that the regulation is estimated to reduce the supply of blood in American hospitals by a substantial amount (roughly 3 to 8 percent), raising the specter of a serious countervailing risk. France has adopted less stringent restrictions on British blood; the UK has undertaken leukodepletion on the theory that the disease agent (the prion) is more likely to be carried by certain blood cells, and it has recently begun importing blood for young children. In short, the US has been more precautionary regarding a risk of much greater impact and public concern in Europe.

Consider a third example: terrorism. In September 2002, President Bush formally announced a new doctrine of American self-defense, promising that "America will act against such emerging threats before they are fully formed. . . . The greater the threat, the greater is the risk of inaction—and the more compelling the case for taking anticipatory action to defend ourselves, even if uncertainty remains as to the time and place of the enemy's attack."[71] Similarly, in a speech at West Point in June 2002, he said: "If we wait for threats to fully materialize, we will have waited too long."[72] This US doctrine of preemptive self-defense against terrorism is, in effect, the precautionary principle applied to terrorism. In advocating precaution regarding the environment, European leaders—especially Greens—invoke the same logic that Mr. Bush has about terrorism: if we wait to confirm that the threat is real, it will be too late. The European Environment Agency advised in January 2002: "Forestalling disasters usually requires acting before there is strong proof of harm."[73] Said the EU's Environment Commissioner, Margot Wallström, in April 2002: "If you smell smoke, you don't wait until your house is burning down before you tackle the cause."[74] Likewise, nongovernmental advocates of the precautionary principle say: "Sometimes if we wait for proof it is too late. . . . If we always wait for scientific certainty, people may suffer and die, and damage to the natural world may be irreversible."[75] These are almost verbatim the reasons given by the Bush administration for its preemptive antiterrorism policy.

In response to the US call for precautionary action against uncertain threats of terrorism, German Foreign Minister (and Green Party member) Joschka Fischer worried aloud on September 14, 2002 in remarks to the UN General Assembly: "To what consequences would military intervention lead? . . . Are there new and definite findings and facts? Does the threat assessment justify taking a very high risk? . . . we are full of deep skepticism regarding military action."[76] Mr. Fischer's call for more evidence of real risk before acting, and his concern about the potential adverse consequences of action, reflect the same objections that industry often raises to calls for precautionary risk regulation.

Hence it is not that the EU endorses precaution and the US rejects precaution. The reality is that the US and Europe both endorse precaution, but regarding different risks; and each side criticizes precaution when applied to risks it discounts. Of course, one good reason for each side's worries about the other side's precautions is that, as noted earlier, there can be real countervailing risks to precaution, whether military or regulatory. Giving airline pilots guns to stop terrorists may lead to inflight accidents, theft, or misuse. Military action causes potentially devastating "collateral damage" (that is, civilian deaths), and also risks inciting reprisals (by both governments and by terrorists). Opposition to a precautionary war on terrorism can be based on these kinds of concerns about countervailing risk.

The same complexity observed with regard to the selection of target risks can be seen from the vantage of concern about countervailing risks. After years of experience with precautionary risk regulations, the US has become somewhat more attentive to the prospect of the countervailing risks that may arise from efforts to reduce target risks.[77] Countervailing risk appears to be a lesser concern in Europe, at least in terms of the official literature.[78] But in another area—the war against terrorism and against drugs—there is a parallel but opposite concern. The EU fears the countervailing risks of intervention, while the US presses ahead notwithstanding (or perhaps neglecting) those risks.[79] This again illustrates the complex pattern of simultaneous precaution but concern about different risks.

What is interesting about this complex pattern is not whether one society is more environmentalist or risk averse or morally upstanding than the other (as is sometimes implied by claims of greater precaution), but why the societies choose to worry about different risks. Several hypotheses can be advanced to answer this question.[80] The choice of which risks to regulate may derive from real differences in the serious-ness of different risks in different places. Or it may arise from different cultures and risk perceptions (including heuristic reactions to recent crises).[81] It may turn on differences in domestic political systems, such as separation of powers versus parliamentary systems, the role of third parties (including the greens), the role of nongovernmental advocacy groups, and industry pressure and rent-seeking (including international trade protectionism and domestic trade rivalry). It may relate to differ-ent background legal systems, including the role of ex post tort law. It may spring from changing positions in global strategy.[82] But to fit the observed complex pattern, any or all of these explanations would have to predict heterogeneous policy choices in both the US and Europe, not a simple contrast between all US and all European policies. Identifying the probative explanatory variables driving the observed complex pattern of relative precaution is a prime question for further research.

Choice of Policy Instruments

In the past there had been some divergence between the US and Europe in the choice of policy instruments, but the future portends increasing convergence. Both the US and Europe had employed best available technology (BAT) approaches for many years. But the US had made increasing use of emissions trading (tradable permit) policies to deal with problems, including lead in gasoline, CFCs, acid rain, land development, and water pollution, while Europe had not; and Europe had made greater use of emissions taxes (charges) than had the US.[83] Of late there appears to have been some convergence, especially as the EU has made greater use of emissions trading—in particular to control greenhouse gas emissions under the Kyoto Protocol.[84] But the US has not yet begun to make widespread use of emissions taxes.

It should be noted here that the use of economic incentives is not a move to favor economic interests over environmental interests. In fact, industry often resists the use of taxes or emissions trading because those instruments (unlike technology standards) force industry to pay for every residual unit of emissions (either as a tax levy or as the earnings forgone from not selling a permit). Nor is the advocacy of market-based instruments based on the premise that the market can solve all environmental problems; it is rather an effort to correct what are recognized to be market failures by adopting government policies that reconstitute incentives in environmentally desirable directions. Moreover, the choice of instruments, such as economic or market-based incentives, is distinct from the choice of the level of environmental protection to be achieved. One can employ economic incentives to achieve quite stringent, precautionary goals.

Information disclosure is an instrument that has been used more frequently in the US than in Europe.[85] In addition to the powerful "discovery" procedures in American civil litigation, the US has enacted several powerful information policies, including the 1966 Freedom of Information Act, the environmental impact statement (EIS) requirements of the National Environmental Policy Act (NEPA) in 1969, the 1986 enactment of the national Toxics Release Inventory (TRI) and of California's Proposition 65, and the facility accident scenario requirements of Clean Air Act section 112r adopted in 1990. In turn, Europe has recently been moving to bolster its information disclosure policies through CEC Directive 1990/313/EEC on access to information from member states, the 1998 Aarhus Convention, Regulation (EC) 1049/2001 of May 30, 2001 on access to information from EU institutions, the new European Pollutant Emissions Registry created in 2000 to be operational by 2003, and the pending Draft Protocol on Pollutant Release and Transfer Registers to be finalized at the United Nations Economic Commission for Europe Ministerial Environmental Conference in Kiev in 2003.[86]

Degree of Integration across Hazards and Media
Environmental regulation in the US is highly fragmented, with many different agencies implementing many different statutes to address

different risks. Even within the EPA, there are separate fiefdoms for air, water, and waste.[87] This fragmentation contributes to cross-media and cross-pollutant shifts, frustrating effective regulation.[88] "Integrated pollution control" (IPC) is the effort to deal with multiple risks more holistically, to ensure actual environmental improvement.[89] Since the early 1990s, the UK has made significant efforts to adopt integrated pollution control, especially in its 1990 and 1995 Environmental Protection Acts and its creation of an integrated pollution control agency.[90] The UK approach has since been borrowed by other countries in Europe and by EU institutions.[91]

Enforcement Mechanisms
The "style" of US and European regulation has long been said to diverge. The US regulatory system is seen as highly legalistic and adversarial, with a strong role for decentralized decision-making in courts (both in the review of regulation and in the application of tort law).[92] Regulatory authority in the US is more fragmented than European regulatory authority, with multiple agencies, courts, committees, and levels of government all having a hand in (and offering opportunities for public input into) policy development.[93] The European regulatory style is seen as more cooperative, hierarchical, and centralized.[94] Even when substantive standards are equivalent, the procedural approaches diverge significantly.[95] American adversarial legalism yields greater opportunities for formal public input and transparency, but also greater delay and antagonism; the European approach invites more negotiation of policy development between government and regulated businesses.[96]

This difference in style reflects the long-standing American mistrust of concentrated power, in both government and business.[97] The US Constitution has few principles obligating the government to act; it speaks of limited government powers and of individual rights to block the government. Mistrust of government power may itself be a reason for American reluctance to embrace the precautionary principle as a formal principle, while European legal culture may be more comfortable with principles of obligatory regulatory action.

The American reliance on courts, both to enforce regulations at the behest of citizen suits and to award compensation to tort victims, may also help explain the disagreement between US and European officials over adoption of an overarching precautionary principle. Knowing that the adversarial US legal system would enforce such a principle more vigorously than European law, US officials may resist agreeing to a principle that would be more costly in the US than elsewhere. And knowing that the US tort system is there to remedy injuries when they occur (and thereby deter future injuries), US officials may feel less need to adopt highly precautionary ex ante regulation. By contrast, European officials may worry less about vigorous and rigid enforcement of precaution, while they may feel they need it more because they lack as robust a tort system.

There are some signs of convergence regarding the style of enforcement. Europe is becoming more formal and legalistic, inviting greater participation by interest groups in policy formulation, in part as a consequence of the integration of European institutions and rise of power in Brussels.[98] European public trust in government and scientists has declined in the wake of several food safety crises, including mad cow disease, thereby prompting greater demands for regulatory transparency and accountability.[99] Meanwhile, American regulation is becoming less adversarial and more cooperative through the use of regulatory negotiation, alternative compliance agreements, habitat conservation plans, and Dutch-style environmental covenants.[100]

Hierarchical Level of Government

There had been divergence between the US and Europe on the hierarchical or vertical level of government responsible for environmental regulation. US policy had moved toward a strong role for the federal government (although federal standards are often implemented by the states), while in Europe the competency of the European Commission to address environmental issues took time to establish, and the principle of subsidiarity still left most decisions in the hands of member state and provincial governments. But now there may be signs of some convergence, as the EU centralizes toward a stronger role for the Commission in Brussels and as the US decentralizes toward a greater role for the states.[101]

Hybridization in Action

The foregoing analysis suggests that one cannot characterize the entirety of US and European environmental policies by either convergence or divergence; both are occurring, but differently in different strata of policy development and implementation. A better model to depict current dynamics, as argued earlier, is hybridization: the exchange of legal concepts across systems.

Examples of such borrowing in environmental policy abound. From the US, Europe has borrowed approaches to emissions trading;[102] cost-benefit analysis, and executive oversight of the regulatory system;[103] product liability[104] and the proposed liability directive; increasing "federal" oversight of environmental policy;[105] information disclosure instruments, including environmental impact assessment (EIA) and toxics release registries;[106] and other measures.

Meanwhile, from Europe, the US has borrowed the Dutch method of environmental covenants and related approaches to voluntary negotiated agreements,[107] and the concept of precaution itself (which originated as *Vorsorgeprinzip* in German law and was later adopted in the noted US case *Ethyl Corp.*).[108]

These examples of hybridization occur in both converging and diverging strata of law. Hybridization is visible in converging strata, such as criteria for standard-setting (benefit-cost), choice of policy instruments (taxes and trading), and hierarchical allocation of authority (federalism/subsidiarity). It is also visible in divergent strata, such as enforcement style (bringing voluntary agreements to the US and litigation to Europe). (See chapters 4–6.)

Additional examples of transatlantic borrowing are undoubtedly under way; for example, Europe may borrow American methods of judicial review and notice and comment rule-making,[109] and the US may borrow from European experience with watershed management and with subsidiarity. Continuing transatlantic dialogue would also be desirable on the meaning, value, improvement, borrowing, and reconciliation of decision-making approaches, such as the precautionary principle, proportionality, and cost-benefit analysis. For example, it would be useful to compare US Executive Order 12866 on Regulatory Planning and Review with the European Commission's Communication on the Precautionary Principle, both on paper and in practice. There may be more room for agreement here than has so far been recognized. Climate change offers another potential arena for hybridization. Judged on cost-benefit criteria, the US should be somewhat more precautionary than its current posture (though not as precautionary as the Kyoto Protocol targets), while Europe should accept US proposals on robust use of market-based incentives and fully global participation.[110] This path would represent a better mixture of the US and European positions than either has advocated to date.

As discussed here, hybridization is not necessarily the same as convergence. Hybridization involves exchange, but it is more complex and dynamic than convergence or divergence. It can yield new offspring that diverge from both parents. And it may be difficult to discern when one is in the midst of its unfolding. Yet it offers both sides an opportunity to reduce acrimony, to study the complex reality, and to learn from each other. We are both observing and shaping the unfolding evolution of our regulatory policies; we can participate in the process of hybridization.

Further research is warranted on why hybridization occurs when and how it does. Why are some legal concepts borrowed and not others? How is this process stimulated or inhibited? How does it relate to convergence and divergence? Hybridization is probably spurred by several factors. The integrating world economy offers greater opportunities for exchange of ideas and counterpart experiences, and at the same time it puts pressure on national regulators to harmonize standards.[111] Transnational networks of environmental NGOs and policy entrepreneurs spread legal ideas,[112] and multinational corporations spread environmental management practices to their foreign operations.[113] Furthermore, government officials, academics, nongovernmental actors, and businesses are all engaged in a process of learning by doing, in which successful innovations in one place can be imitated in other places (and failures can be avoided).

Conclusion

Claims that US and European environmental policies are converging or diverging miss the more complex and more interesting reality. Viewed across several strata of policy development and implementation, there are areas of divergence (such as the issue-framing rhetoric of precaution, the formality of risk assessment, the choice of particular risks to regulate, and the style of legal enforcement), and areas of convergence (such as the substantive criteria for standard-setting, the choice of policy instruments, and the hierarchical level of authority). Viewed across the array of risks, both the US and Europe are precautionary about many risks, but they differ primarily on which risks they select to worry about and regulate most. Neither Europe nor the US appears to be categorically more precautionary than the other across the board. Nor would it be desirable for the US and Europe to race to be ever more precautionary on all fronts, given the costs and countervailing risks of precautionary interventions. The reality is a complex pattern of diverse relative precaution across risks; the interesting question is why different societies are choosing different risks to worry about and regulate most. And the reality is a dynamic pattern of legal hybridization, with interactive exchange of legal concepts occurring continuously among the multiple nodes of these two vast legal system networks. These patterns indicate a process of mutual legal borrowing, from which we can learn a great deal, and to which we can contribute—if we undertake our comparative analyses with seriousness and mutual respect.

Acknowledgments

The author thanks Francesca Bignami, Mark Cantley, Donald Elliott, Elizabeth Fisher, George Gray, James Hammitt Donald Horowitz, Bruce Jentleson, Robert Keohane Andreas Kraemer, Jonathan Losos, Ralf Michaels, Michael Rogers, Peter Sand, Richard Stewart, Cass Sanstein, Norman Vig, David Vogel, Katrina Wyman, and an anonymous referee for helpful comments, Reed Clay, Emily Schilling, and Leah Russin for research assistance, and Joan Ashley and Ann McCloskey for editorial assistance. A substantially expanded treatment of the issues raised in this chapter will appear as Jonathan B. Wiener, "Whose Precaution After All? A Comment on the Comparison and Evolution of Risk Regulatory Systems," *Duke Journal of Comparative and International Law* 13 (2003): 207–262.

Notes

1. See Timothy Garton Ash, "Anti-Europeanism in America," *New York Review of Books*, February 13, 2003 pp. 32–34, also available at http://www.nybooks.com/articles/16059.

2. See also the sources cited in notes 27–34.

3. See Oscar Wilde, "The Canterville Ghost," *Court and Society Review*, February 23, 1887. Reprinted in *The Short Stories of Oscar Wilde* (New York: Heritage Press, 1968), p. 123.

4. See David Vogel, *Ships Passing in the Night: The Changing Politics of Risk Regulation in Europe and the United States*, Working paper 2001/16, Robert Schuman Center for Advanced Studies (Florence, Italy: European University Institute, 2001).

5. John C. Reitz, "How To Do Comparative Law," *American Journal of Comparative Law* 46 (1998): 617–636.

6. See *Whitman v. American Trucking Associations (ATA)*, 531 U.S. 457 (2001).

7. Reitz (1998) in note 5.

8. *Whitman v. ATA*, 531 U.S. 457 (2001).

9. Commission of the European Communities, *Communication from the Commission on the Precautionary Principle*, COM (2000) 1, Brussels, February 2, 2000 (available at http://europa.eu.int/comm/dgs/health_consumer/library/pub/pub07_en.pdf).

10. Peter H. Sand, "The Precautionary Principle: A European Perspective," *Human and Ecological Risk Assessment* 6 (2000): 445, 448.

11. See Nicholas Emiliou, *The Principle of Proportionality in European Law: A Comparative Study* (The Hague: Kluwer Law International, 1996).

12. Henri Tajfel, "Experiments in Intergroup Discrimination," *Scientific American* 223 (November 1970): 96–102; Donald L. Horowitz, *Ethnic Groups in Conflict*, 2nd ed. (Berkeley: University of California Press, 2000), pp. 144–147.

13. See Konrad Zweigert and Hein Kötz, *An Introduction to Comparative Law*, 3rd ed. (New York: Oxford University Press, 1998), p. 32; Hiram E. Chodosh, "Comparing Comparisons: In Search of Methodology," *Iowa Law Review* 84 (1999): 1025–1128.

14. See Gary King, Robert O. Keohane, and Sidney Verba, *Designing Social Inquiry: Scientific Inference in Qualitative Research* (Princeton, N.Y.: Princeton University Press, 1994).

15. David Kennedy, "New Approaches to Comparative Law: Comparativism and International Governance," *Utah Law Review* (1997): 545, 605.

16. Jonathan B. Wiener, "Precaution in a Multirisk World," in Dennis Paustenbach, ed., *Human and Ecological Risk Assessment: Theory and Practice* (New York: Wiley, 2002), pp. 1509–1531.

17. Alan Watson, *Legal Transplants: An Approach to Comparative Law*, 2nd ed. (Athens: University of Georgia Press, 1993).

18. See John Bell, "Mechanisms for Cross-Fertilization of Administrative Law in Europe," in Jack Beatson and Takis Tridimas, eds., *New Directions in European Public Law* (Oxford: Hart, 1998), p. 147 (describing cross-fertilization as a process of external stimulus and responsive adaptation, not necessarily involving full transplantation of a doctrine from one legal system into another).

19. A. J. Toynbee, *A Study of History: Reconsiderations* (London: Oxford University Press, 1961), p. 343.

20. Torsten Hägerstrand, "The Diffusion of Innovations," *International Encyclopedia of Social Sciences*, Vol. 4 (New York: Macmillan, 1968), p. 194.

21. Richard Dawkins, *The Selfish Gene* (New York: Oxford University Press, 1976); Robert Aunger, *The Electric Meme: A New Theory of How We Think* (New York: Free Press, 2002).

22. M. L. Arnold, *Natural Hybridization and and Evolution* (Oxford: Oxford University Press, 1997); Peter R. Grant, *Ecology and Evolution in Darwin's Finches*, 2nd ed. (Princeton, N.Y.: Princeton University Press, 1999); Peter R. Grant and B. Rosemary Grant, "Hybridization in Bird Species," *Science* 256 (1992): 193–197; Peter R. Grant and B. Rosemary Grant, "Speciation and Hybridization in Island Birds," *Philosophical Transactions of the Royal Society of London* 350 (1996): 765–772; Dolph Schluter, *The Ecology of Adaptive Radiation* (Oxford: Oxford University Press, 2000).

23. Hybridization therefore comports better (than convergence or divergence) with models of the "disaggregated state" and transnational and transgovernmental networks for the exchange of policy ideas. See Robert O. Keohane and Joseph S. Nye, "Transgovernmental Relations and International Organizations," *World Politics* 27 (1974): 39, 43; Anne-Marie Slaughter, "The Real New World Order," *Foreign Affairs* (September–October 1997): 183, 184.

24. Mark Tushnet, "The Possibilities of Comparative Constitutional Law," *Yale Law Journal* 108 (1999): 1225–1305.

25. Duke Center for Environmental Solutions, "The Reality of Precaution" project, information available at
http://www.env.duke.edu/solutions/precaution_project.html.

26. Ibid.

27. Jonathan B. Wiener and Michael D. Rogers, "Comparing Precaution in the US and Europe," *Journal of Risk Research* 5 (2002): 317–349.

28. See Sonja Boehmer-Christiansen, "The Precautionary Principle in Germany—Enabling Government," in Tim O'Riordan and James Cameron, eds., *Interpreting the Precautionary Principle* (London: Cameron and May, 1994), p. 33; European Environment Agency, *Late Lessons From Early Warnings: The Precautionary Principle 1896–2000*, Environmental Issue Report No. 22 (Luxembourg: Office for Official Publications of the European Communities, 2001); *Ethyl Corp. v. EPA*, 541 F. 2d 1 (D.C. Cir. 1976); *TVA v. Hill*, 437 U.S. 153 (1978).

29. John S. Applegate, "The Precautionary Preference: An American Perspective on the Precautionary Principle," *Human and Ecological Risk Assessment* 6 (2000): 413–443.

30. *Treaty of Amsterdam Amending the Treaty on European Union, the Treaties Establishing the European Communities and Certain Related Acts*, October 2, 1997, Article 174, OJ (C 340) [formerly Single European Act, Article 130R (1987), as amended by the Maastricht Treaty, 1992], *International Legal Materials* 31: 247 (providing that EU environmental policy shall be "based on the precautionary principle"); see also Commission of the European Communities (2000) in note 9 (elaborating on terms of Article 174).

31. European Environment Agency (2001) in note 28.

32. Suzanne Daley, "More and More, Europeans Find Fault with US: Wide Range of Events Viewed as Menacing," *New York Times*, April 9, 2000, p. A1 [citing widespread European fear of the US as violent (e.g., guns, death penalty), profit-driven, heartless (lets poor go without medical insurance), imperialist (forcing its military, culture, and products on others)]; Donald McNeil, Jr., "Protests on New Genes and Seeds Grow More Passionate in Europe," *New York Times*, March 14, 2000, pp. A1, A10 (quoting Pierre Lellouche, French Parliament environment committee, "The general sense here is that Americans eat garbage food, that they're fat, that they don't know to eat properly."); Stephan-Gotz Richter, "The U.S. Consumer's Friend," *New York Times*, September 21, 2000, p. A31 (lauding "the advantages of European intervention in everything from . . . antitrust policy to food safety" and charging that "the American government is inclined toward allowing industry to regulate itself"); David L. Levy and Peter Newell, "Oceans Apart? Business Responses to Global Environmental Issues in Europe and the United States," *Environment*, November 2000, pp. 9, 10 (describing the "conventional wisdom" that "Europeans demonstrate their considerable concern about environmental issues" while "people in the United States are more individualistic, more concerned about their lifestyles than about the environment, and more ideologically averse to regulation"); Willett Kempton and Paul P. Craig, "European Perspectives on Global Climate Change," *Environment*, April 1993, pp. 16–20, 41–45 (arguing that Europeans are more concerned than Americans about environmental impacts on future generations and on developing countries, and more likely to invoke caution regarding unforeseen risks; and that Americans are more concerned about the economic costs of regulation and more optimistic about future technological solutions to environmental problems).

33. Steve Charnovitz, "The Supervision of Health and Biosafety Regulation by World Trade Rules," *Tulane Environmental Law Journal* 13 (2000): 271, 295 n.181.

34. Jorgen Henningsen, "The Seven Principles of European Environmental Policies," in *Toward a Transatlantic Environmental Policy* (Washington D.C.: European Institute, 1992), pp. 25–26.

35. Ronald Brickman, Sheila Jasanoff, and Thomas Ilgen, *Controlling Chemicals: The Politics of Regulation in Europe and the United States* (Ithaca, N.Y.: Cornell University Press, 1986); David Vogel, *National Styles of Regulation: Environmental Policy in Great Britain and the United States* (Ithaca, N.Y.: Cornell University Press, 1985).

36. Vogel (2001) in note 4.

37. Ibid., p. 1.

38. Ibid., p. 31.

39. Richter (2000) in note 32.

40. John Redwood, *Stars and Strife: The Coming Conflicts Between the USA and the European Union* (New York: Palgrave Macmillan, 2000).

41. Wiener and Rogers (2002) in note 27.

42. Ivo Daalder, "Are the United States and Europe Heading for Divorce?" *International Affairs* 77 (2001): 531–545; Robert Kagan, *Of Paradise and Power: America and Europe in the New World Order* (New York: Knopf, 2003).

43. *Industrial Union Dept., AFL-CIO v. American Petroleum Institute*, 448 U.S. 607 (1980).

44. Jasanoff (1986) in note 35; Sheila Jasanoff, "Contingent Knowledge: Implications for Implementation and Compliance," in Edith Brown Weiss and Harold Jacobson, eds., *Engaging Countries: Strengthening Compliance with International Environmental Accords* (Cambridge, Mass.: MIT Press, 1998).

45. Commission of the European Communities (2000) in note 9.

46. *Commission of the European Communities (CEC) v. French Republic*, European Court of Justice, Case C-1/00 (Failure of a Member State to fulfil its obligations—Refusal to end the ban on British beef and veal), decided December 13, 2001.

47. *Pfizer Animal Health SA v. Council of the EU*, Case T-13/99, 2002 WL 31337, European Court of First Instance, September 11, 2002.

48. *Alpharma Inc v. Council of the EU*, Case T-70/99, 2002 WL 31338, European Court of First Instance, September 11, 2002.

49. See the *Angelopharm*, *Cassis de Dijon*, *German Beer*, and *Danish Bottle* cases, cited in chapter 1.

50. See Stephen G. Breyer, *Breaking the Vicious Circle* (Cambridge, Mass.: Harvard University Press, 1993); Cass R. Sunstein and Timor Kuran, "Avail-

ability Cascades and Risk Regulation," *Stanford Law Review* 51 (1999): 683–768.

51. See Breyer (1993) in note 50; Jonathan Baert Wiener, "Risk in the Republic," *Duke Environmental Law and Policy Forum* 8 (1997): 1–21; Jonathan Baron, *Judgment Misguided: Intuition and Error in Public Decision Making* (New York: Oxford University Press, 1998); Ann Bostrom, "Risk Perception: 'Experts' vs. 'Lay People'," *Duke Environmental Law and Policy Forum* 8 (1997): 101–113; Howard Margolis, *Dealing with Risk: Why the Public and the Experts Disagree on Environmental Issues* (Chicago: University of Chicago Press, 1996).

52. See Frank B. Cross, "The Subtle Vices Behind Environmental Values," *Duke Environmental Law and Policy Forum* 8 (1997): 151–171.

53. See Commission of the European Communities (2000) in note 9; Executive Order 12866: Regulatory Planning and Review, *Federal Register* 58: 51735–51744 (signed September 30, 1993; published October 4, 1993).

54. Emiliou (1996) in note 11.

55. David Vogel, "Risk Regulation in Europe and the United States," *Yearbook of European Environmental Law*, Vol. 3 (New York: Oxford University Press, forthcoming 2003).

56. Vogel (2001) in note 4.

57. Vogel (2002) in note 55.

58. Robert A. Kagan and Lee Axelrad, *Regulatory Encounters* (Berkeley: University of California Press, 2000), pp. 2–3. 376–377. Robert A. Kagan, professor at the University of California at Berkeley, is not the same person as Robert Kagan, author of *Paradise and Power*, cited in note 42.

59. John P. Dwyer, Richard W. Brooks, and Alan C. Marco, "The Air Pollution Permit Process for US and German Automobile Assembly Plants," in Kagan and Axelrad (2000) in note 58, pp. 206–208.

60. Stephen Breyer and Veerle Heyvaert, "Institutions for Regulating Risk," in Richard L. Revesz, Philippe Sands, and Richard B. Stewart, eds., *Environmental Law, the Economy, and Sustainable Development: The United States, the European Union and the International Community* (Cambridge: Cambridge University Press, 2000), pp. 308–309.

61. See Ann P. Bartel and Lacy G. Thomas, "Predation through Regulation: The Wage and Productivity Impacts of OSHA and EPA," *Journal of Law and Economics* 30 (1987): 239–264; Jonathan B. Wiener, "On the Political Economy of Global Environmental Regulation," *Georgetown Law Journal* 87 (1999): 749–794.

62. The argument in favor of precaution on the ground that health matters more than money was made by Talbot Page, "A Generic View of Toxic Chemicals and Similar Risks," *Ecology Law Quarterly* 7 (1978): 207–244. A more recent argument along these lines is Mark Geistfeld, "Implementing the Precautionary

Principle," *Environmental Law Reporter* 31 (2001): 11326–11333; Mark Geistfeld, "Reconciling Cost-Benefit Analysis with the Principle that Safety Matters More than Money," *NYU Law Review* 76 (2001): 114–189.

63. John D. Graham and Jonathan Baert Wiener, *Risk vs. Risk: Tradeoffs in Protecting Health and the Environment* (Cambridge, Mass.: Harvard University Press, 1995).

64. Wiener (2002) in note 16.

65. See Graham and Wiener (1995) in note 63, chap. 1.

66. Wiener and Rogers (2002) in note 27; Jonathan B. Wiener, "Whose Precaution After All?" *Duke Journal of International and Comparative Law* 13 (2003): 207–262; see also Duke Center in note 25.

67. Katharine Q. Seelye, "Administration Approves Stiff Penalties for Diesel Engine Emissions, Angering Industry," *New York Times*, August 3, 2002, p. A9.

68. Diesel Technology Forum, "Demand for Diesels: The European Experience" (July 2001), available at www.dieselforum.org.

69. Wiener and Rogers (2002) in note 27.

70. *Commission of the European Communities v. French Republic* (2001) in note 46.

71. "The National Security Strategy of the United States of America," September 17, 2002 (available at http://www.whitehouse.gov/nsc/nss.html, visited September 22, 2002), introduction and part V.

72. "Remarks by the President at 2002 Graduation Exercise of the United States Military Academy," West Point, N.Y., June 1, 2002 (available at http://www.whitehouse.gov/news/releases/2002/06/20020601-3.html, visited September 22, 2002).

73. European Environment Agency (2001) in note 28, see section 1.2.

74. Margot Wallstrom, "US and EU Environmental Policies: Converging or Diverging?" Speech to the European Institute, Washington D.C., April 25, 2002.

75. Science and Environmental Health Network, "Frequently Asked Questions," at http://www.sehn.org/ppfaqs.html, visited September 22, 2002.

76. Address by Joschka Fischer, Minister for Foreign Affairs of the Federal Republic of Germany, at the Fifty-seventh Session of the United Nations General Assembly, New York, September 14, 2002, available at http://www.auswaertiges-amt.de/www/en/aussenpolitik/index_html.

77. Graham and Wiener (1995) in note 63; Wiener (2002) in note 16.

78. European Environment Agency (2001) in note 28 (reviewing false negatives but neglecting to discuss any false positives).

79. Kagan (2002) in note 42.

80. For an expanded discussion of these hypotheses, see Wiener and Rogers (2002) in note 27.

81. Ortwin Renn and Bernd Rohrmann, eds., *Cross-Cultural Risk Perception: A Survey of Empirical Studies* (Dordrecht, the Netherlands: Kluwer, 2000); Mary Douglas and Aaron Wildavsky, *Risk and Culture: An Essay on the Selection of Technical and Environmental Dangers* (Berkeley: University of California Press, 1982).

82. Kagan (2002) in note 42 (the US and Europe "disagree about what constitutes a threat . . . [they] differ most these days in their evaluation of what constitutes a tolerable versus an intolerable threat").

83. Jonathan Golub, "New Instruments for Environmental Policy in the EU: Introduction and Overview," in J. Golub, ed., *New Instruments for Environmental Policy in the EU* (London: Routledge, 1998), pp. 4–24; Jos Delbeke and Hans Bergman, "Environmental Taxes and Charges in the EU," in Golub (1998), pp. 242–260; Richard B. Stewart, "Environmental Law in the United States and the European Community: Spillovers, Cooperation, Rivalry, Institutions," *University of Chicago Legal Forum* 41 (1992): 75–80.

84. Commission of the European Communities, *Proposal for a Directive of the European Parliament and of the Council Establishing a Scheme for Greenhouse Gas Emission Allowance Trading Within the Community and Amending Council Directive 96/61/EC*, COM (2001) 581 final, Brussels, October 23, 2001.

85. Sand (2000) in note 10; Sand, "The Reality of Precaution: Information Disclosure by Government and Industry," paper presented at the Transatlantic Dialogue on the Reality of Precaution: Comparing Approaches to Risk and Regulation, Airlie House, Virginia, June 15, 2002, available at http://www.env.duke.edu/solutions/precaution_conference.html#Agenda.

86. Sand (2002) in note 85.

87. Alfred Marcus, EPA's "Organizational Structure," *Law and Contemporary Problems* 54 (1991): 5–40.

88. Graham and Wiener (1995) in note 63.

89. Lakshman Guruswamy, "The Case for Integrated Pollution Control," *Law and Contemporary Problems* 54 (1991): 41–56; Nigel Haigh and Frances Irwin, eds., *Integrated Pollution Control in Europe and North America* (Washington, D.C.: Conservation Foundation, 1990).

90. Albert Weale, "Environmental Regulation and Administrative Reform in Britain," in Giandomenico Majone, ed., *Regulating Europe* (London and New York: Routledge, 1996), p. 106; Michael Purdue, "Integrated Pollution Control in the Environmental Protection Act 1990: A Coming of Age of Environmental Law?" *Modern Law Review* 54 (1991): 534–551; Neil Carter and Philip Lowe, "The Establishment of a Cross-Sector Environment Agency," in T. S. Gray, ed., *UK Environmental Policy in the 1990s* (Basingstoke, UK: Macmillan, 1995), p. 38.

91. Chris Backes and Gerrit Betlem, eds., *Integrated Pollution Prevention and Control: The EC Directive from a Comparative Legal and Economic Perspective* (The Hague and Boston: Kluwer Law International, 1999); Johannes Zottl,

"Towards Integrated Protection of the Environment in Germany?" *Journal of Environmental Law* 12 (2000): 281–291.

92. Kagan and Axelrad (2000) in note 58.

93. Ibid., pp. 12–13.

94. Vogel (1985) in note 35; Jasanoff (1986) in note 35; Kagan and Axelrad (2000) in note 58, pp. 11–13.

95. Kagan and Axelrad (2000) in note 58, pp. 3, 23. Compare Donald L. Horowitz, "The Qu'ran and the Common Law: Islamic Law Reform and the Theory of Legal Change," *American Journal of Comparative Law* 42 (1994): 543–580 (observing that procedural law is often slower to converge across countries than substantive law because practicing lawyers cling to the procedure they know).

96. Kagan and Axelrad (2000) in note 58, pp. 23, 404–405.

97. Kagan and Axelrad (2000) in note 58, pp. 10, 13; John C. Reitz, "Standing to Raise Constitutional Issues," *American Journal of Comparative Law* 50 (2002): 437, 457; Richard B. Stewart, "A New Generation of Environmental Regulation?" *Capitol University Law Review* 29 (2001): 21, 85–86.

98. Vogel (2002) in note 55; Kagan and Axelrad (2000) in note 58, pp. 14–15.

99. Ragnar Lofstedt and David Vogel, "The Changing Character of Consumer and Environmental Regulation: A Comparison of Europe and the United States," *Risk Analysis* 21 (June 2001): 399–416.

100. Golub (1998) in note 83; but see Stewart (2001) in note 97 (doubting how far the US will go in this direction).

101. Breyer and Heyvaert (2000) in note 60.

102. Golub (1998) in note 83; Commission of the European Communities (2001) in note 84.

103. Commission of the European Communities (2000) in note 9.

104. Mathias Reiman, "The End of Comparative Law as an Autonomous Subject," *Tulane European and Civil Law Forum* 11 (1996): 49, 62.

105. Eckhard Rehbinder and Richard B. Stewart, *Integration Through Law: Europe and the American Federal Experience*. Vol. 2, *Environmental Protection Policy* (Berlin and New York: de Gruyter, 1985); Breyer and Heyvaert (2000) in note 60.

106. Sand (2002) in note 85.

107. Golub (1998) in note 83; Stewart (2001) in note 97.

108. See Boehmer-Christiansen (1994) in note 28; Sand (2000) in note 10; *Ethyl Corp. EPA*, 541 F. 2d 1 (D.C. Cir. 1976).

109. Martin Shapiro, "The Giving Reasons Requirement," *University of Chicago Legal Forum* 1992: 179–220; Francesca E. Bignami, "The Democratic Deficit in European Community Rulemaking: A Call for Notice and Comment in Comitology," *Harvard International Law Journal* 40 (1999): 451–515.

110. Richard B. Stewart and Jonathan B. Wiener, *Reconstructing Climate Policy: Beyond Kyoto* (Washington, D.C.: American Enterprise Institute, 2003).

111. Kagan and Axelrad (2000) in note 58, pp. 2–3.

112. Mark A. Pollack and Gregory C. Shaffer, eds., *Transatlantic Governance in the Global Economy* (Lanham, Md.: Rowman & Littlefield, 2001); Nicholas A. Robinson, "Introduction," in Nicholas Robinson, ed., *Comparative Environmental Law and Regulation* (Dobbs Ferry, N.Y.: Oceana Publications, 1997), pp. v, xiii.

113. Levy and Newell (2000) in note 32, pp. 17–18; Ronie Garcia-Johnson, *Exporting Environmentalism* (Cambridge, Mass.: MIT Press, 2001).

II

Regulatory Trends: Institutional and Policy Innovations

4

Environmental Federalism in the United States and the European Union

R. Daniel Kelemen

Both the United States and the European Union practice environmental federalism. While the EU lacks important attributes of a fully fledged federal state, it operates as a federal system in the area of environmental regulation. As federal systems, the US and EU have confronted a number of common challenges. Both polities have faced choices regarding the allocation of regulatory authority between the federal and state governments. Similarly, both polities have faced choices regarding the relationships between federal and state governments. In particular, where state governments play a role in implementing federal law, federal authorities have had to establish mechanisms to ensure that states fulfill their regulatory obligations.

At first blush, it appears that the US and EU have addressed environmental policy in extremely different ways. The staff and budget of US federal environmental agencies dwarf those of their EU counterparts; the US government owns huge tracts of land and plays a major role in resource management; and US regulators possess potent regulatory tools that EU regulators lack, such as the power to preempt any state role in a policy area and the power to sue polluters directly. While such differences are indeed significant, they obscure fundamental similarities between patterns of environmental regulation in the US and EU. In both cases, the federal governments have established a powerful role in environmental policy-making, while state governments dominate the funding, implementation, and enforcement of environmental policy. In both cases, federal regulators take a litigious, coercive approach to securing compliance with federal law that places significant constraints on the discretion of state governments. These similarities require explanation.

Why has the US government, which has the power to assume a dominant role, chosen to delegate most implementation and enforcement authority to state governments? Why has the EU, with such a small federal bureaucracy and such a limited base of legitimacy, taken a coercive approach to enforcing member states' compliance with EU environmental law?

This chapter argues that similarities in the institutional structure of the US and EU have led the two polities to adopt broadly similar approaches to environmental regulation. Both the US and the EU combine a vertical division of authority between federal and state governments with horizontal fragmentation of authority at the federal level. The combination of federalism and fragmentation of power at the federal level in the two systems has influenced both the allocation of regulatory authority between federal and state governments and the approach that federal governments take to controlling state governments. In particular, the fragmentation of power at the federal level has encouraged the judicialization of regulatory style, with the enactment of action-forcing environmental laws and a litigious approach to enforcement.

The discussion that follows is divided into three sections. The first section examines the allocation of regulatory authority between federal and state governments in the US and EU. The second examines the means by which EU and US federal regulators attempt to control the regulatory practices of state governments. The final section draws some conclusions.

Allocating Regulatory Authority

The United States

In the United States, the allocation of regulatory authority is determined by politics, not by considerations of efficiency.[1] All rhetoric concerning efficiency and subsidiarity aside, state and federal governments focus primarily on political considerations when struggling with one another to allocate regulatory authority. Typically, these political considerations do not lead to a neat division of authority in the area of environmental regulation, in which some issues are the exclusive province of the states and others the exclusive province of the federal government. Rather, regulatory authority is often shared between the two levels of government, with the federal government playing a powerful role in setting minimum standards and state governments retaining most responsibility for policy implementation.[2]

Political considerations encourage the division of policy-making and policy implementation competencies between federal and state governments. State governments may support a federal policy-making role either because it allows them to resolve collective action problems such as a regulatory "race-to-the-bottom,"[3] or because it allows them to shift blame to the federal government in the event of regulatory failures. However, even states that support a federal role in policy-making are likely to favor a state role in implementation that allows them flexibility in dealing with the regulated entities. Federal governments support this allocation of authority because it allows them to claim credit for addressing issues of great public concern by enacting environmental laws, while shifting blame for much of the cost of these policies to state governments.[4] Finally, any state governments that might attempt to block federal involvement in policy-making are unlikely to succeed because federal courts will adjudicate such disputes and are likely to approve federal assertions of jurisdiction.[5]

In the US, the federal government first became involved in pollution control issues in the late 1940s and 1950s through a series of research and funding programs assisting state and local governments in their air and water pollution control programs.[6] This federal funding came with few conditions attached, and the states remained in control of regulatory policy-making and implementation. In the early 1960s, as public concern with environmental issues mounted, state governments were widely perceived as failing to respond adequately. Where state governments did adopt significant pollution control regulations, differences among state standards threatened to create major problems in interstate commerce.[7]

When public concern regarding environmental issues grew dramatically in the late 1960s, the federal government responded decisively. Political rivals in the House, the Senate, and the White House competed to claim credit as the most ardent advocates of environmental protection.[8] Beginning in 1969, Congress adopted a series of landmark environmental statutes that established federal standards for environmental

impact assessment, air quality, water quality, and control of toxic substances. These legislative developments were coupled with administrative reforms, most prominently President Nixon's creation of the
Environmental Protection Agency (EPA). Together, these legislative and
bureaucratic developments established a major federal role in environmental regulation.

With the growth of the EPA during the 1970s, the US government
came to play a powerful role in implementation and enforcement. Nevertheless, it never sought to supplant state governments as the primary
implementers and enforcers of federal law. Cases of "complete preemption," where the federal government occupies a field of regulation and
prevents states from playing any role in standard-setting, implementation, and enforcement, are extremely rare. Instead, the government
typically relies on a "partial preemption" approach, setting minimum
standards and goals, but allowing states to design and implement their
own regulations and programs aimed at achieving these objectives,
subject to federal approval of a state implementation plan (SIP). A variety
of federal environmental statutes provided that after the delegation of
primacy in enforcement to a state, the EPA could take back responsibility for implementation and enforcement where it found the state was
systematically failing to enforce the statute. However, the EPA almost
never chooses to do so, and the threat of federal preemption is not credible because both states and the federal government recognize that the
EPA has powerful incentives not to preempt the state role completely.[9]

This division of authority suited both the federal government and most
state governments. By playing a visible role in setting minimal standards,
the federal government could claim credit for addressing an area of major
public concern, while shifting blame for most day-to-day enforcement
and much of the cost of regulatory programs to state governments. State
governments, in turn, could use their control over implementation to
maintain some flexibility in dealing with regulated industries. Moreover,
they benefited from substantial federal funding during this period.

By the end of the 1970s, the division of regulatory competencies
between the federal government and the states was well established.
President Reagan came to office calling for a major rollback of federal
environmental regulation. However, Democrats in Congress were able

to block any major legislative rollback. While the Reagan administration did not succeed in repealing major environmental legislation, it did succeed in de-funding environmental programs. Major cuts in federal funding for state programs placed a growing burden on state governments, transforming federal statutes into unfunded mandates.[10] As the financial burden of federal environmental laws increased, the states complained more vociferously about unfunded mandates and demanded an increased voice in federal policy-making. From 1991 to 1993, state and local officials mounted a campaign against federal mandates, in particular environmental mandates, culminating in the nationwide "National Unfunded Mandates Day" protest in October 1993[11] and in the establishment of a new intergovernmental association of state environmental agencies, the Environmental Council of the States.

The Clinton administration responded to these concerns with measures that included an executive order restraining administrative mandates and new EPA programs promising the states more flexibility in implementation, and Congress responded by enacting legislation restricting the adoption of unfunded mandates.[12] Most recently, the Bush administration has promised to grant states increased flexibility in meeting federal environmental mandates.[13] For instance, the EPA has proposed delegating to the states the power to designate "impaired water bodies" under the Clean Water Act and granting them considerable discretion as to how to achieve the "highest attainable" uses of them.[14] Just how far the Bush administration's rollback of federal oversight will go remains to be seen. However, for the time being, the basic division of regulatory authority in which federal regulators play a far-reaching role in policy-making while delegating most implementation to state governments remains firmly in place.

The European Union

The EU entered the field of environmental policy for the same reasons as the US government. In the late 1960s, concern over environmental issues increased across EU member states, and some began adopting stringent environmental regulations. EU officials saw that developing an environmental policy at the EU level could simultaneously increase the EU's popularity with concerned citizens, expand the scope of EU power, and remove distortions to the common market caused by different national standards.[15] At this time, the adoption of EU directives and regulations required unanimous approval of the member states in the Council of Ministers. Strict environmental regulators such as Germany and the Netherlands favored an extensive EU role in policy-making, hoping to pressure their laxer neighbors into raising their regulatory standards.[16] Lax states favored some EU involvement, particularly in establishing common product standards, because they feared that states with high standards might use environmental regulations as disguised restrictions on imports.[17] Generally, the fact that policy measures had to be adopted by unanimity and that Community enforcement was weak increased the willingness of member states to accept EU involvement.[18]

The Community began adopting EU-level environmental directives and regulations in the late 1960s and accelerated the pace of policy-making in the early 1970s. By the mid-1980s, the Community had issued directives and regulations dealing with nearly all major areas of environmental policy. In 1983, the European Commission established a separate Directorate-General (DG XI) to oversee environmental policy-making and enforcement. By the mid-1980s, the EU's jurisdiction in the field of environmental policy was well established, and in the 1987 Single European Act (SEA), the member states added an explicit treaty basis for EU environmental policy (SEA, Articles 130r-t).

Throughout the 1980s, criticism of the EU's "implementation deficit" began to mount, and the European Parliament pressured the commission to increase its enforcement activities.[19] The commission responded by both intensifying its use of the Article 169 (now Article 226) infringement procedure against noncompliant member states and by proposing the establishment of a European Environment Agency (EEA) to enhance the Community's monitoring capacity. As the EU increased its enforcement activities, member states came under increasing pressure to enforce environmental laws, which many had not viewed as strict legal requirements. The EU's poorer member states responded during the Maastricht Treaty negotiations by demanding increased EU funding for the implementation of EU environmental directives.[20] In 1992, the United Kingdom (UK) sought to roll back a number of EU environmental directives that it argued violated the principle of subsidiarity. While the UK did pressure the commission into withdrawing some pending proposals,[21] its effort to return established areas of Community regulation to the member state level failed. In the years since Maastricht, opponents of Community-level regulation have accepted that the Community will not retreat from environmental policy-making. In the 1997 Amsterdam Treaty, no major efforts were made to roll back existing environmental policy; to the contrary, Article 175 (ex Article 130s) of the Amsterdam Treaty extended the use of the codecision procedure in environmental policy-making.

The evolution of the division of competencies in the US and EU followed a common pattern. Growing public concern with environmental issues coupled with the potential for distortions to trade and competitive conditions caused by divergent state environmental regulations led the federal governments to assume a role in environmental policy-making. Although the US government has taken a more direct role in implementation and enforcement than the EU, federal governments in both polities leave implementation and enforcement primarily in the hands of the state governments.

Controlling State Governments

Given the division of regulatory authority in which state governments implement most federal environmental laws, federal regulators in the US and EU face the challenge of controlling state governments. State governments often have incentives to shirk on their commitments to implement and enforce federal law, in order to give their industries competitive advantages over those in neighboring states, or simply to minimize expenditures on environmental programs.[22] The US and EU federal governments use a combination of legal compulsions and fiscal inducements to pressure state governments to implement federal environmental laws. The US government has made much greater use of fiscal inducements than has the EU, which has far more limited resources. As for legal tools, the US has relied primarily on federal enforcement actions against polluters and litigation by private parties, while the EU has relied primarily on enforcement suits against member state governments. While the specific legal tools have varied, both US and the EU federal regulators have taken a coercive, litigious approach to securing state government compliance with federal environmental law. Despite the prevalence of calls for cooperation between levels of government and regulatory flexibility, intergovernmental relations in the EU and US often have been characterized by state government resistance and federal coercion.

Legal Compulsions

The fragmentation of power in the US and EU has encouraged the adoption of strict, inflexible laws and reliance on litigation as a tool of enforcement. First, the fragmentation of power between branches of the federal government creates agency problems because legislative bodies do not trust executives to implement statutes in accord with legislative intent. Second, the multiple veto points in the legislative processes in the EU and US make legislation difficult to enact or amend. Therefore, coalitions that succeed in enacting a piece of legislation can anticipate that it will remain in place for a considerable time and they seek to draft legislation that will lock in their policy victory for the long term. Third, the fragmentation of power in the EU and US provides institutional foundations for judicial independence. Knowing that the fragmentation of power insulates them against easy legislative overrides or other forms of political backlash, the courts may play an active role in the regulatory process. Given agency problems, the durability of legislation, and the likely willingness of courts to challenge the actions of both federal and state executives, drafters of legislation may be tempted to pursue a judicialization strategy.

In the US, when environmental advocates in Congress drafted the landmark environmental legislation of the 1970s, they sought to ensure that federal agencies and state governments would implement the statutes as they intended. They included detailed "action-forcing" requirements aimed at limiting the discretion of federal agencies and state governments, and invited the courts to play an active role in enforceing these requirements.[23] Lawmakers included provisions for citizen suits in statutes, enabling citizens to sue both violators of environmental statutes and the government agencies that failed to perform mandatory duties. The federal courts responded by engaging in active judicial review of executive action on environmental policy, often overturning executive decisions and even forcing federal agencies and states to adopt new regulatory programs.[24] This judicial activism in turn produced an increase in litigation, as environmental advocates attempted to use the courts to influence policy implementation and limit the discretion of federal and state agencies. This pattern persisted throughout the 1980s as the Democratic Congress adopted statutes with highly detailed, justiciable provisions aimed at forcing the intransigent Reagan administration (and later the first Bush administration) to take action on the environment.[25] Judicial supervision of federal and state regulatory agencies diminished somewhat after the late 1970s; however, the federal courts continue to exercise active judicial review and remand many rule-making decisions.[26]

As in the US, the EU's fragmented institutional structure has influenced its approach to controlling member state governments. The separation of executive and legislative power and the lack of trust among the commission, the council, and the European Parliament and among member states within the council, have encouraged the enactment of legislation that often specifies in great detail the goals that member states must achieve, the deadlines they must meet, and the procedures they must follow.[27] The European Parliament distrusts both the commission and the member states, and it favors inflexible, detailed laws that limit member state discretion and encourage the commission to take enforcement actions.[28] Similarly, member states in the council distrust one another, and often favor directives and regulations that spell out legal obligations in great detail, in order to aid enforcement action against noncompliant member states by the commission and the European Court of Justice.[29] The fragmentation of power has also emboldened the court to engage in aggressive judicial review of national administrations.[30] The combination of action-forcing statutes and judicial assertiveness has encouraged the commission to take an active role in pursuing enforcement litigation against noncompliant member states.

While both the US and EU have adopted coercive, legalistic approaches to securing state compliance with environmental law, they have relied on different legal tools. The US government takes legal action directly against polluters and encourages private parties to bring litigation against both state governments and polluters. It does not take legal action directly against state governments; indeed, in most cases it cannot (see later discussion). By contrast, the EU relies primarily on the commission taking enforcement actions against member states. The Community cannot bring legal actions directly against polluters, and litigation brought by private parties has thus far had little role in securing member states' compliance. In the following discussion I examine the use of these enforcement tools in more detail.

Centralized Enforcement   The European Commission's main enforcement tool is the Article 226 (ex Article 169) infringement procedure, which enables the commission to bring member states before the European Court of Justice for their failure to implement EU law. The Treaty on European Union (Maastricht Treaty) strengthened the infringement procedure by amending Article 228 (ex Article 171) to allow the court to impose penalty payments on member states who fail to comply with previous court rulings in infringement cases. Since the mid-1980s, the commission has intensified its application of the infringement procedure. It initiates hundreds of infringement proceedings for suspected breaches of Community environmental law every year and secures compliance in the vast majority of cases.[31] Since 1997, the commission has employed its power under Maastricht's Article 228 (ex Article 171) to request that the court penalize member states that fail to comply with court rulings in infringement cases. Subsequently, the commission has initiated dozens more such cases, and the threat of penalty payments seems to have had a substantial impact, pressuring recalcitrant member states to come into compliance.[32] On July 4, 2000, the European Court of Justice delivered its first ruling on such a case and imposed the EU's first-ever penalty payments, eventually totaling 4.8 million euros, on a member state for failure to implement Community environmental law.[33]

Many infringement cases have severely restricted the discretion of member states, even where environmental directives appeared to afford member states considerable latitude. For instance, infringement rulings have restricted member state discretion in the designation of bird sanctuaries[34] and bathing (swimming) areas[35] under Community directives. The court has even placed great restraints on member state choices of the administrative procedures with which to implement EU environmental directives.[36]

While the commission very often succeeds in using the infringement procedure to secure compliance with Community law, its enforcement efforts suffer from fundamental deficiencies. It has great difficulty identifying failure to apply Community law in practice. Lacking the authority to conduct direct inspections, the commission must rely primarily on individuals and associations using the EU's complaints procedure to serve as its eyes and ears in the member states. The small professional staff in the legal unit of the Directorate-General Environment is overwhelmed by complaints regarding infringements, and the average time between the commission's decision to initiate an infringement procedure on an environmental matter and the actual judgment by the European Court of Justice is nearly 5 years.[37] The procedure fails to provide the private parties that bring complaints with any form of legal certainty, and infringement procedures are not subject to judicial or administrative review. While the threat of penalty payments is likely to make infringement procedures a more effective tool, the problems stemming from the commission's staffing and funding limits persist.

While the EU relies almost exclusively on commandeering the administrative apparatus of member state governments to implement EU policies, the US government is prohibited from doing so. The Supreme Court reaffirmed this prohibition in its recent decision on the "anticommandeering" principle.[38] In *New York v. United States*[39] the court explained that while the federal government may pressure states to implement federal programs, for instance by providing funds to states that do so and denying them to states that do not, it may not directly compel states to administer a federal program. Even before the *New York* decision reaffirmed the "anticommandeering" principle, the federal government avoided enforcement actions against state governments because proving that a state was systematically failing to enforce a federal statute would generate pressure for the EPA to take over enforcement in the state, which it was loath to do.[40]

Instead of suing state governments, in order to secure compliance, the US government relies on two instruments that the EU lacks. First, it prosecutes polluters directly. Even after a state implementation plan is approved and a state assumes primacy in enforcement, the EPA retains the right to bring enforcement actions directly against polluters. It can assess administrative penalties, initiate cases seeking civil penalties, and, in the case of some statutes, seek criminal penalties and jail time. The EPA has brought tens of thousands of administrative enforcement actions and thousands of civil and criminal actions that have forced polluters to pay billions of dollars in fines and even serve jail time. In a practice known as "overfiling," the federal government even brings enforcement actions where states have already initiated an action against the polluter in question. State governments resent this practice because it indicates to regulated industries that the state is not in control of its own regulatory agenda and thus hurts their credibility in negotiations.[41] The threat of federal intervention gives states an incentive to pursue their own enforcement actions vigorously, in the hopes of forestalling federal intervention. Most recently, the George W. Bush administration has signaled its intention to reduce the EPA's role in enforcement and to grant state governments greater discretion. These moves led to the highly publicized resignation of a top EPA enforcement official, who accused the administration of undermining the enforcement of federal environmental law.[42]

Decentralized Enforcement   In the US, federal environmental statutes are designed to encourage decentralized litigation by environmental organizations. The coalitions backing the landmark environmental statutes of the 1970s foresaw that their influence might wane over time. They sought to protect their legislative victories by creating opportunities for environmental advocates to bring legal action should future administrations or state governments fail to implement environmental statutes. For instance, major environmental statutes included "citizen suit" provisions, allowed plaintiffs to request civil penalties, and even allowed successful plaintiffs to recover legal costs from defendants. In the early 1970s, the federal courts also encouraged such litigation by loosening rules governing standing to sue (*locus standi*) for public interest litigants and by demanding that administrative agencies take into account the views of a variety of groups and provide reasons for their decisions.[43]

With Congress and the courts opening up access to the courtroom in the early 1970s, environmental groups increased their use of litigation as a strategy to influence environmental policy. A number of groups were founded with the explicit aim of bringing litigation. Initially, most citizen suits focused on the EPA's failure to properly implement statutes. Some suits brought by environmental organizations led to significant policy victories, forcing the EPA and state environment agencies to initiate new programs.[44] In the early 1980s, when the Reagan administration relaxed federal enforcement efforts, environmental organizations fought back by instigating more lawsuits.

In the 1990s, the Supreme Court's decisions regarding standing and the sovereign immunity of state governments placed new restrictions on the ability of private parties to act as the enforcers of federal law. In three major cases—*Lujan I, Lujan II*, and *Steel Co.*—the Supreme Court restricted the ability of environmental groups to gain standing to sue.[45] While the precedents set in these cases bode poorly for groups with tenuous injury claims who bring litigation under the citizen suit provisions of various environmental statutes, they have by no means slammed the courtroom door shut, and federal circuit and district courts continue to find grounds for standing for environmental groups.

The Supreme Court's recent case law on the doctrine of sovereign immunity also diminishes the ability of private parties to serve as enforcers of federal law. The comparison with the EU is striking in this regard. While the European Court of Justice has been developing a doctrine of member state liability for the nonimplementation of Community law, the US Supreme Court has been doing just the opposite, shielding state governments against liability claims. In a series of recent decisions, the Supreme Court has made it nearly impossible for private parties to seek retrospective relief (i.e., damage awards) from state governments for violations of federal law in federal or state courts.[46]

Taken together, the court's recent rulings on commandeering, standing, and sovereign immunity can be seen as part of an attempt to increase state autonomy and reduce public interest litigation. The court's anti-commandeering decisions reduce the federal government's ability to pressure states into compliance. In this respect, they make it more necessary for the federal government to rely on litigation by private parties to secure enforcement. However, recent Supreme Court rulings on standing for public interest groups deter just such litigation, and the court's sovereign immunity case law shields states against damage claims if they fail to comply with federal law. Together, these legal principles promise to increase the discretion of state governments in implementing federal law and to provide them with protections against federal coercion that even EU member states do not enjoy.

To date, decentralized litigation by private parties has played little role in the enforcement of EU environmental policy. However, while legal and political developments in the US are working to limit such litigation, legal and political developments in the EU are working to expand it. The doctrines of direct effect and supremacy of Community law enable individuals to bring legal action before national courts to defend their rights under Community law in the event of noncompliance by a member state.[47] Cases brought by private parties before national courts can reach the European Court of Justice via the preliminary ruling procedure [Article 234 (ex Article 177)], which provides that national courts may refer questions of EU law to the court for clarification. The preliminary ruling procedure has not yet played a significant role in Community environmental law. From 1976 to 1996 the European Court of Justice made rulings in only thirty-six preliminary environmental ruling cases.[48] Although the pace of referrals from national courts accelerated in the late 1990s and has started to play an important role in such areas as nature conservation policy,[49] overall the impact of the preliminary ruling procedure on EU environmental policy remains limited. One important reason for the infrequency of such cases is that the legal systems in many member states maintain restrictive *locus standi* conditions that prevent environmental organizations from bringing suits before national courts.[50]

Recent commission initiatives and developments in European law promise to create new incentives and opportunities for private parties to initiate litigation on environmental matters before national courts. In the mid-1990s, the commission and the European Parliament began pressuring member states to harmonize their national rules on access of private parties to national courts.[51] In 1998, EU member states and the EU itself signed the UN Aarhus Convention, which includes a set of commitments concerning access to justice in environmental policy-making.[52] Most member states have expressed support for minimum criteria on access to justice. One impediment to the establishment of such minimum standards is the continuing reluctance of the commission and the

European Court of Justice to grant environmental groups standing to challenge commission decisions. The court has consistently denied environmental nongovernmental organizations (NGOs) standing to challenge commission decisions.[53] As long as the court persists in denying environmental plaintiffs standing, it will deter efforts to identify minimum, common standards that would expand access to justice throughout the EU.

The case law of the European Court of Justice concerning the principle of state liability has created the potential for environmental plaintiffs to sue member states for damage they suffer from the nonimplementation of environmental law. In a series of rulings beginning with *Francovich*,[54] the court has developed a doctrine of state liability that provides that under certain conditions member states can be held liable for damage suffered by individuals as a result of the member state's failure to implement Community law. Although there are no reported cases of the application of the state liability principle to environmental matters, considering the criteria for this liability, it is likely that in the future individuals will bring claims for damage relating to environmental directives.[55] Moreover, the mere potential for such suits is likely to have an impact on the implementation practices of member states. Beyond these general principles of state liability, the commission is pushing for the adoption of EU legislation that specifically addresses environmental liability (see chapter 7).[56] Taken together, these developments suggest that the role of decentralized litigation in the enforcement of EU environmental law is likely to increase considerably.

Fiscal Tools

A second set of levers that federal governments can use to secure compliance by state governments relies on financial inducements. The US government has used a variety of such budgetary tools. First, Congress has attached conditions to federal grants-in-aid to state governments, requiring states to adopt specific regulatory measures in order to receive funds.[57] Congress also imposes "cross-cutting" requirements that apply to all recipients of federal funds.[58] Finally, Congress uses "cross-over sanctions," withholding federal funds for particular programs to punish states for violations of other regulatory programs.

To date, the EU's use of fiscal levers has been limited. One reason these levers are weaker than those of the US government is that the EU controls far fewer resources, with a budget of only approximately 1.27 percent of the combined gross national product (GNP) of the member states. Beyond this general limitation, some member states have insisted on limiting the EU's role in funding environmental policy because of fears that the EU would gain undue influence over their domestic policies through Community funding schemes.[59] However, the reluctance of member states to allow a substantial EU role in funding environmental projects appears to be waning. First, the 1988 reform of the structural funds called for an increase in EU spending on environmental activities with an aim to integrating environmental policy into other community policies. EU spending on environmental programs under the structural funds rose dramatically in the 5 years after the reform.[60] In 1992, the EU initiated the LIFE (*L' Instrument Financier pour l' Environnement*) program to help finance implementation of priority areas of EU environmental policy, in particular nature conservation. As part of the Maastricht Treaty negotiations, the EU established a cohesion fund that targeted over 15 billion ECU (European Currency Units—predecessor of the euro) in support between 1993 and 1999 to Portugal, Ireland, Greece, and Spain. Spending on environmental protection constitutes one of the two principal uses of grants from the EU's Cohesion Fund, and since 1997 more than half of this fund has been spent on environmental projects.

Along with increasing spending on environmental measures, the EU has started to attach environmental conditions ("cross-cutting requirements") to other spending programs. In 1993, administrative reforms of the structural funds stipulated that development plans submitted with funding proposals must include environmental impact assessments. The council regulation establishing the guidelines for the 2000–2006 round of structural funds also includes environmental protection and sustainable development as general aims of the structural funds and requires environmental impact assessments for plans and measures receiving financing from these funds.[61] While the approval of development projects is up to national or local officials, the commission can withhold EU financing for particular projects that violate Community environmental law. For instance, since 1999 the commission has started to warn member states that structural funds will be withheld for failures to implement the habitats directive (92/43), which provides for the establishment of special conservation areas to protect wild flora and fauna.

Conclusion

A comparison of environmental federalism in the US and the EU reveals how the institutional structures of the two polities have influenced patterns of environmental regulation and offers surprising insights. First, observers who argue that the EU lacks the regulatory powers of a true federal system often emphasize that member states retain control over most implementation and enforcement of EU policies. However, a comparison with the US reveals that state government control of implementation and enforcement is prevalent even in a system with an extremely powerful federal government. While skeptics of EU power might argue that the EU delegates to member states because it has no choice, the US example suggests that given the choice, political considerations would lead to the same allocation of regulatory authority.

Second, a comparison of the two cases demonstrates how the fragmentation of powers built into the structure of their federal governments encourages a judicialized, litigious style of regulation that places great constraints on state governments. In both cases, the fragmentation of power at the federal level encourages the adoption of detailed, inflexible legislation and encourages the courts to play an active role in the regulatory process. The US experience highlights the importance of decentralized enforcement in environmental law. Even the USEPA with powers of direct enforcement, a staff of over 18,000, and powerful fiscal levers cannot ensure uniform implementation and enforcement of federal environmental laws by state governments. The US government's coercive powers are ultimately limited by its desire to leave states in control of most implementation and enforcement. Given these limits, decentralized enforcement of US federal law by private parties has played a vital role in pressuring recalcitrant states to enforce federal requirements. Decentralized enforcement by private parties promises to play a similar role in the EU, should the EU succeed in expanding opportunities for access to justice across the member states. And therein lies the final irony of the comparison; while the EU is working to expand the ability of private parties to act as enforcers of EU environmental policy, in the US these rights are being restricted.

Notes

1. On the debate concerning what division of regulatory authority serves to maximize social welfare, see Richard Stewart, "Environmental Regulation and International Competitiveness," *Yale Law Journal* 102 (1993):2039; Richard Revesz, "Rehabilitating Interstate Competition: Rethinking the 'Race-to-the-Bottom' Rationale for Federal Environmental Regulation," *NYU Law Review* 67 (1992):1210; Richard Revesz, "Federalism and Environmental Regulation: Lessons for the European Union and the International Community," *Virginia Law Review* 83 (1997):1331; David Vogel, *Trading Up: Consumer and Environmental Regulation in a Global Economy* (Cambridge, Mass.: Harvard University Press, 1995); Peter Swire, "The Race to Laxity and the Race to Undesirability," *Yale Journal on Regulation* 14 (1996): 67; Kirsten Engel, "State Environmental Standard-Setting: Is There a 'Race' and Is It 'To the Bottom'?" *Hastings Law Journal* 48 (1997):271; Daniel Esty and Damien Geradin, eds., *Regulatory Competition and Economic Integration: Comparative Perspectives* (Oxford: Oxford University Press, 2001).

2. Jerry Mashaw and Susan Rose-Ackerman, "Federalism and Regulation," in George Eads and Michael Fix, eds., *The Reagan Regulatory Strategy* (Washington D.C.: Urban Institute Press, 1984): 111–152.

3. For the debate regarding the "race-to-the-bottom," see note 1.

4. John Kincaid, "The New Coercive Federalism," in Franz Gress, Detlef Fetchner, and Matthias Hannes, eds., *The American Federal System: Federal Balance in Comparative Perspective* (Berlin: Peter Lang, 1994); John Dwyer, "The Practice of Federalism Under the Clean Air Act," *Maryland Law Review* 54 (1995):1183.

5. R. Daniel Kelemen, "Regulatory Federalism: EU Environmental Policy in Comparative Perspective," *Journal of Public Policy* 20(2) (2000):133.

6. Charles Jones, *Clean Air: The Policies and Politics of Pollution Control* (Pittsburgh, Pa.: University of Pittsburgh Press, 1975), pp. 29–38; Robert Percival, "Environmental Federalism: Historical Roots and Contemporary Models," *Maryland Law Review* 54 (1995):1141.

7. Donald Elliott, Bruce Ackerman, and John Millian, "Toward a Theory of Statutory Evolution: The Federalization of Environmental Law," *Journal of Law, Economics and Organization* 1(2) (1985):330.

8. Terry Moe, "The Politics of Bureaucratic Structure," in John Chubb and Paul Peterson, eds., *Can the Government Govern?* (Washington D.C.: Brookings Institution Press, 1989), pp. 306–310; Jones, *Clean Air*, pp. 175–210.

9. Hubert Humphrey and LeRoy Paddock, "The Federal and State Roles in Environmental Enforcement: A Proposal for a More Effective and More Efficient Relationship," *Harvard Environmental Law Review* 14(1) (1990):44; Dwyer, "The Practice of Federalism."

10. James Pfander, "Environmental Federalism in Europe and the United States," in John B. Braden, Henk Folmer, and Thomas S. Ulen, eds., *Environmental Policy with Political and Economic Integration* (Northampton, Mass.: Edward Elgar, 1996), p. 88; Kincaid, "The New Coercive Federalism," p. 46.

11. Timothy Conlan, James Riggle, and Donna Schwartz, "Deregulating Federalism? The Politics of Mandate Reform in the 104th Congress," *Publius* 25(3) (1995):26.

12. Kincaid, "The New Coercive Federalism," p. 45.

13. Douglas Jehl, "Whitman Promises Latitude to States on Pollution Rules," *New York Times*, January 18, 2001, p. A18.

14. Eric Pianin, "EPA Seeks Leeway in Rules about Dirty Water," *Washington Post*, August 8, 2002, p. A11.

15. Kelemen,"Regulatory Federalism."

16. States with strict environmental standards share an incentive to see laxer neighbors raise their standards regardless of whether they are motivated by concerns over the environment or economic competition.

17. Giandomenico Majone, "The Rise of the Regulatory State in Europe," *West European Politics* 17(3) (1994), 77–101; Jonathan Golub, "Why Did they Sign? Explaining EC Environmental Policy Bargaining," Robert Schuman Centre No. 96/52 (Florence, Italy: European University Institute, 1996).

18. Joseph Weiler, "The Transformation of Europe," *Yale Law Journal* 100 (1991):2403; Golub, "Why Did they Sign?"

19. Ludwig Krämer, *E.C. Treaty and Environmental Law*, 2nd ed. (London: Sweet and Maxwell, 1995).

20. See the Maastricht Treaty's Article 130s(5).

21. Jonathan Golub, "The Pivotal Role of British Sovereignty in EC Environmental Policy," EUI Working Paper, Robert Schuman Centre No. 94/17 (Florence, Italy: European University Institute, 1994).

22. The strength of these incentives to shirk is subject to considerable debate (see note 1) and Vogel in *Trading Up* has shown that competitive pressures can also give states incentives to raise their environmental standards.

23. Moe, "The Politics of Bureaucratic Structure"; Matthew McCubbins, Roger Noll, and Barry Weingast, "Administrative Procedures as Instruments of Political Control," *Journal of Law Economics and Organization* 3(2) (1987):263; David Vogel, *National Styles of Regulation: Environmental Policy in Great Britain and the United States* (Ithaca, N.Y.: Cornell University Press, 1986).

24. R. Shep Melnick, *Regulation and the Courts: The Case of the Clean Air Act* (Washington, D.C.: Brookings Institution Press, 1983).

25. Richard Stewart, "A New Generation of Environmental Regulation?" *Capital University Law Review* 29 (2001): 56.

26. Susan Rose-Ackerman, *Controlling Environmental Policy: The Limits of Public Law in Germany and the United States* (New Haven, Corn.: Yale University Press, 1995), pp. 151–152.

27. Eckhard Rehbinder and Richard Stewart, *Environmental Protection Policy* (New York: de Gruyter, 1985); Berthold Rittberger and Jeremy Richardson, "(Mis-)Matching Declarations and Actions? Commission Proposals in Light of the Fifth Environmental Action Programme," paper presented at the Biennial Meeting of the European Community Studies Association, Madison, Wisconsin, May 31–June 2, 2001.

28. Renaud Dehousse, "Integration v. Regulation? On the Dynamics of Regulation in the European Community," *Journal of Common Market Studies*, 30(4) (1992):392.

29. Giandomenico Majone, *Regulating Europe* (New York: Routledge, 1996).

30. Geoffrey Garrett, R. Daniel Kelemen, and Heiner Schulz, "The European Court of Justice, National Governments and Legal Integration in the European Union," *International Organization* 52(1) (1998):149; George Tsebelis and Geoffrey Garret, "The Institutional Foundations of Intergovernmentalism and Supranationalism in the European Union," *International Organization* 55(2) (2001):357.

31. European Commission, Communication from the Commission: Implementing Community Environmental Law, COM (96) 500 final (1996); European Commission, "Second annual survey on the implementation and enforcement of Community environmental law," Working document of the Commission Services, Directorate-General XI (2000).

32. European Commission, "Second annual survey"; "EU/Environment," *Agence Europe*, January 30, 1997.

33. Case C-387/97 *Commission v. Greece*, July 4, 2000; "Legal Actions Announced over EU Waste Rules," *Ends Environment Daily*, issue 1038, July 30, 2001.

34. Case C-355/90 *Commission v. Spain* (1993) I ECR 4221; Case 3/96 *Commission v. Netherlands*.

35. Case C-56/90 *Commission of the European Communities v. United Kingdom*. (1993) ECR I-4109; "Bathing Water: Commission Acts Against Several Member States," European Commission press release, IP/00/14, January 11, 2000.

36. See for instance, C-361/88 *Commission v. Germany* (1991) ECR 2567.

37. IMPEL (European Union Network for the Implementation and Enforcement of Environmental Law), *Complaint Procedures and Access to Justice for Citizens and NGOs in the Field of Environment within the European Union* (2000),

pp. 32, 162. Report available at <http://europa.eu.int/comm/environment/impel/access_to_justice.htm>.

38. *New York v. United States*, 505 U.S. 144 (1992); *Printz v. United States*, 521 U.S. 898 (1997).

39. 505 U.S. 144 (1992).

40. Humphrey and Paddock, "The Federal and State Roles in Environmental Enforcement," p. 44.

41. Telephone interview, Environmental Council of the States (ECOS), Washington D.C., March 23, 1998.

42. Katharine Seelye, "Top E.P.A. Official Quits, Criticizing Bush's Policies," *New York Times*, March 1, 2002, p. A19; Faye Fiore, "Top EPA Enforcement Official Quits, Blasts Bush Policy," *Los Angeles Times*, March 1, 2002, p. A17.

43. See notes 23–24. Also see Martin Shapiro, *Who Guards the Guardians?* (Athens: University of Georgia Press, 1988), pp. 36–77.

44. Melnick, *Regulation and the Courts*.

45. *Lujan v. National Wildlife Federation ("Lujan I")* 110 S. Ct. 3177 at 3186–3187 (1990); *Lujan v. Defenders of Wildlife ("Lujan II")* 112 S.Ct. 2130 (1992); *The Steel Co., AKA Chicago Steel and Pickling Co. v. Citizens for a Better Environment (Steel Co.)* 523 S. Ct. 83 (1998). See Phillips, Joseph T. "Comment: Friends of the Earth v. Laidlaw Environmental services: Impact, Outcomes, and the Future Viability of Environmental Citizen Suits," *University of Cincinnati Law Review* 68(2000): 1281.

46. William Araiza, "Alden v. Main and the Web of Environmental Law," *Loyola of Los Angeles Law Review* 33 (2000):1513.

47. See Weiler, "The Transformation of Europe," for an overview of the development of the doctrines of supremacy and direct effect.

48. Rachel Cichowski, "Integrating the Environment: The European Court and the Construction of Supranational policy," *Journal of European Public Policy* 5(3) (1998):396. Categorizing environment cases differently, Krämer reports a figure of only twenty-one cases during the same period. See Ludwig Krämer, "Public Interest Litigation in Environmental Matters Before European Courts," *Journal of Environmental Law* 8(1) (1996):4.

49. Rachel Cichowski, "Litigation, Compliance and European Integration: The Preliminary Ruling Procedure and EU Nature Conservation Policy," paper presented at the Biennial Meeting of the European Community Studies Association, Madison, Wisconsin, May 31–2 June 2, 2001.

50. Han Somsen, "The Private Enforcement of Member State Compliance with EC Environmental Law: An Unfulfilled Promise?" in Han Somsen, ed., *Yearbook of European Environmental Law* (Oxford: Oxford University Press, 2000); IMPEL, *Complaint Procedures and Access to Justice for Citizens and NGOs*.

51. European Commission, Communication from the Commission: Implementing Community Environmental Law; European Parliament, "Report on a Communication from the Commission on implementing Community environmental law," PE 221.176 final, March 21, 1997.

52. However, few member states have yet ratified the Aarhus Convention.

53. See for instance, *P. Stichting Greenpeace Council (Greenpeace International) v. Commission* Case C-321/95 (1998) ECR I-1651 and Krämer, "Public Interest Litigation in Environmental Matters."

54. Joined cases C-6/90 and C-9/90, *Francovich and Others v. Italy* (1991) ECR I-5357; Joined cases C-46/93 and C-48/93.

55. Jürgen Lefevere, "State Liability for Breaches of Community Law," *European Environmental Law Review* 5 (August/September 1996):237.

56. European Commission, White Paper on Environmental Liability, COM (2000) 66 final.

57. Humphrey and Paddock, "The Federal and State Roles in Environmental Enforcement," p. 20.

58. James Lester, "A New Federalism? Environmental Policy in the States," in Norman Vig and Michael Kraft, eds., *Environmental Policy in the 1990s* (Washington D.C.: CQ Press, 1994).

59. Auke Haagsma, "The European Community's Environmental Policy: A Case-Study in Federalism," *Fordham International Law Journal* 12 (1989):311.

60. Andrea Lenschow, "Variation in EC Environmental Policy Integration: Agency Push Within Complex Institutional Structures," *Journal of European Public Policy* 4(1) (1997):109.

61. Council Regulation No. 1260/99/EC of June 21, 1999, Articles 2.5 and 12.

5

Implementation of Environmental Policy and Law in the United States and the European Union

Christoph Demmke

It is no easy task to compare the implementation of environmental policy and law in the European Union and the United States. Ministries and agencies in the EU and the US carry out different laws within different legal systems, address different actors, follow different administrative and criminal enforcement styles, and have vastly different resources.

Another problem is the concept of "implementation," which is difficult to define.[1] Implementation has different meanings in the EU and the US because of the different nature of "federal" relations. In the European Union, implementation of environmental policy and law mainly involves the extent to which the member states of the EU transpose into their own laws and enforce legal acts made at the EU level. In the United States, the states also enforce federal statutes, but only to the extent that specific authority is delegated to them under federal legislation or by federal agencies.

Legislation also takes different forms in the European Union. In the EU, most environmental law is enacted in the form of directives, which, under Article 249 of the EU Treaty, "shall be binding, as to the result to be achieved, upon each Member State to which it is addressed, but shall leave to the national authorities the choice of form and method." This means that member states must "transpose" EU directives into national law, and it allows them considerable flexibility in adapting the implementation of the laws to their national administrative traditions. Other laws take the form of regulations that are directly binding on the states without requiring national legislation, but these are utilized mostly when technical standards must be brought into harmony throughout the EU.

One of the first comparative studies on implementation and enforcement in the US and Europe concluded that there was less variation among countries in institutions and effective policy at the implementation level than the authors expected.[2] The study compared several European states with the US across different policy sectors. Since this study, many developments have taken place. The number of member states in the EU has increased from ten (in 1983) to fifteen (since 1995) and will be up to twenty-five (as of 2004). In addition, the national legal, administrative, and political structures of the member states have changed considerably.

This chapter examines the implementation of environmental policy and law by the US federal government, the institutions of the European Union, and the member states of the EU. (The American states and the member states of the EU also implement their own environmental legislation, but that is beyond the scope of this chapter.) My purpose is to compare trends in innovation and transnational policy in the implementation process in the US and the EU. Since the role of environmental agreements is discussed elsewhere in this book (see chapter 6), I concentrate on new management instruments as well as "new governance" strategies used by the two "green giants." In the conclusions, an answer is given as to whether these new developments might be effective (or not) and lead toward convergence in implementation and enforcement of environmental policy and law.

Role of EU Member States in Policy Formulation and Implementation

The differences in policy formulation are crucial for understanding the differences in policy implementation between the EU and the US. Obviously, implementation can only be effective if there is coherence between policy formulation and policy implementation. In the US, many legislative proposals adopted by Congress are put forward by interest groups, other nongovernmental organizations (NGOs), or agencies of the government itself. After adoption of a law it is (mostly) turned over to an agency for implementation and/or enforcement. Federal agencies such as the Environmental Protection Agency (EPA) may in turn delegate enforcement powers to the states, but they retain supervisory authority over the implementation process.

In the EU, the European Commission formally dominates the policy formulation process since it alone can initiate legislation, but national policies prevail in the implementation phase. Comparative research in the EU has shown that "national administrative traditions and their level of institutionalization influence national implementation of EU legislation. More precisely, national compliance with EU law depends on the level of adaptation pressure perceived in the member states."[3] Therefore, the national administrations try to impose their national regulatory philosophies at the European level. Some, however, exercise more influence than others in convincing the European Commission to adopt their approach.[4] Pedler and Schäfer[5] have found that legislative initiatives usually come from the member states and/or interest groups, and only a small proportion (approximately 6 percent) originate in the commission itself.[6] The function of the commission in developing proposals is thus more that of a mediator than an initiator.[7]

However, analysis of successful national regulatory initiatives requires cautious interpretation. It seems clear that no member state of the EU has set its stamp on any one area of environmental policy. Obviously, some member states concentrate on specific legal acts, instruments, and policy sectors, and thus have little interest in negotiations in other areas. The United Kingdom and France, for example, undertook particularly ambitious (and successful) efforts to "modernize" water protection policy. The Water Framework Directive, which is based on a river basin concept in Article 3, clearly reflects initiatives by the United Kingdom and France, and also to some extent by the Netherlands.[8] Table 5.1 suggests the areas in which member states had especially strong influence on this directive.

The achievement of national interests at EC level is a precondition for, but not a guarantee of, compatibility between European and national law and effective implementation on the national, regional, and local levels. The "export" of national concepts to Brussels does not necessarily preclude later implementation difficulties.[9] One major reason for this is that the interplay among actors, institutions, and interests at the local

Table 5.1
Regulatory Competition and the Water Framework Directive—Influence and Power in the EU Council of Ministers

Who?	Influence	Comments
Presidency of the Council of Ministers	Very strong	British (1998) and German presidency (1999)
Member states	Very strong	United Kingdom—Art. 3 (river basins approach), Art. 8 (combined approach)
		France—Art. 3
		Netherlands—framework concept
		Germany—Art. 3, Annex V, Art. 8
		Spain—Art. 4 (derogations), Art. 9 (water prices)
		Ireland—Art. 9
		Finland—Art. 11.2 (plans)
		Portugal–Art. 13 (coordination)
		Austria–Art. 4 ("heavily modified waters")

level might create a totally different implementation structure than that anticipated by the (mostly) central officials who have negotiated the text over a period of years in the council of ministers. Most observers estimate the EU environmental *acquis* at approximately 300 legal acts. Whatever the exact figure, different experts estimate that the percentage of transposed EU environmental law in national environmental law ranges from approximately 35 percent in Denmark and 80 percent in the United Kingdom to 95 percent in southern member states such as Portugal and Greece.[10] However, EU law has a much deeper impact on the content of national policy and the choice of policy instruments than on policy structures or national policy styles.[11] This impact is "highly differentiated across countries."[12]

Despite this tremendous regulatory impact, the EU has failed to have a significant effect on the fundamental goals and principles of the national environmental policies of its members.[13] This imbalance between strong regulatory importance and weak impact on structures, principles, and content continuously creates difficulties in implementation.

Implementation and Enforcement in the US and the EU and the Changing Role of Government

Unlike the United States, the European Community primarily has legislative and oversight powers. The implementation of environmental policy is left to the member states. As a result, the EC does not have the competence to intervene in the administrative structures of its member states. Neither the European Commission's Directorate-General Environment nor the European Environment Agency thus exercise direct inspection or implementation powers within the member states. Consequently, the European Commission can only indirectly control the application of Community law in the member states. In the United States, the Environmental Protection Agency works through ten regional offices in supervising state implementation of federal environmental statutes and in directly enforcing provisions of the law on individual regulated parties. The states also carry out an impressive array of environmental regulatory activities under their own statutory authority, including both standard-setting and enforcement. Overall, the states carry out almost 90 percent of all environmental enforcement actions nationwide and write more than 90 percent of all permits.[14] Another feature of the US system is that staff and financial resources at the decentralized level far outstrip those of the EPA headquarters in Washington, D.C. (the EPA has approximately 18,000 staff, 12,000 of whom work in its regional offices). By contrast, the Directorate General Environment of the European Commission has a total staff of only about 450, including approximately 20 seconded EU and national officials.[15] (In addition, an unknown number of member state officials enforce EU legislation as part of national law.) Moreover, the EU (generally) does not finance the implementation and enforcement of environmental policies at national level whereas (for example) the average US state relies on the EPA for 26 percent of its budget for pollution control.[16]

Despite these vastly different figures, which are difficult to compare because of the different enforcement structures, personnel shortages are a crucial problem in both the US and the EU. A report of the US General Accounting Office points out that resource shortages at both the federal and state level have been amplified in recent years "by the expansion of the universe of facilities inspected which could be subject to potential enforcement."[17]

The EPA has consistently described its enforcement programs as providing deterrence (among other tools) against noncompliance by regulated facilities. That is, inspections and other forms of compliance monitoring and enforcement are undertaken to deter violators from noncompliance. In the year 2000, the EPA assessed $122 million in criminal penalties, $54 million in civil judicial penalties, and $29 million in administrative penalties. In addition, the EPA together with the states carried out thousands of inspections at regulated facilities.[18]

The use of deterrent instruments is much more limited at the EU level since the EU has—apart from the possibility of judicial penalties against the member states—no competence to impose administrative and criminal law sanctions in the field of environmental policy. Rather, police and inspection activities fall under the competence of the member states of the EU (see later discussion). Because of this and other important differences, there is no doubt that the "environmental regime" prevailing in the United States (especially with the EPA as main actor) has more authority than that of the EU institutions (see also chapter 4).

The Lack of Monitoring Capacities in the EU

One striking difference between the US and the EU models of implementation is the difference in monitoring and controlling the implementation of environmental policy and law. Because of its restricted competence to implement and enforce legislation, the European Commission is forced—more than the EPA—to look for effective methods that are not based on punitive measures. Effective implementation of environmental law is thus more often based on partnership principles and nonhierarchical forms of cooperation because the European Commission does not have the means to operate through authoritative commands. The only authoritative instrument available (the so-called infringement procedure) is very time-consuming and bureaucratic.[19] Consequently, in recent years new informal enforcement networks (such as IMPEL—the European Network for the Implementation and Enforcement of Environmental Law) and new forms of administrative coopera-

tion (consultative forums, working groups, committees, public conferences, etc.) among European, national, regional, and local authorities
have been emerging in the European Union.

The European Commission publishes annual implementation reports
on the basis of information received from the member states. According
to the commission's annual survey issued in 2002 on the control of the
application of Community environmental law,[20] the environmental sector
is the policy field with the highest number of infringements.[21]

The information transmitted by the member states generally contains
legal texts and documents that allow the commission to get a rough
picture of the state of compliance in these states. Reporting requirements
concerning the effect of existing legislation focus on information about
legal transposition, practical compliance, environmental data, and
descriptions of policy measures.[22] The problem, however, is that member
states report very differently (if at all) on the implementation of
measures.

Moreover, a report by the European Environment Agency in 2001
revealed that there is still too little information about the effectiveness
of EU measures. In only 12 percent of all EU environmental legislation
are member states required to provide any evaluative information on the
effects of measures.[23] The report comes to the conclusion that more information and discussion would be needed to assess the effects and effectiveness of EU measures. In the future, the evaluation of the effects of
legislation should become an important requirement in EU environmental legislation. The twin challenge will be to revise the reporting system
to enable us to know more about the effects of EU legislation while at
the same time decreasing the burden of reporting on the member states.[24]
According to the report, the "process of instilling an evaluation culture
in Member States and the European Commission and improving historic
databases and research on environmental and human systems will be a
gradual learning process."[25]

In the United States as well as the EU, central, state, and local administrations are often unable to give a complete picture of the state of
compliance and enforcement owing to "the seriousness of the data
problem."[26] Moreover, the different actors (member states and the commission in the EU and states and regions in the US) apply different

definitions of "enforcement." Consequently, large variations exist concerning enforcement styles, the number of inspections, the number of violations referred to the Justice Department, penalties assessed, availability of resources, etc. For example, a report of the General Accounting Office shows that regional and state inspection coverage (in regard to the Clean Air Act) ranged from a low of 27 percent in the Chicago region to a high of 74 percent in the Philadelphia region.[27]

Despite the problems in getting accurate data, the annual *Measures of Success Management Report* of the EPA[28] contains detailed figures on the number of penalties assessed, the number of facilities inspected, and the results of enforcement actions. However, these statistics (e.g., numbers of enforcement actions and total penalties assessed) do not in themselves indicate enforcement effectiveness and do not say a great deal about trends in implementation issues.[29] In addition, the report says little about the effectiveness of different tools and instruments in policy implementation. Indeed, because of the absence of reliable information, analysis, and data on how the member states in the EU and the states and regions in the US are performing their enforcement responsibilities,[30] it is not possible to say whether US or EU environmental policy is implemented better. Despite all the differences in the US and EU enforcement systems, empirical studies show that enforcement is not very effective in either system.[31] A European-wide study on criminal law enforcement by Faure and Heine shows that sanctions provided in environmental legislation are relatively small and are much more likely to be imposed by the national authorities and courts than severe penalties,[32] although administrative penalties can become very expensive (especially in the Netherlands and in Germany). In addition, Faure and Heine point to the fact that one should take into account "that at a certain level potential offenders cannot be deterred any longer with fines, given the insolvency risk."[33] In the US also, high penalties "fail to produce measurable increases in deterrence or compliance; however, it may be a result not of deterrence not working, but because even the higher penalty amounts remain too low to matter to polluters."[34]

In addition, as a study from the German Environment Agency shows, most violations do not become known to the public authorities.[35] In the case of violations that do become known, most inspection authorities take a cooperative approach toward the violators since they depend very much on the cooperation ("good will"), the technical expertise, and the data provided by those who are regulated. Furthermore, local politicians often have no interest in pursuing important local enterprises for environmental infringements. It seems that the more the inspection authorities are dependent on the regulated parties, the more the authorities seek cooperative behavior and the less likely they are to impose administrative or criminal sanctions.[36]

Because of all these shortcomings in the deterrent approaches, the EPA and the different inspection authorities in the EU will have to concentrate more resources on the design and improvement of performance indicators for enforcement. Initial developments are encouraging. In 2002, a number of EU member states, the EPA, and a number of international authorities started to develop a system for evaluating capabilities for environmental implementation, compliance, and enforcement programs within the International Network for Environmental Compliance and Enforcement (INECE).[37]

Change of Regulatory and Governance Styles and the Impact on the Implementation Process

Traditionally, citizens have played a much more important role in the enforcement process in the US than in the EU. There is no doubt that for a long time EU citizens have had too few control instruments and too little power and information. The principal enforcement means available are a complaint before the European Commission, the direct effect of Community law at national level, the requirement to pay damages if EC law is not or is insufficiently implemented at the national level, access to information, and political pressure on the different institutions.

However, the role of citizens in the EU decision-making and implementation process is about to change. A dialogue-oriented policy, a demand for more openness and clarity, and a change of focus toward the citizens have now also reached the European Union. Article 1 of the Treaty on European Union refers to openness of decision-making at the EU level. Article 255 stipulates the right of access to documents of the European institutions. In regard to secondary law, Regulation 1049/2001 of May 30, 2001 regulates the participation of citizens in the decision-making process. Furthermore, the European Commission's

White Paper on Governance puts concepts such as "openness, transparency, and participation" at the center of future EU governance.[38] Particularly in the field of EC environmental policy, new European Commission proposals regarding the creation of a public emissions register, environmental liability, access to information, and access to national courts, as well as greater involvement of the public in the decision-making process—particularly in the framework of the implementation of the Aarhus Convention—will have significant effects on the role of citizens in the implementation and application of Community law.

This shift toward more openness and transparency is important and is certainly a positive development, but it is nothing special compared with the role that citizens of the United States have had for decades in the implementation process. More important (and much more unique) is a development on the EU level toward the introduction of a "menu" of new management and regulatory instruments. In the past, there was only one way to do things—the classical community method (e.g., Article 251). In the future, new and additional forms of "governance," including nonlegally binding regulatory instruments, will complement the Community method.[39]

The European Commission has the following plans:

▪ to promote greater use of "framework directives" and self-regulation (e.g., use of voluntary agreements, see chapter 6) in addition to legally binding instruments (i.e., regulations, directives, decisions);

▪ to introduce new decision-making procedures (such as open methods of coordination, or coregulation) that may affect the distribution of powers between the EC and the member states and the rights of the European Parliament and the citizens, but have so far been little studied;

▪ to propose more possibilities for derogations instead of detailed rules;

▪ to conclude more so-called "tripartite contracts" with regional and local authorities;

▪ to transfer regulatory activity from the legislative procedure (e.g., via Article 251) to the executive procedure [via Articles 202 or 211, and the Comitology decision (1999/468/EC)];

▪ to set priorities in monitoring the application of Community law and to allow more flexibility, and;

• to apply benchmarking and "good-practice" techniques in the environmental field.

All of these developments and proposals have different and specific positive or negative effects on the role and the involvement of the public and on the effectiveness of the implementation process. Similarly, the classical concepts of "democracy" and "legal certainty" are about to change in the EU.

It is interesting that a number of the new governance methods and instruments proposed in the EC White Paper bear similarities to the "reinvention" strategies that have been adopted during the past decade in the US (see chapter 6). However, it remains doubtful on both sides of the Atlantic whether these new management instruments and regulatory styles are more effective than classical regulatory instruments (see later discussion). It is also not clear yet whether they exhibit more respect for democratic rules, legal certainty, and citizen's rights than did traditional methods of public administration and decision-making. In addition, it is still unclear whether the implementation of environmental law will be improved by new performance-based management methods, nonlegally binding instruments, and new enforcement styles.

To be sure, the new developments might entail advantages as well as disadvantages. For instance, improved access to documents in the EU extends the rights of the citizen. But new approaches do not necessarily strengthen individual rights in the implementation of Community law (e.g., the more flexible the legal instruments, the less they will grant direct rights in the enforcement phase). The tension between flexibility and legal certainty will undoubtedly increase in the future, especially in the EU. However, we are not looking at contradictions here, but rather at conflict-laden developments that will require new solutions in the implementation process.

The Emergence of Compliance Incentives and Compliance Assistance as New Policy Instruments

Because of the inadequate results of traditional means of enforcement, both the US government and the EU have tried to give states and regions more flexibility in implementation and enforcement. Flexibility can be increased by setting result-oriented performance objectives and introducing new (nonbinding) instruments in place of detailed regulations.[40] However, despite the general recognition of the need for goal-oriented approaches, different regulatory instruments, and more flexibility, these new policies have provoked disagreement over roles, duties, and priorities and over how much flexibility the local level should have under different programs. Consequently, a 2001 report of the US General Accounting Office mentions a better working relationship between the EPA and the states and better data and information management systems as the major challenges for the future.[41]

Because of the tremendous societal changes in recent years, it has also become more and more obvious that regulators need to find better ways to help the private sector learn what they must do to comply. For example, rewarding companies for good practices has become an important concept in both the US and in the member states of the EU. If companies achieve environmental objectives and comply with legal obligations, changes can be made in licensing and enforcement by state authorities as compliance incentives.[42] Within a context of mutual trust, detailed and formalistic monitoring could gradually disappear and be replaced by consultation, leading to voluntary agreements on environmental targets (chapter 6). In addition, companies that are doing well in terms of compliance might be less frequently inspected; instead, the authorities could rely more and more on the company's own sense of responsibility.

This model has several advantages since it requires less supervision and coercion, and is based on positive motivation rather than mistrust. It should be noted, however, that because of the limited powers of the EU to intervene in national compliance, inspection, and enforcement policies, the EU cannot—like the EPA—introduce and implement programs on compliance incentives and "regulatory relief." Moreover, if member states adopt more flexible enforcement programs, this could clash with the strict procedural and legal requirements of the European Court of Justice in regard to member state duties to implement and enforce EU environmental policies. Another problem is that relevant laws—such as the Integrated Pollution Prevention and Control (IPPC) directive—still apply and have to be fully and correctly implemented.[43] Thus far the European Court has not allowed the public or private sector to be exempted from legal obligations and procedural requirements, even if the objectives are achieved. Another problem for these incentive policies is that this approach may work only in cultures in which the relationship between the public and private sector is based on communication and mutual trust.

In both the US and the EU one can observe an important shift to new forms of self-regulation in recent years, especially through the introduction of environmental management systems [International Standard on Environmental Management Systems (ISO 14001) and the EU Eco-Management and Audit Scheme (EMAS)] and other voluntary and market instruments. The idea of rewarding companies for good practices has become an important concept, especially in the United States, the Netherlands, the United Kingdom, and Germany.[44] For example, in the Netherlands, regulators acknowledge that companies that participate in the so-called covenanting process are generally subject to a less prescriptive supervision than those that do not. "This means that regulators visit these companies less frequently and are less prescriptive about the technologies and techniques that these companies use to reach their environmental targets. For those companies without an environmental management system, the regulators are much more likely to tell companies what technologies and techniques they must use."[45]

In the Netherlands a government discussion note on the future of Dutch environmental law explicitly calls for more self-regulation policies and more responsibility of the private sector in complying with Dutch environmental law.[46] In the same way, the twelve provinces in the Netherlands are engaged in an enforcement strategy in cooperation with the private sector. This strategy is based on an agreement between the two sides that the inspection authorities will control and inspect less as long as the companies prove that they will reach the fixed objectives. The government of Germany adopted a similar policy in June 2001.[47]

The US Environmental Protection Agency has also developed two strategies that are very similar to the ones that are applied in the EU member states:

- the compliance assistance approach and
- the compliance incentive approach.

The compliance assistance approach includes a number of initiatives such as on-site visit assistance, training of personnel and inspectors, workshops on environmental policy and law, establishment of hotlines for citizens and private firms, setting up assistance centers for enterprises, production of easily understandable guidelines and guidebooks, and invitations to environmental management seminars and/or competitions.

The compliance incentives approach aims at motivating the private sector to comply with legal requirements by offering special advantages to those who comply with the law. Here one can differentiate between leadership programs, innovation programs, and programs that link specific environmental performance by the private sector (for example, carrying out an audit) with the promise not to prosecute the firm if violations are discovered. As in the EU, a number of uncertainties exist with these policies since the relevant laws still apply and companies cannot be sure that they will not be prosecuted, especially under statutes that allow citizen suits.

The policy on compliance assistance is based on the premise that a significant reason for noncompliance is lack of knowledge on the part of the regulated community about their regulatory obligations and how to comply. According to the EPA, "small businesses will do the right thing if they have the information that they need to comply. The key is to get information on environmental requirements into the hands of all business, small, medium and large, who want to comply."[48]

In its broadest sense, the content of compliance assistance can vary greatly, ranging from basic information on legal requirements to specialized advice on what technology may be best suited to achieve compliance at a particular facility. Compliance assistance also may be delivered in a variety of ways, ranging from publications to conferences and computer bulletin boards, and to on-site assistance provided in response to a specific request for help. The EPA has funded nine compliance centers for various industrial sectors, all available through the Internet. The EPA has also developed notebooks for twenty-eight major industries that are intended to help them to understand their regulatory obligations through easily readable guides.

Another model of compliance assistance is used in the United Kingdom, where approximately 100 environmental business clubs were set up that "provide a number of free or low cost services to their members. These clubs are financed by a mixture of private, government and EU funding, and offer information, advice and guidance on environmental matters."⁴⁹ In addition, the UK government runs an environmental helpline. This helpline can be called by any commercial organization and gives free information on a wide range of environmental issues.

The EPA's compliance incentive policy includes two policies that began in 1995 and 1996. The first is simply called the Audit Policy and the second the Policy on Compliance Incentives for Small Business. The purpose of the audit and small business policies, which are available to entities of any size, is to enhance protection of human health, safety, and the environment by encouraging regulated businesses to voluntarily discover, promptly disclose, expeditiously correct, and prevent violations of federal environmental law. Benefits available to businesses that qualify for the audit policy include reductions in the amount of civil penalties and no recommendation for prosecution of potential criminal violations for violations discovered either through regular audits or during government-sponsored on-site compliance assistance activities. In fiscal year 2000 alone, 425 companies disclosed potential violations at nearly 2,200 facilities. Since 1995, approximately 1,150 companies have disclosed potential violations at more than 5,400 facilities.⁵⁰

Given the size of the United States, these figures are encouraging but not overly impressive. However, the case of Germany shows that the policy itself is interesting: "In Germany the Bavarian Industry has committed itself to ensuring that 3,500 organizations conduct environmental reviews. In addition it will ensure that 500 industrial sites are registered under the EMAS—the Eco-management and audit scheme. In return, the government has promised to free companies which are registered under EMAS from various reporting, documentation and control requirements."⁵¹ Despite this success, it is at least questionable whether these policies might conflict with the case law of the Court of Justice and the obligations set by the Integrated Pollution Prevention and Control directive.

The question of whether firms that have introduced environmental management systems perform better than those that have not has—

surprisingly—not yet been answered. One comparative and empirical study by the University of Sussex (UK) revealed that firms with a "certified environmental management system do not appear to perform better than those without."[52] Much more research is needed on this question.

It is also not at all clear whether cooperative and other flexible approaches are cheaper and more effective than traditional approaches. In fact, there is no persuasive argument that deterrent-based approaches are more expensive and time-consuming. Faure and Heine conclude in their study *Criminal Enforcement of Environmental Law in the European Union* that "it is . . . difficult to compare administrative sanctions and criminal penalties at a very abstract level. Both can be effective in specific situations."[53]

Some states in the US generally view compliance assistance policies and working with violators as a cheaper way to achieve environmental results since they perceive civil judicial cases as particularly resource- and time-consuming.[54] However, it is difficult to prove that the cooperative approach will encourage environmental protection more effectively than a deterrent-based approach. Even if one accepts the factual assumption that firms have become more environmentally friendly, this does not mean that replacing a deterrent approach with a more cooperative approach is a logical response. The elimination of deterrent approaches could well encourage attitudes that have changed for the better to change back again. Nobody can guarantee that environmental consciousness will not alter in the future. Because of the growing evidence on the pros and cons of traditional and cooperative enforcement strategies, even critical observers of adversarial approaches agree that deterrent-based approaches must remain important. This is also confirmed by the *Success Management Report* of the EPA for the year 2000.[55] In this report, nothing points to the fact that deterrent approaches will play a less important role in the future than they have so far. On the contrary, the figures presented show that the number of inspections and penalties assessed are rather stable or even increased over the past few years. From this one might draw the conclusion that despite all the public rhetoric, deterrent approaches will continue to play an important role and be complemented—but not replaced—by new compliance incentive and assistance tools.

Conclusions

The US and EU implementation processes are fundamentally different in that the EU can enforce measures only against member states, while in the US federal agencies such as the EPA enforce the law directly against regulated industries. In the EU, the implementation and enforcement of environmental policy is left to the member states. As a result, the European Union does not have the competence to intervene in the administrative structures and the different inspection policies on the national level. The European Commission can only formally monitor the application of Community law in the member states. Unlike the US system (and more specifically the EPA), the EU primarily has legislative and (certain) oversight powers.

Despite all the legal and political differences between the US and the EU, the differences between the regulatory and enforcement styles of the member states of the EU and the US are gradually diminishing. The reasons for this are the growing internationalization of environmental policy, changing environmental problems, and common features of modernization and administrative reform processes. In both the US and the fifteen member states of the EU, there appears to be a trend toward a more sophisticated model of environmental protection that integrates classic regulatory approaches with incentive- and assistance-based methods, provides the public with more information (and the right to know) about environmental problems and (industry) performance, and focuses on a second generation of more diverse environmental problems and sources. In addition, in the enforcement process, deterrent instruments continue to play an important role.

Because of continuing changes in European and American societies, classical models of implementation are coming to seem more and more old-fashioned. New approaches to implementing public policies are being developed that take these changes into account. The concept of implementation is now increasingly seen in a context that encompasses a number of factors. Implementation of a legal act is neither purely a legal process, nor is it only about money and resources. Rather, it involves a complex set of dynamic factors that have to be addressed. Moreover, the state of compliance is continuously changing. New companies are created and others go out of business; new technologies or industries are introduced; markets shift; regulatory provisions are amended or updated; and citizens ask for better governmental services. Therefore effective enforcement requires an intelligent mix of deterrent, assistance, and incentive policies.

The environmental policy toolboxes currently used in the US and the EU member states are surprisingly similar. The role and involvement of citizens, NGOs, and private actors especially has increased substantially in the implementation of US and European environmental policy. The web pages of the EPA and the European Commission now provide a large amount of information to the public on the Internet. Nevertheless, important differences still exist in the administrative, liability, and criminal law systems of the US and the EU and its member states. These are likely to remain since they reflect different legal and political traditions and cultural values.

Despite all the innovations and reform projects that have been applied in recent years, the implementation of environmental policy remains deficient, and there is no empirical evidence that the state of implementation has improved within the past few years. Indeed, it appears that environmental crimes may have increased rather than decreased in some countries.

In addition, too little is still known about the effectiveness and efficiency of new policy instruments compared with the traditional deterrence-based sanctions. One important precondition for effective incentive policies in the future is that experts have a better understanding of what motivates the private sector to improve its environmental performance. In this respect, the concepts of compliance assistance and compliance incentives have significant potential. Still, it is difficult to believe that new consensual approaches will replace the traditional top-down approach. Command-and-control instruments have the advantage that they are predictable.

There has been an extraordinary emergence of new networks dealing with implementation and enforcement issues in recent years. On the European level, INTERPOL and its Working Party on Environmental Crime were founded in 1992. The European Network for the Implementation and Enforcement of Environmental Law (IMPEL) was created in the same period, followed by the Accession States IMPEL (AC-IMPEL). These networks are themselves part of a global network (the International Network for Environmental Compliance and Enforcement). In the Americas, a network was created among Canada, Mexico, and the United States (the North American Commission for Environmental Cooperation, NACEC). The G-8 also started to work on enforcement issues after 1997. Finally, in 1998 the European Commission invited the European Environmental Bureau (the European umbrella organization of more than one hundred environmental NGOs) to become the European secretariat of the Transatlantic Environmental Dialogue (see chapter 13).

Informal exchanges between the EU and the US authorities have also become a part of daily life. The internationalization of enforcement issues and the emergence of networks are so important because growing international trade poses new challenges to classical enforcement methods.

These multilateral and bilateral partnerships and meetings should serve mutual interests by providing more evidence about the effectiveness of different policy instruments. A comparison of enforcement actions, performance measurement systems, compliance assistance and incentive programs (including environmental management systems), and new reporting and data management systems will offer many opportunities to learn from each other. Because of this, transatlantic cooperation in the field of environmental policy implementation is only about to begin.

Notes

1. Paul Berman, "Thinking About Programmed and Adaptive Implementation: Matching Strategies to Situations." In *Why Policies Succeed or Fail?*, Helen Ingram and Dean Mann, eds. (London, Beverly Hills: Sage, 1980).

2. Paul B. Downing and Kenneth Hanf, *International Comparison in Implementing Pollution Laws* (Boston: Kluwer-Nijhoff, 1983), p. 334.

3. Christoph Knill, "European Policies: Impact of National Administrative Traditions," *Journal of European Public Policies* 18 (1999): 24; Christoph Knill, *The Europeanisation of National Administrations* (Cambridge: Cambridge University Press, 2001). See also Matthieu Glachant, ed., *Implementing European Environmental Policy* (Cheltenham, UK: Edward Elgar, 2001).

4. Adrienne Héritier, Christoph Knill, and Susanne Mingers, *Ringing the Changes in Europe. Regulatory Competition and Redefinition of the State* (Berlin and New York: de Gruyter, 1996).

5. Robert Pedler and Günther Schäfer, eds., *Shaping European Policy and Law* (Maastricht, the Netherlands: European Institute of Public Administration, 1996).

6. Sandra Pellegrom, "The Constraints of Daily Work in Brussels: How Relevant Is the Input from National Capitals?" In Duncan Liefferink and Michael Skou Andersen, eds., *The Innovation of EU Environmental Policy* (Oslo: Scandinavian University Press, 1997), p. 49.

7. This assertion does not affect the significance of the commission in the overall legislative process, the administration of Community funds, the implementation of Community law (under Article 211 ECT in conjunction with Article 202.3 ECT), or its supervisory activities. In view of its many powers, the commission is still—in the opinion of this author—underestimated. However, in regard to its capability to fulfill the tasks assigned to it by treaty, the commission is overestimated. Experts thus refer increasingly to a "management deficit" in the commission.

8. These findings are based on interviews the author carried out (in the framework of a research project) with officials in a number of member states. See Christoph Demmke and Martin Unfried, *European Environmental Policy: The Administrative Challenge for the Member States* (Maastricht, the Netherlands: European Institute of Public Administration, 2001). See also Philip Lowe and Stephan Ward, *British Environmental Policy and Europe: Politics and Policy in Transition* (London and New York: Routledge, 1998).

9. Christoph Knill and Andrea Lenschow, "Change as 'Appropriate Adaptation': Administrative Adjustment to European Environmental Policy in Britain and Germany." In European Integration on-line papers, 1998, p. 2, <http://eiop.or.at/eiop/texte/1998-001.htm>.

10. In the academic field much confusion reigns as to the exact number of environmental legal acts and the right method to calculate the "environmental *acquis*." (Should executive acts be included? Should only legal acts based on Article 175 ECT be counted? What about amended acts?) See Mogens Moe, *Environmental Administration in Denmark*, Environmental News No. 17, The Hague, Danish Environmental Protection Agency, 1995; Lowe and Ward, *British Environmental Policy and Europe*, p. 25; Ludwig Krämer, *Focus on European Environmental Law* (London: Sweet and Maxwell, 1997), p. 25.

11. Andrew Jordan, Duncan Liefferink, and Jonathan Fairbrass, "The Europeanization of National Environmental Policy: A Comparative Analysis." In Andrew Jordan and Duncan Liefferink, eds., "The Europeanization of National Environmental Policy: National and EU Perspectives" (photocopy, 2003, p. 22).

12. Ibid.

13. David Liefferink and Andrew Jordan, "An 'Ever Closer Union' of National Policy? The Convergence of National Environmental Policy in the European

Union," paper presented at the Institute of European Studies, University of Belfast, January 31, 2003, p. 16.

14. See http://www.complianceconsortium.org/public/con_0001_00.html. See also National Academy of Public Administration, *Transforming Environmental Protection for the 21ˢᵗ Century* (Washington D.C.: NAPA, 2000). Available at http://38.217.229.6/NAPA/NAPAPubs.nsf/9172a14f9dd0c3668525696700651 0cd/667599b12550441e8525699900753535/$FILE/environdotgov.pdf, p. 136.

15. U.S. Office of Personnel Management, *Federal Civilian Workforce, Statistics, Employment and Trends* (Washington, D.C.: Government Printing Office, 2000). The figures for the Directorate-General Environment were received informally.

16. See National Academy of Public Administration, *Transforming Environmental Protection for the 21ˢᵗ Century*, p. 136.

17. US General Accounting Office, *More Consistency Needed Among EPA Regions in Approach to Enforcement*, GAO-Report, GAO/RCED-00-108 (Washington, D.C.: Government Printing Office, 2000), p. 41.

18. Environmental Protection Agency, Office of Compliance, *FY 2000 Measures of Success Management Report* (Washington, D.C.: EPA, 2001). Available at http://es.epa.gov/oeca/main/fedgov/npms.html.

19. In regard to the recent plans to reform the procedure see European Commission, COM (2002) 725 final of December 13, 2002.

20. European Commission, Third annual survey on the implementation and enforcement of Community environmental law, SEC (2002) of October 1, 2002; see also XVIIth Report on monitoring the application of Community law, COM (2000) 92 final of June 23, 2000.

21. See Annex II in XVIIIth Report on monitoring the application of Community law, COM (2001) 309 final of July 16, 2001. See also the Annual survey on the implementation and enforcement of Community environmental law (1996/7) (SEC 1999/592), adopted on April 27, 1999. Article 226 of the Treaty of Amsterdam stipulates: "If the Commission considers that a Member State has failed to fullfil an obligation under this Treaty, it shall deliver a reasoned opinion on the matter after giving the State concerned the opportunity to submit its observations. If the State does not comply with the opinion within the period laid down by the Commission, the latter may bring the matter before the Court of Justice." Article 228 allows the Court of Justice to impose a lump sum or penalty payment on member states who have not fulfilled their obligations.

22. See for this classification, European Environment Agency, ed., *Reporting on Environmental Measures: Are We Being Effective?* Environmental issue report No. 25 (Copenhagen: EEA, 2001).

23. Ibid., p. 15.

24. See foreword by Nigel Haigh in European Environment Agency, *Reporting on Environmental Measures*, p. 6.

25. Ibid., p. 26.

26. Obtaining and managing environmental information has been a long-standing challenge for the EPA, and the lack of consistent, comprehensive, and accurate data is still considered one of the major problems in US environmental policy. This is remarkable because this issue is much less discussed in the EU. See also US General Accounting Office, *More Consistency Needed among EPA Regions*, p. 44.

27. This figure, however, should not be interpreted in the sense that enforcement is better in Philadelphia than in Chicago. Ibid., p. 23.

28. See note 18.

29. Commission for Environmental Cooperation, *Indicators of Effective Environmental Enforcement*, Proceedings of a North American Dialogue (Montreal: CEC, 1999), p. 43.

30. At present, the ENVIRONET system in the United States (http://es.epa.gov/oeca/main/compdata/index.html) provides an enormous amount of compliance data (to anyone interested) and can be compared to the Finnish VAHTI system or the German GEIN (which functions as an information broker and less for the purpose of compliance).

31. See Deutsches Umweltbundesamt, *Umweltdelikte 1999* (Berlin: Umweltbundesamt, 2000); Werner Rüther, *Die behördliche Praxis bei der Entscheidung und Definition von Umweltstrafsachen* (Bonn: Kriminologisches Seminar der Universität Bonn, 1991), p. 126; A. B Blomberg and F. C. M. A Michiels, *Handhaven met effect* (The Hague: Vuga, 1997).

32. Michael G. Faure and Günther Heine, *Criminal Enforcement of Environmental Law in the European Union*. Report for the IMPEL Network and the Danish Environmental Protection Agency, Maastricht University, July 2000, p. 54.

33. Ibid., p. 63.

34. John D. Silberman, "Does Environmental Deterrence Work? Evidence and Experience Say Yes, but We Need to Understand How and Why," *Environmental Law Reporter* 30 (July 2000): 10523.

35. Deutsches Umweltbundesamt, *Umweltdelikte 1999*.

36. Ibid.

37. See under performance indicators: http://www.inece.org.

38. European Commission, *White Paper on European Governance*, COM (2001) 428 final, July 25, 2001, pp. 19 and 20.

39. European Commission, Report from the Commission on European Governance, COM (2002) 705 final of December 11, 2002.

40. In the US, a National Environmental Performance Partnership System (NEPPS) was established in 1995. Similar "governance" concepts are included in the *White Book on European Governance of the European Commission*, which was published on July 25, 2001 COM (2001) 428 final.

41. US General Accounting Office, *Major Management Challenges and Program Risks*, Report GAO-01-257 (Washington, D.C.: Government Printing Office, 2001), p. 30. US General Accounting Office, *Status of Achieving Key Outcomes and Addressing Major Management Challenges*, Report GAO-01-774 (Washington, D.C.: Government Printing Office, 2001). See also Environmental Protection Agency, "Key Open GAO Recommendations," at http://www.gao.gov/transition/trans_epaopenrec.htm.

42. See, for example, Interprovincial Overleg (IPO), *Changing Role of Licensing and Enforcement* (Rijswijk, the Netherlands: IPO, 1998).

43. OJ L 257 October 10, 1996, 26.

44. Interprovincial Overleg, *Changing Role of Licensing.*

45. Richard Starkey, ed., *Environmental Management Tools for SMEs: A Handbook* (Copenhagen: European Environment Agency, 1998), p. 60.

46. Ministerie van Volkshuisvesting, Ruimtelijke Ordening en Mileubeheer, *Met Recht Veraantwoordelijk!* (The Hague: VROM, 2001).

47. German Ministry of Environment, "Nature Protection and Nuclear Safety," press release No. 182/01, Berlin, September 19, 2001.

48. Environmental Protection Agency, *Protecting your Health and the Environment through Innovative Approaches* (Washington, D.C.: EPA, 1999), p. 6.

49. Richard Starkey, *Environmental Management Tools*, p. 29.

50. Environmental Protection Agency, *Audit Policy Update*, Vol. 5, No. 1 (spring 2001) (Washington, D.C.: EPA), p. 1.

51. Richard Starkey, *Environmental Management Tools*, p. 24.

52. University of Sussex (SPRU) et al., *Measuring the Environmental Performance of Industry (MEPI)*, Final Report (February 2001), p. 211.

53. Faure and Heine, *Criminal Enforcement*, p. 85. Another problem is that environmental crime is often corporate crime that cannot be reached by traditional criminal law, which deals with natural persons.

54. See US General Accounting Office, *More Consistency Among Regions Needed*, p. 42.

55. See note 18.

6

Convergence or Divergence in the Use of "Negotiated Environmental Agreements" in European and US Environmental Policy: An Overview

David J. E. Grimeaud

As a result of a greater understanding of environmental degradation by European and US policymakers, numerous related policies, laws, regulations, codes, and measures have been adopted both at the European Community and at the national level by member states and by US authorities. Environmental regulations that set, inter alia, conservation principles and objectives, environmental quality standards and emissions limits, or product and process-related requirements, are also the result of the emergence of resourceful and informed "environmental stakeholders," including civil society and nongovernmental organizations (NGOs), and the progressive acceptance by business sectors of the need to address environmental externalities.

However, despite regulatory efforts, scientists and administrative agencies claim that the global environment is still degrading. Causal factors include the ongoing development and emissions of new and still-unregulated industrial processes and substances, the inadequacies of laws and standards based primarily on command-and-control (CAC) regulations, the regulatory bodies' insufficient knowledge and resources, and the capture of the regulatory process by economic lobby groups. Also, poor compliance records often may be a result of the unwillingness or inability of industries to undertake the necessary investment and managerial measures, combined with the absence of adequate monitoring and enforcement actions by regulators or by civil society, and pressure groups who might not always be entitled to initiate legal proceedings before judicial and administrative courts.

Consequently, numerous stakeholders have called for the development of new instruments, including "negotiated environmental agreements"

(NAs).[1] They argue that a cost-effective, efficient, less burdensome, and appropriately targeted environmental policy that is better complied with requires the participation of all stakeholders in the design and implementation of innovative prevention and mitigation and tools, as opposed to "classic" regulatory means.

This chapter examines, in accordance with the core objectives of this book, the sources of and the extent to which such instruments have developed on both continents, whether common patterns can be identified, and whether mutual learning and convergence might take place between the US and European experiments. The first section describes the environmental policy context within which NAs have emerged, as well as the main incentives and alleged benefits that provide the foundations for their development. The second examines the guidelines and criteria that are generally identified as being the key components of effective, efficient, and transparent NAs. In light of the findings in the first two sections, the third provides a comparative survey of European and US experiments, including an examination of the type of NAs that have developed in EU member states, in the Netherlands (the so-called Dutch model of environmental covenants) and in the US (the Project XL initiative). Conclusions are drawn in the final section.

Policy Context and Related Incentives as a Basis for the Emergence of Negotiated Environmental Agreements in Europe and the US

Negotiated environmental agreements respond to concerns raised among policymakers, industries, and to a lesser extent by environmental groups as to whether the classic set of regulatory tools is appropriate for addressing complex pollution issues in an efficient and cost-effective manner. In particular, critics point to the extensive use of CAC regulations as being too prescriptive, detailed, ineffective, inflexible, and adversarial, as well as not being site-specific enough and not allowing sufficient public participation.[2] Some argue that market forces should be permitted to jointly define environmental targets more often or, at the very least, the means of addressing environmental externalities, while public authorities should simply set general environmental goals, act as an enforcer, and monitor the state of the environment. Accordingly, the following section examines the relevant policy and legal context as well as the main arguments, incentives, or presumptions that European and US stakeholders have relied upon to promote the adoption of more collaborative approaches as opposed to top-down government intervention.

An Overview of the European and US Policy Context
Many claim that the emergence of NAs in Europe and in the US reflects the long-awaited instrumental "revolution" in which environmental laws and/or implementing measures will no longer be designed solely by regulatory public authorities, but also by those in charge of realizing them, those whose activities have a direct bearing on the environment. In fact, on both continents, policy and legal contexts are characterized by, on the one hand, the predominance of the top-down CAC regulatory model, and on the other hand, by the non- or partial achievement of environmental objectives.

The European Policy Context A distinction must be drawn between NAs developed at the EU level and those that are designed and implemented by member states. EU interest in using NAs as a tool of EC environmental policy was formally expressed in the 1992 EC Fifth Environmental Action Program (EAP), "Towards Sustainability."[3] It insisted on the need for shared responsibility in the design of relevant policies between national and Community institutions and industry, which should also be part of solutions to environmental problems.[4] In this regard, the use of a broad mix of Community environmental instruments, including voluntary agreements, is called for as a complement or possibly an alternative to prescriptive EC directives and regulations so as to achieve EU-wide sustainable development. In turn, this would allow all sectors of the society to have a say in policy-making processes since they are the ones who possess expertise and have to implement and comply with environmental measures.

In the same vein, the sixth EAP, "Environment 2010: Our Future, Our Choice" (2002), stipulates that nonregulatory methods may be the most appropriate and flexible means, as an alternative to traditional regulation, to address certain environmental issues.[5] In practice, provided that the EC Treaty does not allow EC institutions to resort to NAs as a legislative instrument along with directives, regulations, and decisions, it is a regulatory tool that has so far been only randomly used at the Community level and primarily as nonlegally binding early experiments. Yet, if one considers the recent EC commitment to simplify and improve the Community regulatory environment, one may then expect the EU to make more use of EC-wide voluntary agreements in the form of self-regulation and coregulation. However, at the time of writing, the EC is still debating how such tools may eventually be used as a complement or an alternative to traditional EC environmental law.[6]

In sum, as in the US, there is a growing concern about the inability of regulators, including those at the EC level, to bear the sole responsibility for designing environmental regulations and directives that are both environmentally sound and cost-effective. Yet, since the use of NAs at the EC level is still in a very early experimental stage, the third section of this chapter will focus on NAs that have emerged in the member states, particularly in the Netherlands. However, the EU interest in voluntary approaches is derived not only from the need to simplify and improve EC environmental law but also from the necessity to regulate a practice that had already developed at the domestic level, namely, the implementation of more than 300 voluntary environmental agreements.[7] The EC Commission adopted specific guidelines that define the conditions within which NAs might be concluded, not only by EC institutions as a replacement for Community directives and regulations, but also at the national level to implement either national environmental policies or Community measures.[8]

The US Policy Context   The 1995 Clinton administration's "Reinventing Environmental Regulation" program formed the official basis upon which NAs have emerged as an alternative and/or a complement to traditional US environmental regulations.[9] Similarly to the EU, the call for the reforming of the environmental regulatory system responded to the need to address significant environmental problems that had not been solved by traditional CAC regulations and to improve compliance. The "Principles for Reinventing Environmental Protection" do not seek to abandon the predominant "end-of-pipe" regulatory model, but to "reinvent" it through the design and implementation of more flexible, site-specific, efficient, effective, and cheaper "commonsense" solutions. Environmental policies and regulations should be based on more collaborative as opposed to adversarial approaches and be characterized by the broader participation of all stakeholders in decision-making and the use of new instruments such as NAs. Accordingly, the Environmental Protection Agency (EPA) has promoted new policy tools, including the 1994 Common Sense Initiative (CSI) and the 1995 Project XL (eXcellence for Leadership).[10]

An Overview of the Incentives and Alleged Benefits of Negotiated Agreements

As seen earlier, European and US policy and legal contexts show that the motives for using nontraditional environmental instruments share many similarities. In this framework, four main categories of incentives may be distinguished.

First, those who advocate NAs argue that a proactive attitude on the part of industry is crucial to ensure "realistic" and appropriately targeted environmental norms and standards and compliance with them. Policymakers should not be the only ones to determine "what to do" and "how to do it"; there should be shared responsibility in designing environmental measures in order to benefit from the expertise and experience of all concerned stakeholders. Such a call for dialogue, mutual understanding, and trust between, in particular, regulators and industries, contrasts with the adversarial feature of the US environmental regulatory system. In the same vein, the EC Commission states that industry is often consulted at a late stage, leading to a defensive attitude toward regulations, which it perceives as a threat to economic profitability and competitiveness. Also, by encouraging collaboration, the objective is to obtain a consensus on environmental goals, which may lead to better compliance records. However, environmental groups claim that well-resourced and staffed industry groups already influence the content and adoption of regulations. Thus, if NAs are to be used as a policy tool, which may increase industry inputs or industry capture, room must also be made for noneconomic actors to participate in drafting processes.[11] In addition, the search for consensus may lead to the quest for the lowest common denominator because industries may negotiate for low environmental standards on technological and economic grounds.[12]

A second incentive concerns the adoption of cost-effective measures. Indeed, on both continents, many claim that environmental failure is partly due to the predominance of CAC regulations. While they impose technology design and environmental performance on a whole range of industries and economic sectors, they may be inefficient and an obstacle to innovation.[13] Uniform standards do not take account of the fact that the benefits and costs of pollution abatement vary from one facility to another and that environmental conditions differ according to locations. Traditional regulations may not reward those companies that achieve higher environmental objectives than are prescribed by law unless it leads to economic gains or better reputation.[14] Conversely, NAs may have the advantage of providing industries with the freedom and flexibility to determine at the sector, company, or site level how they will best achieve relevant national environmental policy objectives using tailormade environmental planning and decisions. In turn this may encourage the development of new pollution reduction technologies and managerial innovations. In fact, as explained later, flexibility is one of the core aspects of the US Project XL initiative (regulatory flexibility) and of the Dutch covenants (company-specific environmental plans).

Third, NAs may result in higher environmental standards. For instance, the US Project XL aims at encouraging participating companies to achieve greater environmental performance than required by existing laws, in exchange for which they are provided with "regulatory relief." Moreover, NAs promote the adoption of integrated permits that allow all interactions among air, water, and soil to be taken into account in a holistic manner.[15] However, where NAs preempt future legislation, doubts may remain as to the ambitiousness of agreed-upm environmental objectives, which may be undermined by the search for consensus in the agreement drafting process and the lack of representativeness of third-party interests in negotiations.

A fourth alleged benefit relates to the ability to speed up the process by which environmental objectives are adopted and implemented. In this context, the EC Commission states that more than 2 years are usually needed between the date of a proposal of an EC environmental directive and its adoption, and another 2 or 3 years are needed for its transposition into national legal orders. However, one may also argue that time gained with an NA may well be lost where, for instance, an agreement fails to be reached, which may then require a whole new policy-making process.

Thus, both European and US NAs have emerged on the basis of the these alleged benefits, according to which a collaborative approach may lead to better-designed, flexible, and targeted environmental policies that provide companies with the opportunity to adopt cost-effective and site-specific measures. However, in the particular case where NAs replace legislation, despite the fact that they may reduce the regulatory burden, concerns may then be raised about the level of environmental protection that consensus-based agreements set and the involvement of environmental groups. In this context, beyond the identification of European and US environmental policy contexts and incentives, a comparative examination of NAs on both continents also requires us to look at whether these agreements have developed according to specific guidelines, what form they take, and whether they have achieved their alleged benefits.

Guidelines and Criteria for Designing and Assessing Negotiated Environmental Agreements

To accompany the emergence of European NAs, nonlegally binding guidelines were adopted by the EC Commission. These guidelines lay down a general framework to ensure that NAs are designed in an appropriate fashion to bring about environmental benefits, transparency, monitoring, accountability, and compliance with legal obligations.[16] Similarly, with regard to Project XL, the USEPA published a guidebook to help firms draw up proposals and pass the EPA review process.[17] More particularly, beyond classic contractual provisions on the duration, revision, and termination of NAs, five broad categories of key guidelines may be identified, which may then serve as criteria for assessing how NAs have developed.

First, all relevant stakeholders need to be involved. Public authorities will have to make sure that in the case of sector-based agreements, they negotiate with the business or industry sectors that are directly and indirectly responsible for the concerned environmental issues. In this regard, NAs concluded at the EC and member state level often have involved trade associations rather than individual companies so as to ease and speed up negotiations and to control free riding, which is essential when the agreement replaces legislation.[18] Along with industry, other stakeholders, including local public authorities, communities, and environmental groups, should also participate, in particular where no regulations complement an NA. However, the problem is how to determine when and to what extent public participation should take place. While the EC Commission simply encourages European NAs to provide nonindustry actors with a right to be consulted on draft texts, to make comments, and to be informed on the content of the agreement and on reported data,[19] it seems as if public involvement would be greater under the Project XL initiative in which communities near the concerned facility, local governments, and environmental or public interest groups may be involved as direct participants in the framework of a stakeholder group.[20] In sum, although stakeholders' participation is essential to ensure that NAs are transparent, "democratic," and socially acceptable, European and US practice differs. In the former, third-party interests are often granted only a right of passive participation, whereas Project XL requires a more active involvement, which may derive from the fact that the US initiative leads to regulatory relief.

Second, NAs must clearly determine the environmental objectives to be achieved. The EC Commission's guidelines insist on the necessity of including quantitative objectives expressed in absolute terms, such as a percentage or quantity of emission reductions, using an appropriate and unequivocal base year, as opposed to "best efforts" provisions. In the same vein, the EPA's *Best Practices for Proposal Development* for XL projects provides that the obligation to reach "superior environmental performance" at the facility level should also relate to quantitative and qualitative data and figures and to comparisons in terms of emission levels achieved with and without the project. In addition, the EC Commission stipulates that noncompulsory intermediary objectives should also be set, together with a timetable, to enable the monitoring of progress and the eventual adoption of an amending calendar.[21]

Third, NAs are to include provisions on the monitoring, reporting, and verification of results. Both the EC Commission and the EPA's guidelines hold that a monitoring methodology should be determined in advance and be sufficiently reliable to enable the assessment of environmental performance. Whereas the guidelines call for regular reporting, transparency may also be essential, which would imply that stakeholders should have access to reported information. One may refer to the 1990 EC Council directive on access to environmental information, which provides citizens with the ability to ask public authorities for relevant information, which may include that related to NAs.[22] In the US, the 1986 Emergency Planning and Community Right-to-Know Act (EPCRA) and the Toxics Release Inventory give the general public access to information on emissions of toxic chemicals that may concern data on the implementation of Project XL.[23] For verification of results, the EC Commission proposes the creation of an independent body that would collect and assess the reported information. However, European companies usually reject external verification of their environmental performance. This raises concerns, in particular where NAs replace existing or future laws, since self-reporting and verification may be clearly insufficient and socially unacceptable.

The fourth key element relates to the legal status of NAs. As opposed to Project XL, European NAs are mostly not legally binding and thus provide no sanctions in case of a breach of the agreement. Accordingly, the EC Commission not only calls for agreements that have legal force, but also for each national legal system to define NAs as either private or public law contracts so as to determine precisely what legal regime will be applicable, including the terms of sanctions, liability, and competent jurisdiction. In fact, as an alternative to regulation, most European NAs are linked to the threat of the introduction of new binding laws, which provides both a collective incentive for compliance and an enforcement mechanism.[24]

The guidelines indicate the minimum key elements that all NAs should have to make them collaborative, ambitious, effective, efficient, transparent, socially acceptable, and enforceable instruments. Yet, as seen in the next section, European NAs seem to be generally still far from complying with these criteria. In fact, apart from the Flemish case, there have not been any specific national policy or legal frameworks adopted in member states, which has inevitably affected the design, scope of application, legal status, and other features of NAs. The EC Commission's guidelines simply seek to, among other things, influence the practices of member states in the absence of relevant EC harmonizing legislation. Accordingly, the next section attempts to provide a survey of the main tendencies that have characterized European NAs and to draw a comparison with the Project XL initiative.

A Comparative Survey of European and US Negotiated Environmental Agreements

In light of the policy and legal contexts, alleged benefits, criteria, and guidelines, discussed have, this section provides a critical examination of NAs as they are developing in member states and in the US. A distinction is made between those that have emerged in the Netherlands, where they have become a key instrument of environmental policy, and those that have been concluded in other European countries. US Project XL will then be looked at in a comparative perspective.

European Negotiated Environmental Agreements: A Piecemeal Development

In the absence of EC harmonizing rules and converging national legislative frameworks, NAs have developed on an ad hoc and piecemeal basis.[25] However, whereas one may not talk of a "European model," Dutch environmental covenants may still be distinguished from NAs concluded in other European states.[26]

Except for the Netherlands, NAs concluded in member states address mainly the energy and industry sectors, including agreements on waste management, cholrofluorocarbon (CFC) abatement, and reduction of greenhouse gas emission.[27] In fact, in light of international competitiveness interests, public authorities may be reluctant to impose far-reaching and inflexible energy efficiency, consumption, and emission requirements unless regional or global standards are enacted. Thus, apart from the

Dutch covenants, European NAs are concerned only with a narrow set of environmental issues.[28] Regarding their link to other policy tools, most NAs have been concluded as a substitute for future laws. The primary aims are to prevent the adoption of new CAC regulations and reduce regulatory intervention and the administrative burden. The secondary aims are to allow site-specific environmental measures, to benefit from companies' knowledge of "what to do" and "how to do it" at plant level in a cost-effective and flexible manner, and to promote shared responsibility.[29]

The great majority are "gentlemen's agreements" because national executive government branches are usually not entitled to conclude NAs as an alternative to existing EU or national decrees, laws, or regulations. Accordingly, as nonlegally binding instruments, they do not contain legal sanctions in case of a breach. In practice, compliance incentives derive from the public authorities' threat to adopt new CAC regulations when the agreement is not realized. Yet, while NAs often replace regulations, concerns may be raised about their democratic accountability and the level of public participation. In fact, the parties that participate in the design, monitoring, and signing of NAs are often national, regional or local public authorities and, in sector-wide agreements, trade associations. Conversely, apart from environmental agreements to preserve particular neighborhoods, the active participation of third-party interests, including the public and environmental and consumer groups or local communities, is not a common feature.[30] The involvement of nonindustry actors is often limited to a right to comment on the draft or final text of the agreement and access to the monitoring and verification reports, which may weaken the social acceptability of the NA.

A 1998 study showed that European NAs scored rather low on their environmental performance.[31] While they generally improve the state of the environment, they hardly achieve ambitious protection levels. In most cases the targets are not significantly superior to a business-as-usual scenario or to the "no-regrets" option and do not reach higher environmental performance unless they are combined with an existing ambitious regulatory framework. Furthermore, in contrast to Project XL, most of them are focused on a single environmental medium and do not impose integrated pollution prevention. Thus it may prove difficult to determine whether those agreements have led to cost-effective measures and technological innovations. Then too the lack of environmental ambitiousness raises concern about the degree of capture by industry participants and the need for enlarging negotiations to include other stakeholders, including environmental groups and local communities.

In sum, despite the great diversity of European NAs, they share common features that do not respond to the earlier-mentioned incentives, alleged benefits, and evaluation criteria. They generally are nonbinding, do not achieve higher environmental performance and related innovations, and do not allow comprehensive and meaningful public participation despite the fact that they are usually concluded as an alternative to CAC regulations. In fact, one may argue that the development of NAs in most member states is still in a transitional experimental period. However, should NAs be given further impetus in the future, European countries would have to make sure that they contain ambitious environmental objectives, transparency, participation, and enforcement provisions. The NAs may well be best used within a mixed policy framework where they would be combined with other policy instruments, such as permits, economic instruments, or backup CAC regulations.[32]

In this context, the Dutch covenants may deserve special attention because they are usually considered a workable and effective European NA model.[33] The use of covenants as a proper environmental policy tool followed the adoption in the Netherlands of the first 1989 National Environmental Policy Plan (NEPP), which set ambitious emission reduction targets and timetables for more than 200 substances. Accordingly, industry claimed that the realization of these goals would require a new policy approach based upon environmental covenants because they would need flexibility in defining at company and plant level the best-suited implementation measures.

In practice, negotiations first take place between public authorities and each industry sector. They result in the signing of a sector-wide nonbinding "declaration of intent" that sets long-term emission objectives and timelines. Second, each firm that belongs to the concerned sector and that adheres to the terms of the declaration concludes an individual agreement (private law contract) with the government. The latter includes a company environmental plan (CEP) that is reviewed every 4 years, emission reduction targets, a description of the implementing measures, and a time schedule. When the CEP is approved by the competent licensing authority, it is translated into permit requirements.[34] In fact, such a scheme, which has a high rate of participation by firms, is considered successful in avoiding the shortcomings that often characterize other European NAs. In particular, beyond the mere fact that those agreements fit within a policy mix (the NEPP in which targets are set and the permitting system) and are binding, they give companies flexibility in designing, together with the licensing authorities, the most appropriate and cost-effective implementing measures at plant level, as opposed to having detailed and highly prescriptive permit conditions imposed on them. In addition, each CEP requires yearly reporting, whereas monitoring is ensured within the licensing framework. As private law contracts, any breach of the agreement may lead to a company's civil liability. Also, since agreement provisions are translated into permit requirements, licensing authorities and third parties may lodge administrative actions when license conditions are violated.

Thus, because they are developed within a regulatory framework; ensure democratic accountability and transparency (targets are set in a legislatively adopted NEPP); and provide reporting, monitoring, and enforcement mechanisms, the Dutch covenants are often seen as a potential model for European NAs. However, one has to remember that their objective is not to achieve higher environmental performance than existing targets or to replace regulation, but to reduce government intervention and promote shared responsibility in designing cost-effective and efficient measures.[35] It is precisely within the context of this "limited" aim that NAs are most likely to be socially acceptable.

US Project XL: An Attractive Initiative

As already mentioned, the EPA launched fifty pilot Project XL initiatives in the wake of President Clinton's 1995 "Reinventing Environmental Regulation" program.[36] Whereas the latter aims to encourage industry to produce "cleaner, smarter and cheaper," Project XL relies on four key elements: a site-specific approach, regulatory flexibility, achievement of "superior environmental performance," and the active involvement of stakeholders, including local communities and environmental groups.

For a company to be awarded a Project XL initiative, it must first make a project proposal, which is then subjected to EPA screening and review.[37] A final project agreement (FPA) is then signed, which determines what is to be achieved (environmental performance), how it is to be achieved (which production process and on the basis of what kinds of regulatory flexibility), the timetable of implementation, and the way progress will be monitored and reported and the eventual enforcement actions taken. Whereas the FPA is not in itself a legally binding document, it contains a binding permit or site-specific rule-making (the Project XL requirements) that replaces conflicting existing CAC provisions in accordance with the rule on regulatory relief.[38] Thus, Project XL seeks to identify how regulatory standards may be best adjusted to the features of a specific site so as to implement tailor made environmental requirements.

Another key aspect is the degree of participation of nonindustry stakeholders in the drafting process. Indeed, although they are not signatories to the FPA, local communities and environmental groups may play a significant role because the EPA only awards Project XL proposals that have the support of those who may be affected. Stakeholders can work either as a commentator, but more important, also as a direct participant. In the latter case, they then form a stakeholder group that is consulted by the company at every stage of the FPA development process. Thus they may directly influence the content of the FPA and the scope of regulatory flexibility that would be provided to the firm. Indeed, it is precisely by providing regulatory relief that environmental impacts may be shifted from one environmental medium or from one resident group to another. In this context, all stakeholders (EPA, the firm, and other actors) would have to agree on an appropriate tracking and reporting scheme. At a minimum, data must include information on environmental performance and evaluation methods, their link to the particular regulatory flexibility provided, the involvement of stakeholders in the implementation of the project, and the cost savings.[39]

A third key component of Project XL is the "superior environmental performance" requirement compared with what would have been achieved through "compliance with otherwise applicable requirements." Worded differently, as a result of regulatory flexibility, the concerned firms must achieve greater environmental performance than that prescribed in existing CAC regulations. In 1997 the EPA issued a *Federal Register* notice, which specifies precisely how this requirement should be interpreted.[40] In particular, it stipulates that such a requirement would be evaluated on the basis of a comparison between the actual environmental loadings or the future allowable environmental loadings, whichever is more protective, and the environmental performance achievable under Project XL. In addressing tradeoffs between environmental media and the achievement of "superior environmental performance," the EPA ruled that tradeoffs that may threaten ecological health or risk-based environmental standards (e.g., water quality standards) would be prohibited, as well as those that would result in a shifting of risk burden among local communities.

A final attractive factor is the possibility for firms to benefit from regulatory flexibility. In fact, while participants must achieve "superior environmental performance," they may be allowed to not comply with certain existing regulatory requirements.[41] As an illustrative example, the XL Project for Merck & Co. gave the firm the right for one of its plants not to have to apply for a new permit, contrary to Clean Air Act provisions, every time its production process changed and resulted in increased volatile organic compound (VOC) emissions, but stipulated that such loadings must remain under certain limits. In this context, it was found that those emissions were not leading to more ozone, given the specific meteorological conditions that prevailed at the location of the facility. In exchange for this permitting flexibility, Merck agreed to reduce its emissions of sulfur dioxide and oxides of nitrogen beyond what is currently prescribed. This Project XL was thus a win-win initiative in which both parties were achieving greater environmental performance by taking into consideration the features of the plant and the characteristics of the surrounding environment. Nevertheless, as mentioned earlier, pollutant tradeoffs must not lead to shifting risk onto local communities or plant workers or to violation of public health standards.[42]

In sum, it seems that Project XL may effectively "reinvent" environmental protection, at least to the extent that it leads to flexible, site-specific, cost-effective, transparent, and enforceable environmental measures. Contrary to most European NAs, XL projects are plant-level agreements, are binding, and are not concluded to replace future legislation, but to adjust existing regulatory requirements to the characteristics of a company's production processes and surrounding environment. In addition, one may welcome Project XL's scope of public participation and the fact that its predevelopment costs are likely to be offset by the savings that will result from regulatory flexibility, technological innovations, and tailormade environmental measures. Thus, while many see the Dutch covenants as the NA model, one should not underestimate the Project XL initiative. However, one should also keep in mind that in practice, Project XL has so far had a rather low rate of participation because relatively few company agreements have been concluded and implemented. In this respect, its long-term success and social acceptability may depend on the overcoming of barriers, including clarification of the scope of regulatory relief, the capacity of stakeholders to control pollutant tradeoffs, and the ability to establish criteria for evaluating environmental benefits.[43]

Conclusions

Negotiated environmental agreements in the US and in Europe share both similarities and differences. Their use on both continents is still at an early stage. On the European side, the emergence of EC-wide voluntary environmental agreements is yet to be formalized, while their development as a regulatory environmental tool at the level of member states has been mostly confined to the Netherlands. In the US, the relatively low numbers of XL projects shows, as is the case in Europe, that legal, technical, and negotiating barriers need to be overcome. Another similarity refers to the arguments of US and European legislators on the need to reinvent environmental regulation, namely, the necessity to make it more efficient, specific for a site and local conditions, cost-effective, flexible, and participatory.

In terms of transatlantic differences, European NAs have mostly developed as an alternative to future legislation, and they generally contain sector-wide, unambitious environmental objectives and lack transparency and enforceability. Thus, they would inevitably have to be combined with other binding policy instruments such as the Dutch covenants that are linked to policy-set targets and the permitting system. In sum, if European and national public authorities wish to develop them as a key policy tool, they will have to make sure that the NAs are but one element of an instrument mix, fulfill certain criteria so as to benefit all stakeholders (civil society, the environment, and industries), and contain safeguards (public participation, bindingness, enforceability). In this respect, considering the features of the US XL projects (site-specificity, stakeholders' participation, the requirement for "superior environmental performance," and regulatory flexibility), attention would certainly have to be paid to the outcome of future assessments because these projects may provide a convergence model for a limited use of well-framed NAs.

Notes

1. Negotiated environmental agreements are one of several so-called "voluntary approaches," which can be divided into three broad categories. They include unilateral commitments (voluntary environmental standards designed by the industry), public voluntary schemes (voluntary environmental standards designed by government authorities where industry is free to apply them), and voluntary environmental agreements or negotiated environmental agreements that refer to voluntary *or binding* rules that are agreed upon both by industry and government authorities and eventually by other stakeholders as a complement or an alternative to traditional environmental agreements. I discuss negotiated environmental agreements only as a comparative case illustrative of innovations in environmental regulation that are taking place in both the US and Europe. This focus is based on the assumption that they reflect the objectives of "reinventing" regulatory policy; that is, the enacting and implementation of cost-effective, site-specific, flexible, participatory, environmentally friendly, and accountable forms of management.

2. This chapter holds that negotiated environmental agreements are mainly a response to the partial failure of command-and-control instruments (permits, environmental quality standards, emissions limits, product standards, and production and process methods) to bring about satisfactory environmental outcomes. They should be viewed as one component, among others, of US and European environmental policies that are characterized by a regulatory mix in which other types of regulatory tools are also used or being developed, including, e.g., pollution taxes or emissions trading schemes.

3. See Resolution of the Council and the Representatives of the Governments of the Member States meeting in the Council of February 1, 1993 on a Community program of policy and action in relation to the environment and sustainable development, OJC 138, May 17, 1993, pp. 1–4 and Decision No. 2179/98/EC

of the European Parliament and of the council of September 24, 1998 on the review of the European Community program of policy and action in relation to the environment and sustainable development "Towards sustainability," whereby voluntary agreements are defined as one of the five key priorities of EC environmental policy, OJ L 275, October 10, 1998, pp. 1–13.

4. "Whereas previous environmental measures tended to be prescriptive in character with an emphasis on the 'thou shalt not' approach, the new strategy leans more towards a 'let's work together' approach. . . . The new approach implies, in particular, a reinforcement of the dialogue with industry and the encouragement, in appropriate circumstances, of voluntary agreements and other forms of self-regulation." "Towards Sustainability—A European Community Programme of Policy and Action in Relation to the Environment and Sustainable Development," OJ C 138, May 17, 1993, p. 68.

5. See Decision No. 1600/2002/EC of the European Parliament and of the council of July 22, 2002 laying down the Sixth Community Environmental Action Program, OJ 242, September 10, 2002, p. 61.

6. The EC Commission has committed itself to encouraging the reforming of the European regulatory framework with the objectives of making it more simple, effective, participatory, cost-efficient, complied-with, and better attuned to local conditions in particular sectors of industries or areas. See Communication from the European Commission Action Plan, "Simplifying and improving the regulatory environment," COM (2002) 278 of June 5, 2002. Available at http://www.europa.eu.int/eur0lex/en/index.html. Upon this background and alongside the Sixth Community Environment Action Program, the EC Commission has also adopted a communication on environmental agreements at the EC level. It proposes that EC institutions make use, in particular, of self-regulation. It refers to nonlegally binding unilateral commitments initiated and negotiated only by industry sectors (and in some cases with other economic and social actors), but which would eventually be "formally acknowledged" by the EC Commission. Self-regulation may concern areas of environmental policy where the commission has neither already proposed legislation nor expressed its intention to do so, or conversely, where it has announced its will to regulate. Whereas such voluntary agreements have no legal binding force at the EU level, the commission specifies that its eventual acknowledgment would not only be based on a set of criteria but might also be complemented by the adoption of a binding EC decision that would set monitoring procedures and reporting duties. Also, the commission held that the mere existence of a self-regulatory agreement would never alter its right to initiate EC regulation, all the more so if the agreement failed to bring about its environmental objectives. As examples, one can cite the EC-wide environmental agreements (as self-regulation) concluded with European, Japanese, and Korean auto manufacturing associations on reduction of carbon dioxide emissions from new passenger cars. See Communication from the EC Commission, "Environmental Agreements at the Community Level Within the Framework of the Action Plan on the Simplification and Improve-

ment of the Regulatory Environment," COM (2002) 412 of July 17, 2002. Available at http://www.europa.eu.int/eur-lex/en/index.html.

7. The legal context of European law inevitably affects the scope of national NAs since member states have to make sure that they comply with existing EC environmental law, case law, and the EC Treaty, including Article 28–30 EC on the free movement of goods; Articles 81(1,2,3) and 82 EC on competition law, and Article 87 EC on state aid. In addition, when EC directives intend to create rights and obligations to individuals, NAs cannot implement them because they would be no guarantee that the beneficiaries would be able to ascertain their rights before national courts and that free riders would be avoided. Conversely, where directives only require the establishment of general environmental protection programs, a binding NA can be used if it is combined with an implementing and binding national law. See Commission Recommendation 96/733/EC of December 9, 1996 concerning Environmental Agreements implementing Community directives, OJ L 333, December 21, 1996, pp. 59–61. Note also that EC environmental law may expressly provide the possibility of using binding NAs as an implementing tool in limited and specific circumstances. In that regard see Directive 2000/53/EC of the European Parliament and of the Council of September 18, 2000 on end-of-life vehicles—Commission Statements, OJ L 269, October 21, 2000, pp. 34–43 [Article 10(3)].

8. COM (96) 561 final "Communication from the Commission to the Council and the European Parliament on Environmental Agreements," November 27, 1996, OJ L 333, p. 69 and Council Resolution of October 7, 1997 on environmental agreements, OJ C 321, October 22, 1997.

9. See "Reinventing Environmental Regulation: Clinton Administration Regulatory Reform Initiatives," March 16, 1995 at www.epa.gov.opei.; W. Clinton and A. Gore, "*Reinventing Environmental Regulation,*" USEPA, Office of Policy Analysis and Review, Office of Air and Radiation, Washington, D.C., 1995. See also A. A. Marcus, D. A. Geffen, and K. Sexton, *Reinventing Environmental Regulation: Lessons from Project XL* (Washington, D.C.: Resources for the Future, 2002), pp. 1–9.

Note that the 1990 Negotiated Rulemaking Act of November 29, 1990 (title 5, US Code, 581–590) establishes an administrative collaborative process by which draft environmental regulations may be subject to comment by interested parties. However, it does not lead to enforceable contracts against public authorities and has been used in less than 2 percent of legislative proposals. See G. C. Hazards and E. W. Orts, "Environmental Contracts in the United States," in *Environmental Contracts—Comparative Approaches to the Regulatory Innovation in the United States and Europe*, E. W. Orts and K. Deketelaere, eds. (The Hague: Kluwer Law International, 2001), pp. 71–91.

10. See the 2000 EPA report, "Innovation at the Environmental Protection Agency—A Decade of Progress," at www.epa.gov/opei. This chapter examines the Project XL initiative only as it results in binding agreements. As stated earlier, this chapter aims at examining voluntary approaches that stand as legal

regulatory tools per se. See USEPA, *Regulatory Reinvention (XL) Pilot Projects*, 60 Federal Register 27,282 (1995). Note that the Project XL initiative consists of Project XL for sectors, Project XL for communities, Project XL for government agencies, and Project XL for facilities. This chapter discusses Project XL for facilities because it is the most developed category. See Marcus, Geffen, and Sexton, *Reinventing Environmental Regulation*, pp. 10–23.

11. The need for public participation in (environmental) decision-making has been acknowledged in several international and regional instruments, including the 1992 Rio Declaration on Environment and Development (Principle 10) (at http://www.unep.org/unep/rio.htm) and the 1998 United Nations Economic Commission for Europe Aarhus Convention on Access to Information, Public Participation in Decision-making and Access to Justice in Environmental Matters (Article 8) (at http://www.unece.org/env/.)

12. See C. Coglianese, "Is Consensus an Appropriate Basis for Regulatory Policy?" in Orts and Deketelaere, *Environmental Contracts*, pp. 93–114.

13. See D. D. Hirsh, "Understanding Project XL: A Comparative Legal and Policy Analysis" in Orts and Deketelaere, *Environmental Contracts*, pp. 116–117.

14. Note that numerous countries have developed strict liability regimes for environmental damage that do not provide a compliance-with-regulation defense.

15. Note that a large range of European industries are subject to an integrated permit system under Council Directive 96/61/EC of September 24, 1996 concerning integrated pollution prevention and control, OJ L 257, October 10, 1996, pp. 26–40.

16. See note 8.

17. See EPA, "Project XL: Best Practices for Proposal Development and Principles of Development of Project XL Final Project Agreements" at www.epa.gov.

18. Trade associations must be representative of the industry sectors in terms of their membership or the contribution of their members to the particular environmental problem. Also, the allocation of responsibility between trade associations and their individual members should be addressed via burden-sharing provisions, especially where there are no obligations imposed at the company level. In such a case, the whole sector may be held jointly liable for failure and be sanctioned collectively through the adoption of legally binding laws or regulations. A related issue concerns the need to provide nonmembers with the ability to join the agreement in order to prevent potential competitive distortions.

19. The 1994 Flemish decree on environmental agreements provides for such a right, but draws a distinction between comments made by public environmental authorities that are added as an annex to the final text of the agreement and comments made by other stakeholders that are simply taken into account but not published. See *Belgian State Gazette*, July 8, 1994.

20. See 1999 EPA, "Project XL Stakeholder Involvement: A Guide for Project Sponsors and Stakeholders" at www.epa.gov.

21. Recall the distinction between target-based NAs and implementation-based NAs, which only address implementing measures to meet targets set in legislation or policies.

22. Council Directive 90/313/EEC of June 7, 1990 on the freedom of access to information on the environment, OJ L 158, June 23, 1990, pp. 56–58 and COM (2000) 402, Proposal for a Directive of the European Parliament and of the Council on public access to environmental information where Article 2 (1) (d) defines "environmental information" by referring to, among others, information on environmental agreements.

23. The Emergency Planning and Community Right-to-Know Act (EPCRA); 42 U.S.C. 11011 et seq. (1986).

24. See P. M. Bailey, "The Creation and Enforcement of Environmental Agreements," *European Environmental Law Review* 8(6) (June 1999): 170–179.

25. In fact the 1994 Flemish decree on environmental covenants is still the only European legislative framework. It is concerned with legally binding sector-wide agreements concluded between the Flemish Region and trade organizations. They do not intend to replace existing laws and cannot depart from them in a less strict sense. As part of the agreement, the Flemish Region cannot adopt, at the time of its application, regulations that would contain stricter requirements, unless prescribed by new international or European legal requirements. Infringement cases may give rise to claims for specific performance or damages. For full text and details on the Flemish legislation, see M. G. Faure, "Environmental Contracts: a Flemish and Economic Perspective," in Orts and Deketelaere, *Environmental Contracts*, pp. 167–178.

26. More particularly, whereas 300 NAs had been identified in 1996, Germany and the Netherlands alone accounted for two-thirds of these. See European Commission, *Study on Voluntary Agreements Concluded between Industry and Public Authorities in the Field of the Environment* (Copenhagen: Enviroplan, 1996); P. Borkey, M. Glachant, and F. Lévêque, "Voluntary Approaches for Environmental Policy in OECD Countries: An Assessment (2000)," at www.cerna.ensemp.fr.; and Öko-Institut report "New Instruments for Sustainability—The New Contribution of Voluntary Agreements to Environmental Policy" at http://www.oeko.de/elni/index.htm.

27. See for instance the 1996 German agreement on the reduction of greenhouse gas emissions and the 1996 French agreement on these emissions from the aluminum industry. (See Borkey, Glachant, and Lévêque, *Voluntary Approaches for Environmental Policy*.)

28. Waste-related NAs respond to the need for industry's expertise in setting realistic recycling and management targets (within the limits imposed under EC law) whereas climate-oriented NA are mainly the result of industries' refusal to be taxed for carbon dioxide emissions.

29. The 1996 German agreement on the reduction of greenhouse gas emissions is linked, for instance, to the German authorities' commitment not to adopt new laws relating to the use of waste heat and energy audits.

30. Regional or local public authorities may also be involved in neighborhood or regional environmental agreements such as the nonlegally binding French "rivers or bay contracts" concluded between national and local public authorities, industry, tourism businesses, and riverside owners. The 1992 French water law provides them with a legal basis as a planning instrument. Similar "river contracts" have been developed in Wallonia, Belgium, including the 1996 River Contract of the Upper Meuse. Conversely, neighborhood agreements are rare in other member states.

31. *Environmental Agreements—The Role and Effects of Environmental Agreements in Environmental Policies*, edited by ELNI (Environmental Law Network International) (London: Cameron May 1998).

32. The 1996 Danish agreement on greenhouse emission reductions is linked to the national CO_2 tax system. Note also that several agreements concluded in Luxembourg, Portugal, and Spain complement existing command-and-control regulations. In such a case, agreements provide flexibility only in terms of implementing measures, whereas targets are set in legislation.

33. Early environmental agreements developed in the 1980s were product-related agreements, including the 1987 Environmental Agreements on Detergents and the 1988 Environmental Agreements on Crop Production Products. They were "gentlemen's agreements" with no explicit goals or provisions for monitoring and enforcement.

34. Note that where the industry sector is homogeneous (the same operating conditions apply in all concerned companies), a standardized CEP is adopted at the branch level.

35. Recall that the use of Dutch covenants may nevertheless be limited by the application of EC law. See note 6. For further details on Dutch covenants and related legal issues, see R. Seerden, "Legal Aspects of Environmental Agreements in The Netherlands, in Particular the Agreement on Packaging and Packaging Waste," in Orts and Deketelaere, *Environmental Agreements*, pp. 179–197.

36. See J. Mazureck, "The Use of Voluntary Agreements in the United States: An Initial Survey," 1998, ENV/EPOC/GEEI(98)27/FINAL, posted at www.oecd.org; EPA, "Project XL 2000 Comprehensive Report—Volume 2: Directory of Project Experiments and Results," at www.epa.gov/projectxl/xlcompreport00.htm.

37. The proposal provides information on, among other things, the plant's production, production processes, location and proximity to local communities and residential areas, and the state of the surrounding environment. Information must also address, inter alia, the achievement of a "superior environmental performance," the expected benefits from regulatory flexibility, the involvement of stakeholders, the innovative pollution prevention strategies, and the firm's past record of compliance with existing EPA regulations.

38. In case of a breach, the EPA or private citizens may sue the firm either on the basis of the violation of the command-and-control provisions that the firm has been exempted from, or for violation of the site-specific agreement.

39. See EPA, "Project Tracking, Reporting and Evaluation: A Guide for XL Project Teams," posted at www.epa.gov.

40. See http://www.epa.gov/projectxl/eval2.htm. In fact, disagreements occurred regarding the meaning and scope of the "superior environmental performance" requirement, as shown by the failure of the 3M Project XL proposal. In short, 3M had proposed that one of its facilities be awarded a Project XL 10-year permit that would have allowed it to release 4,500 tons of volatile organic compounds per year, which was below the Clean Air Act targets. However, the EPA and stakeholders rejected this proposal on the ground that 3M's facility had released less than 2,400 tons per year for the past 9 years. The parties were unable to agree on the exact meaning of the terms of "superior environmental performance relative to what would have been achieved through compliance with otherwise applicable requirements."

41. By contrast, the Flemish decree on environmental covenants stipulates that they must not depart from existing regulations.

42. In fact, regulatory relief has led to controversies because of the legal uncertainty regarding the EPA's authority to provide regulatory flexibility. One major problem relates to the fact that Project XL agreements do not prevent legal actions by citizens, under, for instance, the 1994 Clean Water Act or Clean Air Act where the provisions of these acts might be violated by a firm that would be awarded regulatory relief in a Project XL agreement. See Hirsch," *Understanding Project XL*" and A. A. Marcus, D. Geffen, and K. Sexton, "The Quest for Cooperative Environmental Management: Lessons from the 3M Hutchinson Project XL in Minnesota," in Orts and Deketelaere, *Environmental Agreements*, pp. 143–164. See also D. D. Hirsh, "Bill and Al's XL-ENT Adventure: An Analysis of the EPA's Legal Authority to Implement the Clinton Administration's Project XL," *University of Illinois Law Review* 1 (1998): 129–172.

43. See Marcus, Geffen, and Sexton, *Reinventing Environmental Regulation*, pp. 159–196.

7

What Future for Environmental Liability? The Use of Liability Systems for Environmental Regulation in the Courtrooms of the United States and the European Union

Timothy Swanson and Andreas Kontoleon

The proposed EU Environmental Liability Directive has recently been published. After evolving for more than a decade, it has taken a dramatic change of direction at the last hurdle. The White Paper on Environmental Liability had proposed a wide-ranging framework for civil and environmental liability, introducing the prospect of nongovernmental organizations (NGOs) bringing suits for environmental harm. The proposed directive no longer provides for such extended liability, but instead creates a system of environmental restoration and cost recovery resembling that in the US Comprehensive Environmental Response, Compensation and Liability Act (CERCLA or "Superfund" legislation). At least for the time being, the EU is retreating from the prospect of empowering individuals and organizations, and instead is focusing its efforts on creating state-level "qualified entities" responsible for retrieving the costs of contamination cleanups.

This is a fundamental decision for the EU. At present the EU is reluctant to enlarge the set of individuals and organizations authorized to monitor and regulate environmental harm. The EU is also hesitant about extending the net of liability beyond specifically designated "operators."[1] The commission's explanation for these limitations is a statement that "national legal systems are quite developed with respect to traditional damages [i.e., personal injury and damage to goods], which constitute their subject matter by excellence."[2] This explanation for the retreat on environmental liability is unconvincing and indicates the reluctance with which the individual states are willing to surrender their authority over environmental management to entities such as nongovernmental organizations.

Should the EU act to require devolution of authority from the state level to the citizen level? Is empowerment of individuals conducive to better environmental management? These are the general questions that we address in this chapter.

An important subsidiary question concerns the use of valuation methods in the courtroom. This is because the empowerment of nongovernmental organizations to pursue civil liability actions implies the lodging of suits for very general environmental harms. Many times these suits will have to be based on generally incurred natural resource damage that might not impair any specific uses (e.g., the loss of some little-known species in a little-used wetland). Then the courts in a civil liability action are left with the question of how to value an actionable harm that has no perceptible impact on a given set of individuals. Who should be able to bring such an action? What damages might be assessed?

This chapter seeks to assess the US experience using the liability approach in its courts to determine whether this approach should have been included within the EU liability regime. In general we are sympathetic to the concept of environmental liability; in theory, liability systems can play an important role in the regulation of environmental problems by empowering individuals to monitor environmental harms and by authorizing courts to charge polluters for the damage caused to environmental resources.[3] In practice, however, it is far more difficult to implement environmental liability for the reasons set out earlier.

We undertake our study of the practical difficulties in implementing environmental liability by examining the US experience with environmental liability and courtroom valuations. We conclude that the US experience resembles an attempt to "make the foot fit the shoe," i.e., the problem of environmental harm does not easily fit into the paradigm of civil liability. Our review raises many issues concerning the relatively large costs and low accuracy of valuation methods, but most of all our analysis focuses on the problematic issue concerning the standing to claim damages for environmental injuries. Despite more than 10 years of experience with these problems, the US courts have made little real progress toward the resolution of this issue, partly because (we believe) it is not resolvable in this context.

Does this mean that there is no role for civil liability within environmental management? We argue that environmental liability has made its way into the courtrooms because of dissatisfaction with alternative political mechanisms for controlling environmental harms, and thus a residual role for this approach continues to exist. However, environmental liability is a cumbersome way of providing governmental assurance that "something will be done" about environmental problems. It will be more or less effective, depending on how responsive a system of representative government is to the concerns of its public. For that reason, different jurisdictions might reasonably take different approaches on the issue of liability, depending on how responsive they believe their other methods of governance to be.

Our discussion proceeds as follows. The next section outlines the US experience with using environmental liability and valuation methods in the courts. The third section discusses the EU White Paper on Environmental Liability and the important changes that occurred in the proposed directive. The fourth section presents certain issues regarding valuation that have occupied both academics and the courts in the US. We conclude with a discussion about the future role of environmental liability.

Environmental Valuation in US Courtrooms

In the US public, natural resources such as the atmosphere, oceans, estuaries, rivers, and plant and animal species often hold the legal status of "public trust resources." The main federal statutes that contain provisions establishing management agencies as trustees of natural resources are the Comprehensive Environmental Response, Compensation and Liability Act of 1980, the Oil Pollution Act of 1990 (OPA), and the National Marine Sanctuaries Act of 1996 (NMSA).[4] Under these acts, designated trustees are to assess and recover damages resulting from injury to natural resources (such as an oil spill or the release of a hazardous substance). Federal trustees include the Department of Interior (DOI) and the National Oceanic and Atmospheric Administration

(NOAA). The statutes also acknowledge various state or local governments and Native American tribes as trustees.[5]

Under all three statutes, natural resource damage claims are based on the restoration of public resources and have three basic components. The measure of damages is (1) the cost of restoring, rehabilitating, replacing, or acquiring the equivalent of the damaged natural resources (primary restoration); (2) the diminution in value of the natural resources pending recovery of the resource to the baseline prior to the injury (interim lost value); and (3) the reasonable cost of assessing those damages. The first component provides for restoring injured resources to their baseline level. The second component compensates the public for reductions in the value of resource services pending recovery of the injured resources.[6]

The important questions for our purposes are the extent to which these measures of damages include nonuse values (NUVs) and the extent to which these NUVs (if allowed) may be estimated by the use of individual preference-based techniques. NUVs are those values of resources that are attributable to an individual's abstract concern or caring about the existence or quality of a resource, irrespective of any physical interaction with the resource. An example of an NUV would be the personal loss an individual feels with the extinction of a species of which he or she has no personal experience. NUVs are crucially important to the use of environmental liability because they are capable of empowering the widest possible constituency regarding harm to environmental resources. Once NUVs are included in the set of actionable harms, it is possible for environmental groups, and others interested but not directly harmed, to become involved in bringing environmental actions.

Since NUVs do not involve actual or direct harm to a personal interest, other methods for valuing these impacts must be used. An example of an individual preference-based technique for estimating an NUV would be the contingent valuation method (CV). The CV is based on the idea that the individual with a personal loss of welfare will be able to report a value for that loss if an artificial market is created within which that loss is valued. A CV is conducted through surveys across a sample of the affected population to assess individual values and the results are then aggregated to determine the total value of the resource to the population.

The issue of the applicability of NUVs and individual preference-based techniques was resolved in the 1989 case of *Ohio v. US Department of Interior* (motivated by the *Exxon Valdez* oil spill), in which the court granted equal weight to use and nonuse values in the damage assessment. The allowance of NUVs in the scope of damages implies the use of stated preference techniques since these are currently the only feasible method for estimating such values. Furthermore, individual preference-based valuation techniques were given a "rebuttable presumption of validity."[7] The defendants can appeal the specific application of these methods, but not the methods in general.[8]

Various industries and stakeholder groups fiercely opposed the use of preference-based techniques and especially the use of the CV method for estimating nonuse values. This criticism manifested itself in academic journals and also in the courts.[9] As a response to these attacks, the Department of Commerce convened a panel of leading economists to assess the validity of the CV method in the measurement of nonuse values. The resulting "NOAA panel" cautiously supported the use of stated preference techniques in damage assessment cases.[10] They concluded that the information provided by stated preference techniques is as reliable as that derived in the marketing analyses of new products, and other techniques of damage assessment normally allowed in court proceedings. A stringent list of guidelines from the National Oceanic and Atmospheric Administration ("NOAA guidelines") was recommended to ensure reliably and validity, but from this point on, NUVs and CV techniques have been allowed in US court proceedings.

Courtroom Experience with Economic Valuation Techniques
Probably the most publicized case using the CV methodology concerned the *Exxon Valdez* oil spill off the shores of Prince William Sound in Alaska. The damage was estimated to lie between $3 and $15 billion.[11] Exxon settled out of court by agreeing to pay a total of $1 billion. In the Montrose damage assessment, which was settled recently, the trustees used a CV to assess the cost of the impacts of DDT contamination off the coast of California, and recovered the value of interim losses.[12] Other examples of the successful use of CV techniques to estimate environmental damage include the state of Colorado's case quantifying the

damage caused to watersheds by the Eagle Mine,[13] and the state of Washington's case quantifying the damage from an oil spill that soiled the coastline of the state.[14] In both these cases the trustees estimated both use and nonuse values. Finally, the *American Trader* case is one of the few examples of the application of these valuation techniques that was not settled out of court; there the trustees used the benefit transfer method to estimate the damage to the affected coastline from an oil spill.[15]

Evolving Approaches in the US to Economic Valuation of Environmental Damage

The implementation of the NOAA natural resource damage assessment (NRDA) guidelines has altered significantly over time. In particular, a shift in emphasis occurred in the mid-1990s, with respect to approaches to determining the scale of compensatory restoration. In the early 1990s, economic assessments of natural resource damage were conducted with the objective of determining a monetary value of damage that, if paid as compensation, would make the public whole again. Since the mid-1990s, the procedures for NRDA, and the applicable legislation, have shifted toward resource compensation and the resource-to-resource (or service-to-service) approaches to determining the scale of compensatory restoration. The guidelines suggest that the service-to-service approach is used when the injured and replacement resources and services are of the same type and quality and of comparable value. It is similar to in-kind trading between the injured and replacement resources and services. The defendant is allowed to substitute "equally valued" replacement resources for the injured ones.[16]

The scaling analysis (i.e., the determination of the size of compensatory restoration) simplifies to selecting the scale of a restoration action for which the present discounted quantity of replacement services equals the present discounted quantity of services lost as a result of the injury.[17] Also, monetary valuation procedures are still to be used when there are no appropriate compensatory restoration options and when the injured and restored resources and services are of comparable type and quality, but not comparable value.[18] Finally, the latest NRDA guidelines allow the use of valuation techniques in order to show that the costs of primary restoration may be grossly disproportionate to the benefits. If this is shown, then incomplete primary restoration may be permitted. The responsibility for demonstrating this rests with the party responsible for the damage.[19] Therefore, although the movement within the US has been toward the substitution of replacement resources for injured ones, there still remains a role for valuation, if only to allow an assessment of the extent of replacement or restoration to be required.

Environmental Valuation in the National Courts of the EU

The environmental liability regimes within EU member states make very limited provision for assessment of environmental damages, and few of them have made any progress in delineating the role of individual preference-based techniques in estimating these. Most liability-type legislation found in member states deals with traditional legal forms of damages, such as personal injury or property damages, rather than with environmental damages per se. Moreover, such damages usually have been assessed using techniques based on market prices and costs, and not on broader approaches that encompass nonmarket values (such as those that are the focus of the CV methodology). For example, the German Environmental Liability Act of 1990 and the Danish Compensation for Environmental Damage Act of 1994 are drafted in this spirit. In Belgium, the courts are using a concept of "collective goods" similar in spirit to that found in the US NRDA so that ecological and aesthetic loss can be compensated. Though in some other national laws impairment of the environment is also covered, few rulings have been made to clarify this notion. Also, there has been no ruling on the appropriate role of valuation techniques in the assessment of environmental damages. In short, the national experience with environmental liability across the EU, although no doubt excellent, is extremely limited.

The Proposals within the White Paper on Environmental Liability

The original proposals within the EU White Paper on Environmental Liability[20] sought to fill this legislative vacuum and to broaden the notion of damage to cover loss of biodiversity (in addition to damage in the form of contamination of sites and traditional damage, which is covered by the environmental liability laws in most member states). The main features of the White Paper liability regime are listed here only briefly; they included (1) the absence of retroactive application; (2) coverage of both environmental damage (site contamination and damage to biodiversity) and traditional damage (harm to health and property); (3) scope of application linked to preexisting EC environmental legislation (contaminated sites and traditional damage to be covered only if caused by an EC-regulated hazardous or potentially hazardous activity, and damage to biodiversity only if the resource is protected under the Natura 2000 network); (4) strict liability for damage caused by inherently dangerous activities; (5) fault-based liability for damage to biodiversity caused by a nondangerous activity; (6) liability focused on the operator in control of the activity that caused the damage; (7) economic valuation of environmental harms would be allowed; and (8) compensation received from the polluter would have to be spent on environmental restoration.

Most important, the White Paper proposed that the EU liability system should allow environmental organizations and other interested parties to act as "trustees" and pursue legal actions themselves [see section 4.7 of COM (2000) 66]. This was to allow challenges to environmental harms that were going unchallenged on account of dilatory or negligent state authorities. Thus, the explicit aim of the White Paper provision was to enable interested citizens to take action when governments were unresponsive. This goal had been the subject of discussion since the initial drafts of the proposed directive on civil liability more than a decade ago.

The White Paper would have paved the way for using valuation methods both when damage is irreparable and also when damage is reparable, but the costs of restoration are disproportionate to the damage. In cases where the costs of restoration are considerably higher than the estimated value of the damaged natural resource, the compensation to be paid should amount (at least) to the value of the damaged natural resource, while the damages awarded must be utilized for providing environmental services of a quality and quantity equivalent to those lost.

The document endorsed the use of techniques such as CV, but it was cautious about the costs involved in undertaking original on-site studies. The White Paper encouraged the use of estimates derived in similar studies to determine the damage in other cases: the so-called benefit transfer method. To this end, the White Paper stressed the importance of developing databases [such as the Environmental Valuation Resource Inventory (EVRI)] of economic valuation studies to be used for inferring natural resource damage.[21] Such an approach would involve the accumulation of information from all previous valuation studies in the hope that the incremental cost of future valuation exercises would be reduced on account of the continuing usefulness of earlier ones. Thus, the White Paper advocated the broad-based use of, and exclusive reliance upon, valuation methods to an extent that would have exceeded its level of application even in the US.

The Proposed Directive on Environmental Liability

The Proposed Directive on Environmental Liability retreated from the White Paper's reliance on citizen suits and valuation methods. Under Article 2, section 1, the proposed directive allows but does not require the empowerment of groups other than the state itself:

(14) "qualified entity" means any person who, according to criteria laid down in national law, has an interest in ensuring that environmental damage is remedied, including bodies and organizations whose purpose, as indicated by the articles of incorporation thereof, is to protect the environment and which meet any requirements specified by national law.

Thus, under this provision it is national law that determines the identities of those able to bring actions for environmental harms. As indicated earlier, at present there are few states in the EU that possess laws enabling civil actions for environmental harms, and almost no experience with these sorts of actions. The EU has allowed this state of affairs to continue.

Similarly, the proposed directive makes little use of the concept of valuation, even though the definition of "value" under Article 2 is based on the economist's concept of willingness to pay: "value means the maximum amount of goods, services or money that an individual is willing to give up to obtain a specific good or service" [Article 2, section 1(14)]. Instead, the proposed directive has changed to almost exclusive reliance on restoration.

"Indeed, the Commission's proposal relies more on restoration, whose costs are easier and cheaper to estimate than monetary estimates of the value of natural resources. Unlike [the US] Superfund, the Commission's proposal also gives explicit preference to least cost options."[22] This is indeed the case. The terms of the proposed directive now make little provision for claiming damages, and instead focus on restoration and remediation costs (Articles 5, 7). Prevention is now phrased in injunctive language rather on the basis of "making the polluter pay" (Article 4). In general, the objective of the proposed directive is now to enable "qualified entities" to pursue injunctive or restorative activities, with the ability to reclaim costs from responsible parties after the event.

This is of course a pronounced change of direction from the one laid out within the White Paper, with its explicit empowerment of environmental groups and civil liability actions. In the remainder of this chapter, we examine whether this last-minute change of course is the result of well-considered difficulties with environmental liability systems, or merely an example of a loss of nerve on the part of the commission.

Role of Liability and Valuation in the Courts

It is clear that the proposed directive has withdrawn from the approach of citizen action adopted within the White Paper. The issue at hand concerns whether this is a useful retreat or a costly one for the environment. We believe that the most useful way to frame the issues concerning civil liability for environmental harms is to focus on the problems with using the valuation methods that they imply. We turn now to these problems.

There are four issues that are most problematic regarding the use of valuation methods in courtrooms: (1) Are valuation methods sufficiently accurate for use in courts? (2) Are valuation methods consistent with the compensatory objectives of liability? (3) Can valuation methods be applied at reasonable expense? (4) Who should be counted as part of the affected population in assessing damages? These issues have been the focus of the continuing debate in the US and are reviewed here as an introduction to the complexities of the problems that would be faced by the EU under a regime of environmental liability similar to that outlined in the White Paper.

The Debate over Accuracy

Many objections to the use of valuation in courts have focused on measurement issues. Measurement issues involve two aspects of the problems concerning the accuracy of stated preference studies (such as CVs). One aspect is the credibility of the stated preferences, i.e., how well do the surveys create incentives for the truthful revelation of preferences? For example, if an individual wishes to skew the results of the exercise, does the methodology create incentives or mechanisms that will constrain this sort of behavior? These are problems of survey design that exist in all sorts of similar exercises (such as marketing studies). The NOAA panel found that properly constructed surveys could in fact produce incentives for truthful revelation, and that there existed additional methods by which the results of the survey might be checked. For example, individual bids are usually checked against the salient characteristics of the bidder (such as income level, interest in the issue, family status) to determine whether the bid is consistent with the character of the bidder. Thus, the credibility of the results of a survey is a function of the quality of the survey design.

The other problem of accuracy concerns the margin of error surrounding the valuation. This variance will depend to some extent on the size of the sample and the nature of the good being valued, but it will necessarily remain fairly large and uncertain on account of the technique that is used. This is of course true when valuation is used in cost-benefit analyses generally, and not just in courtrooms. However, some have argued[23] that damage assessment in courtrooms requires a much higher degree of accuracy than that required for policy and regulatory reviews. Errors in welfare estimates for policy purposes may or may not influence the realized outcomes, and (if they do) the realized benefits and costs are usually distributed widely across many gainers and losers in the population. In contrast, the damage estimated in a court proceeding might be borne by a single or a few responsible parties. This concentration of impact renders the range of variability, and its relative uncertainty, more objectionable in the case of courtroom applications. This is a cause for concern about the use of liability and valuation, but it seems reasonable to us to conclude (as did the NOAA panel) that the method is allowable in courtrooms when the methodology is carefully regulated.

Costs of Valuation

The second point of concern has to do with the costs required to undertake a state-of-the-art cost-benefit analysis. Some have argued[24] that in many cases the cost of undertaking the study may exceed the damage itself, and thus the conduct of a valuation may not pass a cost-benefit test! The White Paper recognized that original valuation studies may be too costly, and strongly endorsed the use of benefit transfer techniques, as described earlier. However, economists have stressed that the benefit transfer method, even if suitable for policy decisions, may lack the accuracy required for awarding damages.[25] Moreover, the White Paper suggested that only "significant damages should be covered" under this new regime. This suggests that there should be a *de minimis* standard before economic valuation is applied. But how would it be known in advance if the nonuse value is significant? Since most nonuse values will appear inconsequential from the standpoint of all marketed or direct-use values (by definition), it is difficult to know when a valuation approach should be applied, or how to create a standard that would authorize one. Most likely, the use of liability is something that will have to be allowed overall or not at all.

Is Valuation Consistent with Compensation?

Several legal theorists in the US have examined the extent to which damages calculated using CV techniques corresponds to ordinary legal definitions of compensable damage and loss. They argue that although the ex ante use of preference-based values for the determination of benefits may be valuable for policy decisions, it does not follow that it is equally useful or desirable to use these methods ex post for the measurement of damage. According to Daum,[26] the model of damage calculation embedded in tort law for determining compensation is not compatible with the types of damages that are derived from (stated) preference-based techniques because such studies are always carried out after the damage has occurred and cannot reflect preexisting values independent of the accident and of the valuation process.

Economists do recognize that statements concerning the willingness to pay to avoid damage are a different welfare concept than the valuation of damage to an environmental resource after it has been harmed. This simply means that stated preference techniques should be designed to capture the change in the value of the asset as a result of harm as opposed to estimating willingness to pay to avoid harm. This is an important design issue that must be incorporated into courtroom-directed valuation, but it is not a serious reason for avoiding the use of liability.

The Debate over Standing

Finally, the most critical issues concerning the use of valuation techniques in court revolve around the question of whose preferences matter. Generally, we can say that we should count whoever has suffered a *real* loss. Determining this population is relevant for both sampling and aggregating. Sampling will produce an estimate of an average unit of damage. Aggregation will produce the total amount of damage. The choice of the relevant affected population will affect the estimated shape of the demand function, but more important, the choice of population will have an even greater effect on the estimated level of damage. Hence, if we were merely interested in unit mean values, then the problems of defining the relevant population are not so severe. However, in environmental damage assessment, aggregate values are what matter and hence determining who should be included in the aggregated population can have profound consequences for the outcome of the litigation process.[27]

The economic conception of standing is much broader than the legal definition. It implies that everyone who experiences a real welfare loss should be included in the aggregated population.[28] Legal standing has been traditionally defined as a much less inclusive concept and includes only those individuals who have experienced a compensable injury. A categorization and clarification of some of the issues concerning standing follows. Only a selection of the issues involved is presented.[29] The issue of standing is still very much open both in the courts and in the academic journals.[30] The discussion here highlights some of the misunderstandings and disagreements between economists and lawyers rather than purporting to offer a definitive resolution.

In practice, the courts in the US have been inconsistent in defining the relevant population of nonusers. In the *Nestucca* oil spill case, for example, the populations of Washington state and British Columbia were used for estimating damage, while in the case of the *Exxon Valdez* spill, the population of the entire United States was held to be the potentially affected population. In a more recent case, *Montrose Chemical Corp. v. Superior Court,* the trustees defined the potentially affected population as the English-speaking households in California.[31]

The recognized rights of the claimant do not constitute a sufficient basis for delineating between those within and those outside the affected population. This is because NUV has been defined as the value one obtains from a natural resource when no present or future direct personal use is realized or intended. However, an NUV cannot reasonably be claimed with regard to a resource if it never existed. From this it can be inferred that NUVs are derived from the individual's knowledge of the existence of a natural resource. Hence, human perception or some knowledge about the resource is an important part of the definition of NUVs and has been the basis for the debate over standing.

Dunford et al.[32] and Johnson et al.[33] have argued that a demand for knowledge about the resource and/or its injury is required for one's NUV to have legal standing. The lack of such a demand for information tells the court something about the true preferences of these individuals. Information acquisition activities involve opportunity costs and thus are indicators of one's interest in a particular natural resource. Respondents in CV studies that have not (endogenously) acquired such information nevertheless receive (exogenous) information from the study itself. The authors claim that expressed nonuse values obtained from individuals with no prior or no intended demand to acquire information are somehow "induced" preferences and that the subsequent estimated losses would not have occurred if the respondent had not been sampled.

This raises the familiar issue of the role of the person conducting a study in providing information.[34] It has long been emphasized in the economic literature that the sampled population requires full knowledge about a policy problem in order to make an informed judgment.

However, these are ex ante studies of proposed policy changes, and thus none of the population can have knowledge of the proposed changes. It does not necessarily follow, however, that supplying information to respondents is also appropriate when assessing ex post compensation for actual welfare losses from a sample of respondents that represents the general population.[35] Hence, attempts to aggregate losses over informationally unrepresentative subsamples of larger populations may be inconsistent with the revealed knowledge and concerns of that population.[36] Some sort of prior knowledge of the resource might be made a prerequisite to claiming standing, and thus to taking part in a survey regarding NUVs.[37]

But what sort of "prior knowledge" of the resource is required? It has been argued that individuals have preferences over general classes of environmental goods (not specific forms of environmental resource) and thus they would suffer a legitimate loss in NUV from damage to a particular environmental asset even if they had no prior knowledge of the asset and/or the injury.[38] Randall describes the existence of such preferences as a form of heuristic to deal with the realities of an overwhelmingly complex world. People care about a class of things, which implies that they care about particulars in that class. People might then claim a compensable injury if they are informed about damage to a member of a class about which they care.

However, accepting that individuals care about classes of environmental resources poses problems in interpreting how people make choices about specific resources when asked to do so. That is, if, for example, people care about "all species," does this mean that individuals would have the same value for any member of the class of resources? How would they trade off the loss of any species within that class against the acquisition of a good from another class (such as a road or a hospital project)? The purpose of a legal action is to obtain compensation for injuries to specific natural resources and this requires a well-defined position regarding that resource. General knowledge of "the environment" is more of a political position than a justification for a legally recognized right.[39]

Under the terms of the White Paper, the environmental liability regime would have allowed nongovernmental organizations to pursue actions and to claim damages on "behalf of society." It is likely that a large proportion of such sought damages would have been of the NUV type and hence the issues discussed in this section would have been raised within the courtrooms of EU states. These are complex issues that continue to perplex academics and practitioners alike; no real resolution of them is yet in sight. The delineation of the group eligible to claim standing for damages under environmental liability regimes is the most serious problem facing those who advocate the use of such regimes.

Conclusion: A Future Role for Environmental Liability?

This chapter has sought to assess the future of environmental liability within the US and the EU. We have addressed this question in light of the removal from the EU proposed directive of the provisions enabling suits by nongovernmental organizations for environmental harms. Was this a good thing? Should environmental liability be used to empower a broader set of social activists on behalf of the environment?

In theory this would appear to us to be a good thing because it allows broad-based monitoring for environmental harms and provides a mechanism for assessing environment-sensitive measures of damages against the polluter. In practice there are substantial difficulties with any sort of attempt to realize these benefits. The US experience highlights the issues that are likely to occupy the courts of the EU if environmental liability is extended. These were identified as (1) accuracy of valuation studies, (2) the cost of valuation studies, (3) consistency of valuation with the compensatory objective of a liability regime, and (4) standing and aggregation regarding nonuse values.

In our review of these issues we believe that we have identified two or three crucial problems in the extension of environmental liability. There are, for example, reasonable concerns about the cost and accuracy of valuation methods used to implement such systems of liability. The most significant issue concerning the extension of environmental liability is the problem of who is harmed by damage to an environmental good or system. If the nature of the harm we are assessing concerns "nonuse," how do we decide which "nonusers" to exclude from the assessment? This is the most serious problem that undermines the usefulness of environmental liability with respect to environmental goods or services, such as biodiversity. Courtrooms have usually been reserved for use by those who are individually and directly affected by others' actions, while the legislatures have been reserved for abstract policy debates.

This raises a general point: Do nonuse values attempt to translate an abstract policy issue into a personal injury context and forum? If so, why don't we trust our legislatures to deal with harms to general public goods, both through ex ante regulation and ex post punishment? The pressure for the extension of environmental liability is a by-product of this general problem. It is an attempt to provide the public with a right of redress when its regulatory institutions are perceived to be unresponsive to the public will. Attempts to overcome the problems with valuation tools and standing are understandable when seen from within this broader context.

What future is there for environmental liability within the US and the EU? Concerns about the extension of environmental liability have recently moved the courts in the US toward "in kind" substitution of resources, and away from damage assessment. When injured resources are replaced by reasonable substitutes, all of the valuation problems listed here are avoided. The problem with this approach is that it depends on the availability of reasonable replacement resources for its existence. This is difficult in any circumstances, and even more difficult in the EU (where the only resources for which biodiversity damages may be claimed are the relatively unique ones on the Natura 2000 list). So, valuation might need to remain a tool of last resort in Europe even if it is possible to move away from it in the US.

In sum, it is clear that the use of economic valuation in courtrooms is a poor substitute for adequate environmental regulation ex ante and ex post, and the costs and complexities of dealing with environmental harms within the courtroom environment are substantial. This is a case of attempting to make the problem fit the process, rather than the other way around. Despite this fact, there is the clear perception that there are instances in which damage to important and valued environmental resources (such as biodiversity) is not being dealt with adequately, and recourse for the worst transgressions must be afforded. In this light, the role of environmental liability is to provide an outlet for a statement of public values in a few extreme cases where it appears to the public that they are being entirely overlooked. The White Paper probably served this purpose better than does the proposed EU directive.

Acknowledgments

This work was partly funded by the European Commission's Energy, Environmental and Sustainable Development Program (under the EMERGE and Biodiversity and Economics for Conservation projects). The authors would like to thank professors Richard Macrory, David W. Pearce, Richard B. Stewart, and Richard O. Zerbe for helpful comments on this paper.

Notes

1. "Operator" is defined in the proposed directive [Directive 1(10)] as any person who operates an activity covered under the directive (as specifically designated in Annex I to the directive). Liability is limited to those operators carrying out designated activities that attract responsibilities under already-existing EU environmental obligations (listed in Annex I). In addition, the proposed directive also disallows retroactive liability.

2. COM (2002) 17 final, p. 17.

3. T. Swanson, "Environmental Liability and Environmental Regulation," in O. Lomas, ed., *Frontiers in Environmental Law* (London: Chapman and Hall, 1992).

4. Apart from the CERCLA, OPA, and NMSA, trustees can currently sue for environmental damages under the Clean Water Act of 1972, the Superfund Amendments and Reauthorization Act of 1986, the Deepwater Port Act of 1996, the Trans-Alaska Pipeline Act of 1973, and the Outer Continental Shelf Lands Act of 1953. Some state laws also allow damage recovery and provide various types and levels of coverage. See H. Breedlove, *Natural Resources: Assessing Non-market Values Through Contingent Valuation* (Washington, D.C.: Congressional Research Service, 1999), for more details.

5. T. A. Penn, *Summary of the Natural Resource Damage Assessment Regulations under the United States Oil Protection Act*, NOAA Report (Washington, D.C.: National Oceanographic and Atmospheric Administration, 2000).

6. Ibid., p. 1.

7. John B. Loomis, "Contingent Valuation Methodology and the US Institutional Framework," in Ian Bateman and K. G. Willis, eds., *Valuing Environmental Preferences:Theory and Practice of the Contingent Evaluation Method in the US, EU, and Developing Countries* (Oxford: Oxford University Press, 2000), pp. 613–628.

8. The rebuttable presumption status of preference-based techniques was attacked by industries, yet both the US Court of Appeals and Department of Interior found preference techniques to be reliable for estimating both use and nonuse values.

9. See, for example, the debates in J. A. Hausman, ed., *Contingent Valuation: A Critical Assessment* (Amsterdam: Elsevier, 1993); and between P. Diamond and Hausman, W. M. Hanemann, and P. Portney in a special issue of the *Journal of Economic Perspectives* 9(4) (1994).

10. The panel concluded "that CV [contingent valuation] can produce estimates reliable enough to be the starting point of a judicial process of damage assessment, including lost passive-use values." K. R. Arrow et al., "Report of the NOAA Panel on Contingent Valuation," *Federal Register* 58: 4601–4614.

11. R. T. Carson et al., "Contingent Valuation and Lost Passive Use: Damages from the *Exxon Valdez*," RFF Discussion Paper 94-18 (Washington, D.C.: Resources for the Future, 1994).

12. Penn, *Summary of the Natural Resource Damage Assessment Regulations under the United States Oil Protection Act.*

13. R. Kopp and V. K. Smith, "Benefit Estimation Goes to Court: The Case of Natural Damage Assessments," *Journal of Policy Analysis and Management* 8 (1989): 593–612.

14. See R. Rowe, D. Shaw, and W. Schulze, "Nestucca Oil Spill," in J. Ward and J. Duffield, eds., *Natural Resource Damages: Law and Economics* (New York: Wiley, 1992).

15. See D. Chapman and W. M. Hanemann, "Environmental Damages in Court: The American Trader Case," University of California, Berkeley, Department of Agricultural and Resource Economics, Working Paper No. 913. See also Loomis, "Contingent Valuation Methodology and the US Institutional Framework"; W. M. Hanemann, "Natural Resource Damages for Oil Spills in California," in Ward and Duffield, *Natural Resource Damages*, pp. 555–580; and Breedlove, *Natural Resources: Assessing Non-market Values Through Contingent Valuation* for more examples of the use of preference-based techniques in US legal damage assessment cases.

16. R. E. Unsworth and R. C. Bishop in "Assessing Natural Resource Damages Using Environmental Annuities," *Ecological Economics* 11 (1994): 35–42, have proposed a variant of the service-to-service approach for assessing natural resource damage. The habitat version of the approach, habitat equivalency analysis, has been applied in a number of damage assessment cases and has been largely accepted by the responsible parties. This approach is particularly suitable when dealing with modest injuries to homogeneous resources and thus scaling is a relatively straightforward matter. In dealing with acres of damaged wetlands, Unsworth and Bishop assume that restored wetlands will be homogeneous with the injured wetlands and, from that point, scaling is largely a matter of determining the time-path of resource recovery and applying the appropriate discount

rate. For larger and more complicated injuries, methods such as choice experiments are appropriate. However, it has been recognized [e.g., E. MacAlister et al., *Study on the Valuation and Restoration of Biodiversity Damage for the Purpose of Environmental Liability*. Final Report of project B4-3040/2000/265781/MAR/B3 (Brussels: EU Commission, 2001)] that such methods, while promising, have yet to be validated in large-scale application under litigation conditions.

17. To determine the scale of compensatory restoration in practice, a number of parameters have to be identified. The services lost as a result of the injury are quantified by defining the time of the injury, the extent of the injury, the reduction in resources and services from baseline, and the path of recovery back to baseline. The parameters that define the benefits of restoration include when the restoration project begins, the time until the project provides full services, the productivity of the project through time, and the relative productivity of the created or enhanced resources and services compared with the injured resources and services. A discount rate is applied in quantifying the lost and replacement services because the services occur in different time periods and they are not comparable otherwise. Without identifying these parameters, it would not be possible to determine how much compensatory restoration is required to make the public whole.

18. MacAlister et al., *Study on Valuation and Restoration of Biodiversity Damage*. For full details of the NRDA process recommended by the OPA, see http://www.darcnw.noaa.gov/opa.htm.

19. Penn, *Summary of the Natural Resource Damage Assessment Regulations under the United States Oil Protection Act*; MacAlister, *Study on Valuation and Restoration of Biodiversity Damage*. In some circumstances, the "value-to-cost" variant of the valuation approach may be employed. Value-to-cost is only appropriate when valuation of the lost services is practicable, but valuation of the replacement natural resources and services cannot be performed within a reasonable time frame or at a reasonable cost. With this approach, the restoration is scaled by equating the cost of the restoration plan with the value (in dollar terms) of losses due to the injury. The value-to-cost approach is equivalent to the framework for compensation prescribed by the CERCLA damage assessment regulations.

20. COM (2000) 66. For an analysis of the White Paper, see background papers commissioned by the EU at http://www.europa.eu.int/comm/environment/liability/background.htm and http://europa.eu.int/comm/environment/liability/followup.htm.

21. COM (2000) 66.

22. COM (2002) 17 final, 2002/0021(C0D), p. 13.

23. See, e.g., W. H. Desvousges, F. R. Johnson, R. W. Dunford, K. J. Boyle, S. P. Hudson, and K. N. Wilson, "Measuring Natural Resource Damages with Contingent Valuation: Tests of Validity and Reliability," in Hausman, *Contin-*

gent Valuation, pp. 91–156; F. R. Johnson, M. R. Banzhaf, R. W. Dunford, and W. M. Desvousges, "Informed Concern: The Role of Knowledge in Assessing Compensation for Natural Resource Damages," *Growth and Change* 32 (winter 2000): 48–68.

24. For example, S. Shavell, "Contingent Valuation of the Non-use Value of Natural Resources: Implications for Public Policy and the Liability System," in Hausman, *Contingent Valuation*, pp. 371–388.

25. For example, S. Navrud and G. J. Pruckner, "Environmental Valuation—To Use or Not to Use? A Comparative Study of the United States and Europe," *Environmental and Resource Economics* 10 (1997): 1–26.

26. J. F. Daum, "Some Legal and Regulatory Aspects of Contingent Valuation," in Hausman, *Contingent Valuation*, pp. 389–411.

27. The Eagle Mine case is typical of the relative importance of the issue of standing in estimating average unit damage. In this case the state of Colorado sought damages for the release of hazardous substances into groundwater. What is interesting is that although both the trustees and the defendants' estimates of unit average damage coincided, their estimates of aggregate damage differed by several orders of magnitude (see Kopp and Smith, "Benefit Estimation Goes to Court," for more details).

28. D. Whittington and D. Macrae Jr., "The Issue of Standing in Benefit-Cost Analysis," *Journal of Policy Analysis and Management* 9 (1986): 201–218.

29. For a more comprehensive view of the debate, see R. W. Dunford, F. R. Johnson, and E. S. West, "Whose Losses Count in Natural Resources Damages?" *Contemporary Economic Policy* 15 (1997), pp. 77–87; Alan Randall, "Whose Losses Count? Examining Some Claims about Aggregation Rules for Natural Resources," *Contemporary Economic Policy* 15 (October 1997): 88–97; Johnson et al., "Informed Concern"; R. O. Zerbe, "Comment: Does Benefit-cost Analysis Stand Alone? Rights and Standing," *Journal of Policy Analysis and Management* 10(1) (1991): 96–105; R. O. Zerbe, "Is Cost-Benefit Analysis Legal? Three Rules," *Journal of Policy Analysis and Management* 17(3) (1998): 419–456; R. O. Zerbe, "Can Law and Economics Stand the Purchase of Moral Satisfaction?" paper presented at a symposium on law and economics, University College London, September 5, 2001; W. N. Trumbull, "Who Has Standing in Cost-Benefit Analysis?" *Journal of Policy Analysis and Management* 9 (1990): 201–219; Whittington and Macrae, "The Issue of Standing in Benefit-Cost Analysis"; and Kopp and Smith, "Benefit Estimation Goes to Court."

30. "Of all the issues of CBA few are misunderstood more," Trumbull, "Who Has Standing in Cost-Benefit Analysis?" p. 201.

31. Zerbe, "Is Cost-Benefit Analysis Legal? Three Rules."

32. Dunford et al., "Whose Losses Count in Natural Resources Damages?" p. 77.

33. Johnson et al., "Informed Concern."

34. For an overview of these issues, see A. Munro and N. D. Hanley, "Information, Uncertainty, and Contingent Valuation," in Bateman and Willis, *Valuing Environmental Preferences*; S. M. Chilton and W. G. Hutchinson, Exploring Divergence Between Respondent and Researcher Definitions of the Good in Contingent Valuation Studies," *Journal of Agricultural Economics* 50(1) (January 1999): 1–16; G. C. Blomquist and J. C. Whitehead, "Resource Quality Information and Validity of Willingness to Pay in Contingent Valuation," *Resource and Energy Economics* 20 (1998): 179–196; K. J. Boyle, M. P. Welsh, R. C. Bishop, and B. M. Banmgartner, "Validating Contingent Valuation with Surveys of Experts," *Agricultural and Resource Economics Review* 2 (October 1995): 247–253; J. C. Whitehead and G. C. Blomquist, "Measuring Contingent Values for Wetlands: Effects of Information about Related Environmental Goods," *Water Resources Research* 27 (1991): 2523–2531; and J. C. Bergstrom, J. R. Stoll, and A. Randall, "The Impact of Information on Environmental Commodity Valuation Decisions," *American Journal of Agricultural Economics* 72 (1990): 614–621.

35. Dunford et al., "Whose Losses Count in Natural Resources Damages?"

36. Johnson et al., "Informed Concern."

37. Economists are divided over the necessity of positive (actual or potential) information demand as a precondition for real compensable losses in NUVs. For example, see Zerbe, "Can Law and Economics Stand the Purchase of Moral Satisfaction?"; Zerbe, "Is Cost-Benefit Analysis Legal?"; and Randall, "Whose Losses Count?" arguing against its necessity, and D. Moran, "Accounting for Non-use Value in Optics Appraisal: Environmental Benefits Transfer and Low Flow Alleviation," in *Economic Valuation of Water Resources: Policy and Practice* (London: Chartered Institution of Water and Environmental Management, 2000) arguing in favor of it.

38. Randall, "Whose Losses Count?"; Zerbe, "Can Law and Economics Stand the Purchase of Moral Satisfaction?"

39. Johnson et al., "Informed Concern." Note that there is also ample empirical evidence that willingness to pay by nonusers declines and eventually is reduced to zero when demand for information is absent. Various studies have shown that NUVs have declined with distance and familiarity with the resource. See Bateman and Willis, *Valuing Environmental Preferences*; Moran, "Accounting for Non-use Value in Options Appraisal"; and R. J. Sutherland and R. G. Walsh, "Effect of Distance on the Preservation Value of Water Quality," *Land Economics* 61 (1985): 281–291.

III

Policy Divergence on Global Issues

8

The Climate Change Divide: The European Union, the United States, and the Future of the Kyoto Protocol

Miranda A. Schreurs

A decade after the United States and the European Union first agreed to work together to address climate change at the first Earth Summit in 1992, they are struggling to find ways to talk with each other about the issue. The EU and the US have reluctantly agreed to disagree on their climate change strategies. The fifteen member nations of the EU strongly support the Kyoto Protocol to the Framework Convention on Climate Change (FCCC). The United States, in contrast, has rejected the Kyoto Protocol and instead is advocating what it calls a voluntary, science-based approach to reduction of greenhouse gas emissions. Neither the EU nor the US appears willing to abandon the course it has set out on, and this has complicated their ability to work together in addressing climate change.

The Kyoto Protocol is an international agreement formed in 1997 that if enacted will commit developed countries to reduce their greenhouse gas emissions by an average of 5.2 percent of 1990 levels by 2008–2012.[1] Under the agreement, the EU is committed to reducing its joint greenhouse gas emissions by 8 percent of 1990 levels by 2008–2012. Under EU rules, ratification of an international environmental agreement requires the consent of all member state parliaments as well as of the EU. The EU Council of Environmental Ministers gave its stamp of approval in March 2002. In the following 2 months, the parliaments of all fifteen member nations agreed to ratification. On May 31, 2002, the EU delivered the ratification documents to the United Nations, bringing to sixty-nine the number of nations that had ratified the agreement as of that date.[2] A formal ceremony was held at the United Nations in New York City.

Ratification of the Kyoto Protocol by the EU attests to the importance that Europeans place on taking immediate action to slow down climate change by reducing anthropogenic emissions of greenhouse gases. The EU decision to ratify the agreement is also a direct challenge to the US. In March 2001, shortly after taking office, US President George W. Bush announced in a letter to four conservative senators that he opposed the Kyoto Protocol, and a few days later, Environmental Protection Agency Administrator Christine Todd Whitman announced that as far as the administration was concerned, Kyoto was "dead."[3] Although President William J. Clinton had signed the agreement in 1998, it had not been presented to the Senate for ratification.

Bush could have simply let the protocol die by not presenting it to the Senate for ratification. Many international agreements fade into oblivion in this way. Instead, Bush took the issue head on and simply announced his intentions to withdraw the US from the agreement. In the mind of Bush and several of his key advisors, including Vice President Richard Cheney, the Kyoto Protocol is an unworkable agreement that would impose unacceptable costs on the US economy while exempting developing countries from taking action. The White House position is that the best way to deal with climate change is through additional scientific research and long-term support of technological change rather than committing to a complex and "flawed" international protocol.

The EU decision to move forward with ratification despite the US rejection of the Kyoto Protocol is a historic event. Under the formal rules of the Kyoto Protocol, fifty-five states representing 55 percent of the developed countries' carbon dioxide emissions in 1990 must ratify the agreement before it can go into effect. The United States alone accounts for approximately one-third of CO_2 emissions by developed countries. Thus, without US participation in the agreement, almost all other major emitters of CO_2 in the developed world (the EU, Japan, Canada) plus the transition economies (Russia and Central Europe) would have to ratify the agreement for it to go into effect. Presumably, this is why the Bush administration referred to the protocol as "dead" when it pulled out of the agreement. It seemed highly improbable that the industrialized nations of the world would have the will to move forward in cutting back greenhouse gas emissions if the US did not join them in this effort.

Certainly, European industry would recognize that this would put them at a competitive disadvantage. The EU, however, surprised the Bush administration. Rather than joining the US in abandoning the agreement, the EU's resolve to move forward with ratification and win the support of other nations to do the same intensified. This was significant since the EU member states combined accounted for close to a quarter of 1990 CO_2 emissions. Prior to the Bush announcement, cracks were evident in the EU regarding climate change. The Bush decision, however, so angered the Europeans that they were able to overcome internal disagreements and present a strong united block supporting the Kyoto Protocol. They called upon the Bush administration to do the same. Seldom in history has the EU criticized the US as forcefully as it has over this issue.

For several years now, the EU and the US have been at loggerheads over how to manage climate change. The differences across the Atlantic reflect disagreements regarding the seriousness with which the climate change threat is perceived. They also reflect sharp differences in the political cultures of the EU and US and the ability of different actor constellations to influence the policy process. The future of the Kyoto Protocol remains precarious (at the time of this writing, it is uncertain whether Russia will ratify the agreement; its failure to do so would make it next to impossible for the agreement to go into effect), but the EU action may be what it takes to persuade other nations to move forward with Kyoto. Japan's cabinet agreed to ratify the agreement on June 4, 2002. By the end of the year, 100 countries had ratified the agreement.

This chapter explores why the EU and the US have diverged so sharply in their reaction to the Kyoto Protocol. It begins with a brief historical overview of the international climate change negotiations involving these two major players. The starting point is the preparations for the 1992 United Nations Conference on Environment and Development (UNCED) and the formation of the UN Framework Convention on Climate Change. The chapter then examines EU and US efforts to shape the 1997 Kyoto Protocol and its implementation mechanisms. The reasons for the near collapse of the negotiations in 2000, the subsequent US withdrawal from the agreement, and the historic decision by the EU to move forward on the Kyoto Protocol even without the US are the main focus of the analysis. The chapter concludes by considering what the consequences of the climate change divide may be for international efforts to mitigate climate change and describes recent challenges within the US to the Bush administration's voluntary approach to addressing climate change.

The Climate Change Negotiations

The United Nations Conference on Environment and Development

Following the successful completion of the Montreal Protocol on Substances that Deplete the Ozone Layer in 1987, the international community turned its attention to the threat of global warming. By the late 1980s, there was growing concern that average global temperatures were rising as a result of anthropogenic emissions of greenhouse gases, and especially CO_2. Other global environmental issues, such as loss of biodiversity and increasing deforestation, were also on the agenda. Efforts began within the United Nations to shape an international consensus on the need for action to address global environmental threats and sustainable development. The United Nations General Assembly established an intergovernmental negotiating committee to work out plans for the FCCC. The political negotiations among member states were to occur at the UNCED.

Differences between the EU and the US were already apparent at this early stage of the international climate change negotiations. As early as 1990, the EU established a goal to stabilize its CO_2 emissions at 1990 levels within the course of a decade. Some individual member states established even deeper cuts as goals. Germany, for example, announced in the summer of 1990 that by 2005 it intended to reduce its CO_2 emissions by 25 percent of 1987 levels.[4]

The EU was embracing the precautionary principle, the idea that when there are serious threats or the potential of irreversible damage, measures to mitigate the problem should be taken even where there is a lack of complete scientific consensus on the nature of the problem (see chapter 1). The precautionary principle is embodied in the European Community Treaty, which states: "Community policy on the environment shall aim at a high level of protection taking into account the diversity of situations in the various regions of the Community. It shall be based on the precautionary principle and on the principles that preventive action should be taken, that environmental damage should as a priority be rectified at source and that the polluter should pay."[5]

The domestic political climate was different in the US. The US was shifting toward the use of market-based mechanisms for pollution control and cost-benefit analysis, in which the costs of taking action are weighed against the benefits to be derived from that action. The George H. W. Bush administration pushed through the 1990 Clean Air Act Amendments strengthening domestic controls on acid rain-producing sulfur oxide emissions (based on an emissions trading system), but took a much more cautious stance when it came to controlling CO_2 emissions. The US did not share the EU's embrace of the precautionary principle and refused to establish an emissions stabilization target. Instead, the White House promoted what came to be known as a "no regrets" policy—the idea that the US would limit its climate change mitigation initiatives to actions that would be beneficial for other reasons as well (such as reducing costs to industry by improving energy efficiency).[6] The Bush administration was reluctant to take any actions that might have serious economic repercussions for the US economy.

The EU went to the UNCED in Rio de Janeiro, Brazil, urging nations to agree to the establishment of binding reduction targets for greenhouse gas emissions. They also wanted countries to agree to a set of dates by which those cuts were to be made. The White House, in contrast, argued that more scientific and technological research was needed before nations should commit to establishing any binding targets or timetables. The US argued, for instance, that there was still no reliable global inventory of greenhouse gas emission levels or of the extent of carbon-absorbing sinks (forests, agricultural land, or other areas). The US also was concerned about the role to be played by developing countries and expectations regarding financial transfers from the developed to the developing world. The EU's position was that the developed countries had the responsibility to act first, although they shared US concerns regarding developing country demands for financial aid and technology transfers.

A framework convention for dealing with climate change was established in Rio, but because of the disagreements between the US and the EU, no binding emissions reduction targets were set. Instead, the FCCC simply called upon nations to work to stabilize greenhouse gas concentrations at a level that would not interfere with the natural climate system. The developed countries did agree "to adopt national policies and take corresponding measures" to limit their greenhouse gas emissions and to enhance their greenhouse gas sinks, but action was simply to be voluntary.[7]

The FCCC established two groups of nations, Annex 1 and Annex 2 countries. Annex 1 countries agreed to take measures to mitigate national climate change and conduct regular inventories of greenhouse gas sources and sinks. Annex 2 countries were a subset of the Annex 1 countries who also agreed to provide financial and technical assistance to developing countries. The EU, its member states, and the US fell into both categories.

It is interesting that although George H. W. Bush had expressed considerable ambivalence about attending the Rio Conference, the US was among the first nations in the world to ratify the FCCC. EU ratification, which first required ratification by each of its member states, followed.

Establishing the Kyoto Protocol

It is common in international environmental law that once a framework convention that lays down broad principles and goals is formed, a protocol, stipulating more specific obligations and the technical measures to be followed in the implementation of the framework convention, be established. Thus, after the FCCC was negotiated, the real work of coming to international agreement on specific goals and objectives and implementation mechanisms had to begin. The FCCC, which went into force in 1994 after fifty nations had ratified it, established an annual Conference of the Parties (COP) for this purpose. Eventually it was agreed that by the time of the third COP in 1997 a protocol should be formed.

The division between the EU and the US that was evident at the UNCED persisted in the following years, although at times the distance between them did narrow and there were signs of a growing consensus on the need for action on both sides of the Atlantic. Indeed, had there been no narrowing of the divide, the Kyoto Protocol would not have been signed.

The election of William J. Clinton and his running mate, Albert Gore, to the White House in 1992 changed the US negotiating stance, taking it a step closer to that of the Europeans. Vice President Gore was well versed on climate change and supported a stronger US position.[8] The Clinton–Gore administration announced that in principle it accepted the idea of quantifiable emissions targets in an international agreement. They also tried to pass an energy tax based on British thermal units (Btus) that would have hit fossil fuels, and especially petroleum, particularly hard. The Btu tax could have been a powerful economic tool for US efforts to mitigate climate change, but Congress blocked it, arguing that it would be too expensive for individual households and would hurt US industry, costing hundreds of thousands of jobs.[9]

This foreshadowed a domestic political divide that was to hamper the Clinton administration's efforts to assume any kind of leadership role in the international climate change negotiations. In fact, the divide between the White House and the Congress on how to address climate change was to grow even wider when the Democratic Party lost control of Congress in the 1994 landslide electoral victory by the Republican Party. The Clinton White House found itself caught between European and congressional demands. The Europeans were calling for an international agreement based on binding emissions targets and timetables and premised on the understanding that the industrialized states were to act first by cutting domestic emissions. The US Congress, in contrast, was skeptical of climate change science, was opposed to an agreement that would hurt the US economy, and was unwilling to take action unless developing countries were required to make some commitments as well. The administration's hands were tied.

Within the EU, Germany was a particularly strong advocate of action. Chancellor Helmut Kohl offered to host the first COP in Berlin in 1995. Out of this conference emerged what came to be known as the Berlin Mandate, an agreement to negotiate a protocol that would "set quantified limitation and reduction objectives" for the Annex 1 countries of the FCCC within specified time frames. At the second COP in Geneva, Switzerland, Timothy Wirth, head of the US negotiating team, announced that the US agreed to the idea of "verifiable and binding medium-term emissions targets."[10] He did not go so far, however, as to accept EU calls for mandatory policies and measures to be assumed by all industrialized nations. Wirth had signaled the Clinton administration's willingness to work with the EU to find a middle ground.

Under the US Constitution, the Senate must vote by a two-thirds majority in favor of an international agreement signed by the president before it can be ratified. The Senate was wary of the direction the White House was moving in the international negotiations. The Senate's unease was made clear in July 1997 when it voted 95–0 in support of the Byrd–Hagel Resolution, which stated that "the United States should not be a signatory to any protocol to, or other agreement regarding, the United Nations Framework Convention on Climate Change of 1992, at negotiations in Kyoto in December 1997, or thereafter, which would (A) mandate new commitments to limit or reduce greenhouse gas emissions for the Annex I parties, unless the protocol or other agreement also mandates new specific scheduled commitments to limit or reduce greenhouse gas emissions for Developing Country Parties within the same compliance period, or (B) would result in serious harm to the economy of the United States."[11]

While there were many environmental groups in the US (such as the World Wildlife Federation, the Natural Resources Defense Council, the Climate Action Network, the Environmental Defense Fund, and the Union of Concerned Scientists) urging the Clinton and Gore administration to join the Europeans in forging a strong agreement, there were also many opponents to the emerging agreement. The oil, coal, and automobile industries, among others, had made it clear they were not supportive of the agreement. Indeed, the Global Climate Coalition (GCC), a powerful group of hundreds of industries and companies that were opposed to the Kyoto Protocol, began a $13 million lobbying campaign in an effort to derail the negotiations. The GCC had the ear of the Senate.[12]

Thus, going into the third COP, the Clinton administration was in an awkward position. Unlike the case in Europe where the European public and industry were generally supportive (if in some cases only mildly supportive) of taking action on climate change, the US public and industry were highly divided on this issue.

The third COP began on December 1, 1997 in the ancient capital of Japan. There were still many contentious issues that needed to be resolved by the negotiators from more than 150 nations. The EU and the US were key players in these negotiations. One major issue that divided them because it would influence how easily they could meet emission reduction goals was which gases were to be covered by the agreement. The EU wanted the agreement to address only three greenhouse gases: CO_2, nitrous oxide, and methane. The US argued that three additional gases should be included in the mix: hydrofluorocarbons, perfluorocarbons, and sulfur hexafluoride.

A second point of contention concerned the emissions target to be achieved. The EU went into the negotiations calling for a 15 percent reduction target for all developed nations. It found itself in an awkward position on this point, however, because it was not requiring a uniform 15 percent reduction for each of its member states. Instead, it had established a "bubble" approach for the EU as a whole. This "bubble" created a mix of targets for EU member states that would add up to a 15 percent reduction for the EU, but would allow some of the less-developed member states (e.g., Spain, Portugal, Greece, and Ireland) to substantially increase their emissions while others made deep cuts in their emissions to make up the difference (Germany, UK, Denmark, Austria). The EU argued that this was fair because the EU is a political unit with a parliament, but the US and other nations dismissed this argument because it would put them at a competitive disadvantage. In response to the opposition from the US and Japan, the EU quickly gave in on its demand for uniform targets and accepted the idea of differentiated targets for countries based on their economic and energy structures. The US proposed a stabilization target for itself at 1990 levels by the year 2012.[13]

A third very important difference between the EU and the US reflected differences between them in the extent to which they felt developed nations should be required to cut domestic emissions. The EU strongly supported the idea that the main responsibility for global warming lay with the developed states and that therefore they should set an example for the rest of the world by making sharp emissions cuts at home through a mix of regulations, incentives, and voluntary measures. The US, in contrast, argued that developing countries needed to take meaningful measures as well.

Finally, the US called for maximum flexibility in how nations could meet their targets. The US embraced the idea that states should be allowed to employ various "flexible mechanisms," including emissions trading, joint implementation, and the clean development mechanism. In emissions trading, pollution is given a value so that the right to pollute can be bought and sold, but also controlled through the establishment of emissions ceilings. If the price is right, holders of pollution permits may have an incentive to reduce their pollution in order to be able to sell their pollution rights. Depending on where a ceiling is set, this can result in a decrease in pollution as firms try to cut their costs. The US support of a carbon dioxide emissions trading system was premised on its successful experience with reducing sulfur oxide emissions through an emissions trading scheme under the 1990 Clean Air Act Amendments. Joint implementation and the clean development mechanism are ways for developed countries to offset their own emissions by taking action to reduce emissions, respectively, in transition economies or in developing countries. The EU, which had no experience with emissions trading, was initially very skeptical of this idea. The EU argued that the majority of a nation's emissions reductions must occur domestically. The US remained firm in its position that no limit should be placed on the use of flexible mechanisms.

The stalemate between the EU and the US on these points was partially dealt with in the last days of the COP 3 in Kyoto after Vice President Gore made a last-minute appearance at the negotiations. A compromise was reached, making it possible to establish a protocol. The EU accepted the idea that six rather than three greenhouse gases should be controlled. Because the inclusion of the three additional gases would have made it virtually impossible for the EU to obtain a 15 percent reduction in emissions, they agreed to a much lower 8 percent cut in their greenhouse gas emissions from 1990 levels by 2008–2012. The US negotiating team responded by agreeing to a 7 percent cut in its greenhouse gas emissions over the same period (see table 8.1). Finally, they agreed to work out additional technical issues in subsequent COPs.[14]

Table 8.1
Emission Targets for 2008–2012 Set by the Kyoto Protocol

Country	Percent change from 1990 level
Austria	−13
Belgium	−7.5
Denmark	−21
Finland	0
France	0
Germany	−21
Greece	+25
Ireland	+13
Japan	−6
Luxembourg	−28
Netherlands	−6
Portugal	+27
Spain	+15
Sweden	+4
United Kingdom	−12.5
EU total	−8
United States	−7

From Kyoto to Marrakech

The important question of how these cuts were to be achieved was left for future negotiations. Thus, the question remained open as to what extent countries would be allowed to meet their targets through flexible mechanisms. Also left unresolved was how these flexible mechanisms were to be structured and how compliance with the agreement was to be ensured. Indeed, no industrialized nation was willing to ratify the agreement until these remaining issues were addressed. These issues were taken up at subsequent COPs.

Particularly important in these negotiations was the sixth COP, which was held in The Hague in November 2000. The negotiations collapsed because the EU and the US could not find a middle ground on the issue of whether a cap should be placed on the use of flexible mechanisms. The EU was adamant that at least 50 percent of emissions reductions must occur at home. The US held to its position that no limit should be placed on the use of the flexible mechanisms. The meeting dragged on and the EU refused to budge, claiming that it had already given in too much. Rather than allow the meeting to end in failure, the chairman suspended the conference. An agreement was reached to hold a COP 6-Part II in June of the following year to continue the discussions after additional rounds of preparatory negotiations.

In the meantime, George W. Bush won the 2000 election. One of his first major decisions as president was to reject the Kyoto Protocol. The Bush administration's unilateral decision to withdraw from the protocol without first informing or negotiating with the states of Europe, Japan, and other industrialized countries caught the world by surprise. The EU responded angrily, calling the US decision irresponsible and wrong. EU Environment Commissioner Margot Wallström made repeated visits to Washington to urge the White House to change its mind. Leaders of various EU member states joined in pressuring Bush to rejoin the agreement. German Chancellor Gerhard Schröder asked Bush on behalf of the EU to reconsider his decision. Swedish Environment Minister Kjell Larsson criticized the Bush administration's position, responding that "The Kyoto Protocol is still alive," and French President Jacques Chirac called the US move "a worrying and unacceptable challenge to the Kyoto Protocol."[15]

When these efforts failed to persuade Bush, the EU chose to move forward with the COP anyway, hoping that international and domestic pressure would build and the Bush administration would change its mind. At the COP 6-Part II held in the summer of 2001 in Bonn, the EU and Japan were the main players. A US delegation attended the meeting, but only as a means of protecting US interests. The US negotiating team was on the sidelines; it had been instructed to remain outside of the agreement. In the negotiations, Japan, a close ally of the US, took over many of the negotiating positions that had been held by the Clinton administration's negotiating team. Japan was eager to have the US reenter the negotiations and thought that by shaping a treaty favorable to the US, it would be easier for the US to return to the multilateral framework in the future. The EU needed Japan. Thus, in Bonn, the EU had little choice but to give in to almost all of the demands made by Japan. No cap was placed on the amount of emissions reductions a state could achieve through the use of flexible mechanisms, and countries were allowed to count the development and management of carbon sinks domestically and in developing countries toward their emission targets.

In September 2001 the European Parliament voted 398–9 with three abstentions in favor of the decision reached at COP 6-Part II. The parliament urged quick ratification of the agreement and the initiation of an EU emissions trading system. It even went so far as to criticize the agreement reached in Bonn for not going far enough to address climate change by failing to include stringent sanctions for noncompliance.[16]

Final technical details for the political agreement reached in Bonn were worked out at COP 7 in Marrakech, Morocco, in November 2001. At the end of the Marrakech round of negotiations there was widespread celebration among delegates present at the meeting. The *Washington Post* dubbed Marrakech "an important victory for European and environmental leaders."[17]

US Efforts to Reframe the Climate Change Debate

Stung by the domestic and international criticism that his withdrawal from the Kyoto Protocol elicited, Bush has tried very hard to reframe the climate change debate and portray his decision as forward looking. The first such effort was to put the climate change problematique into the context of the California energy crisis and US energy security. These efforts are embodied in the administration's energy policy plan. In May 2001, an energy task force headed up by Vice President Cheney presented the nation with a national energy plan that described the growing US demand for energy as a result of increased population and changing lifestyles. It noted that despite enhanced energy efficiency over the past decades, the US appetite for energy was strong. To meet the US need for more energy in the future, the plan argues, new energy sources, including oil, coal, and possibly nuclear energy, will have to be developed. There is also some mention in the report of the need for energy conservation and the potential for renewable energies to contribute to meeting future energy demands, but the report's emphasis is clearly on developing traditional energy sources.[18]

This effort to refocus the debate has proven highly contentious. The energy policy plan was lambasted by the environmental community as being beholden to the fossil energy industry. The Natural Resources Defense Council and the Sierra Club successfully sued the vice president's office under the Freedom of Information Act and won the release of thousands of pages of documents pertaining to the development of the report. Vermont Senator James Jeffords bolted from the Republican Party over the administration's energy and environmental policies, turning control of the Senate over to the Democrats (until the election of November 2002 once again gave the Republicans a majority). Although Jeffords registered as an Independent, the Democrats elected him chair of the Senate Environment and Public Works Committee as a reward for his defection. In this position, Jeffords worked with many other colleagues in the Senate to defeat Bush's proposal to drill for oil in the Arctic National Wildlife Refuge (ANWR). With the Senate back in the hands of the Republicans, however, the administration is again pushing for oil drilling in ANWR in pending energy legislation.

Another effort by the administration to reframe the climate change debate has emphasized the uncertainty of climate change science and the importance of long-term technology development. In June 2001 at a press conference called by the White House, Bush announced plans for a US Climate Change Research Initiative to support scientific research on climate change and to determine priority areas for investment and a National Climate Change Technology Initiative to enhance research at universities and national laboratories related to technology that could reduce greenhouse gas emissions.[19] Related to these initiatives and in response to Marrakech, in February 2002 the Bush White House announced its own plans to address climate change domestically through voluntary conservation measures, some added support for renewable energy technologies, and additional research into the science of climate change and mitigation technologies.[20] Consistent with this orientation, in his State of the Union address in January 2003, Bush announced that the government would provide $1.2 billion in support for hydrogen fuel initiatives toward the development of commercially viable hydrogen-powered fuel cells.[21]

The Bush administration also shifted away somewhat from its initial portrayal of climate change science as being highly uncertain and recognized that climate change is partially manmade and will have impacts on the US and other regions of the world. In fact, paralleling the EU rat-ification of the Kyoto Protocol, the US submitted its national climate report to the United Nations (as required by the United Nations FCCC, to which the US is still a party).[22] This document, "U.S. Climate Action Report 2002," indicates that the US will be significantly affected by climate change. According to the report, the US is likely to suffer more frequent and intense heat waves and to lose some ecologically sensitive natural areas. The report does not, however, call for immediate action. Instead, it concludes that regardless of what is done to cut emissions, it is too late to address several decades' worth of greenhouse gases that have already been emitted into the atmosphere. The report states: "Climate change is a long-term problem, decades in the making, that cannot be solved overnight. A real solution must be durable, science-based, and economically sustainable. In particular, we seek an environmentally sound approach that will not harm the US economy, which remains a critically important engine of global prosperity. We believe that economic development is key to protecting the global environment."[23]

The report goes on to note that based on a cabinet-level review and recommendations regarding climate change, President Bush announced that the US would commit to reducing its greenhouse gas intensity by 18 percent over the course of the next decade through voluntary measures, incentive schemes, and existing mandatory measures. According to the report: "This represents a 4.5 percent reduction from forecast emissions in 2012, a serious, sensible, and science-based response to this global problem—despite the remaining uncertainties concerning the precise magnitude, timing, and regional patterns of climate change."[24] Yet both US environmentalists and the EU were quick to point out that this policy is too little, too late, and falls far below the commitments the US made in Kyoto. It should be recognized that reducing greenhouse gas intensity is not the same as reducing greenhouse gas emissions. Greenhouse gas intensity is the amount of greenhouse gases produced per dollar of gross domestic product. According to the Pew Center on Global Climate Change, reducing greenhouse gas intensity by 18 percent equates to allowing total US emissions to *climb by 12 percent* over the same period because of expected growth in the economy![25]

The administration is adamant that the Kyoto Protocol is not the right approach. In a mid-May meeting in London, the newly appointed chief climate negotiator for the US, Harlan Watson, stated that the US will not participate in negotiations set to begin in 2005 to establish emissions targets for the second commitment period under the Kyoto Protocol (the period after 2008–2012). "We want no part of that. . . . The next time we take stock on climate change has been set by the president at 2012."[26] The administration also has worked to win developing country acceptance of its approach and has done this with some success. At COP 8 in New Delhi, the US, China, and India formed a coalition that argued that it would be unfair to expect developing countries to adopt targets (see chapter 10). To the EU's dismay, little reference was made to the Kyoto Protocol.[27]

Comparing the Responses of the EU and the US to the Climate Change Negotiations

The US was considered an international leader in the development of the 1987 Montreal Protocol addressing stratospheric ozone depletion caused by the release of chlorofluorocarbons and other manmade substances. In contrast, it has been the EU that has been the champion of the Kyoto Protocol and international efforts to address climate change.

The differences between the EU and the US in the climate change negotiations reflect the different political cultures that have taken root in Europe and the US. Environmental protection has become a major issue for EU foreign policy, and the EU has increasingly come to see itself as an international environmental leader. The EU rejects the US line of reasoning on climate change. At the UN ratification ceremony, EU Environment Commissioner Wallström urged the US to reconsider: "The United States is the only nation to have spoken out against and rejected the global framework for addressing climate change. The European Union urges the United States to reconsider its position."[28] The statement of an EU delegation to the US sums up well the EU position: "Although hope exists that technological fixes will materialize, they are likely to be difficult and costly to implement. Existing 'climate-friendly' technologies are already facing difficulties penetrating the market, and it is difficult to predict by when new break-through technological solutions would become effective. Immediate action to reduce consumption and to increase the share of those energy products which are less carbon-intensive is necessary because greenhouse gases such as carbon dioxide (CO_2) are long-lived and emissions today have a lasting effect on climate."[29]

Environmental interests are quite strong on both sides of the Atlantic. In the US, many large environmental groups have campaigned for the Kyoto Protocol. However, their efforts have been stymied by strong industrial lobbies opposed to the agreement. In Europe, environmental interests have won direct representation in many national parliaments as well as in the European Parliament. As a result, they have been able to green European politics in a way that is difficult for US environmental groups to achieve (see chapter 12). Even more important, while European industry, like US industry, is averse to energy taxes, it has generally supported government initiatives to address climate change far more than has been the case in the US. Industries in Europe have not openly opposed the Kyoto Protocol as some of their American counterparts have; instead, they have worked to try to influence domestic and EU implementation programs. It is noteworthy that American multinational corporations doing business in Europe play a very different tune on climate change there than they do in the US.

In many EU member states, a social-democratic politics prevails. Throughout the European continent there is greater support for government involvement in economic decision-making than is the case in the US. Government, industry, and nongovernmental organizations work more cooperatively on environmental policy matters there than they do in the US, where relations among these actors are often adversarial (see chapter 13).[30]

The US supports a neoliberal laissez-faire model of limited government intervention in the economy far more strongly than is the case in the EU. While it has proven difficult to introduce carbon and energy taxes in both Europe and the US, the push to get the government out of the economy is far stronger in the US than in Europe. Consistent with efforts to reduce the size of government, the US is moving away from heavy use of regulation to protect the environment and is experimenting more and more with the use of voluntary measures in pollution control. There is also greater use in the US of cost-benefit analysis than in Europe, where the precautionary principle holds greater sway.

Still, despite the wide gulf that separates the EU and the US on climate change, there are some signs of convergence across the Atlantic. There has, for example, been some convergence on the importance of using a mix of policy measures to reduce emissions. Of particular interest is the change in the EU's attitude towards emissions trading. Although initially skeptical of emissions trading, the EU has come to embrace it as a strategy for reducing emissions from power plants, steel producers, and oil refiners. In fact, in December 2002 the EU Council agreed unanimously on the establishment of a carbon dioxide trading system beginning in 2005 that will encompass the fifteen EU member states plus the ten EU accession countries. The plan, approved by the EU Parliament in July 2003, will affect some 10,000 installations and cover 46 percent of the EU 15's total CO_2 emissions in 2010.[31]

It is also important to recognize the role that federalism plays in climate change policy (chapter 4). The European Climate Change Program notes the need for a multistakeholder and twin-track approach at the EU and national government levels to dealing with climate change. At the EU level, the commission is proposing directives to promote generating electricity from renewable energy, voluntary commitments by automobile manufacturers to improve the fuel economy of cars by 25 percent, and energy taxes. National governments, however, have considerable autonomy regarding the policies and measures they adopt to reduce their greenhouse gas emissions.[32] In the US too we are seeing a dualistic approach, with some US states favoring the EU's stance over that of Washington, D.C.

A substantial segment of the US public is not pleased with the White House's lackluster approach to dealing with climate change. A Zogby international poll conducted in June 2002 of 1,008 likely voters chosen at random from across the nation found that only 21 percent of the respondents agreed with Bush's voluntary approach to reducing global warming pollutants. More than three-quarters (76 percent) of the survey respondents said the US should set emission standards on fossil fuel power plants and other industries.[33] Frustrated by Washington's inaction, and responding to the demands of their citizens, some states are beginning to move on their own.

Challenging the Bush administration, in the summer of 2002 California Governor Gray Davis, a Democrat, signed a bill into law that will require manufacturers to reduce CO_2 emissions from vehicles.[34] The law requires that the California Air Resources Board establish emission standards for the "maximum feasible reduction" of greenhouse gases from vehicles by 2005. Manufacturers will then have until 2009 to come into compliance with the standards. This move is highly significant since California accounts for approximately one-tenth of the cars in the United States. Through its proactivism, California has forced the automobile industry to introduce catalytic converters, seat belts, unleaded fuel, clean diesel fuel, reformulated gasoline, and electric and hybrid cars. Other states could follow California's lead in reducing greenhouse gas emissions from automobiles as well.[35] In describing his decision to sign this important environmental legislation, Governor Davis stated: "The federal government and Congress by failing to ratify the Kyoto Treaty on global warming have missed the opportunity to do the right thing. . . . We can now join the long-standing and successful effort of European nations against global warming, learn from their experience and build upon it."[36]

In testimony before the Senate in January 2003, Eileen Claussen, president of the Pew Center on Global Climate Change, noted that "Other countries are moving forward to address climate change, and in the United States, states and companies are exercising leadership to fill the void left by inaction at the federal level." She gave several examples. The New England states have joined five eastern Canadian provinces in agreeing to reduce their regional greenhouse gas emissions to 1990 levels by 2010 and more thereafter. New Hampshire and Massachusetts are now regulating CO_2 emissions from power plants, and several midwestern states are allowing agricultural interests to sell their sequestered carbon as a commodity. She also pointed out that forty companies based in the US or with substantial US operations had voluntarily committed to reducing their greenhouse gas emissions.[37]

Table 8.2
Trends in Greenhouse Gas Emissions

Country	Percent change 1990–2000
Austria	+2.7
Belgium	+6.3
Denmark	−9.8
Finland	−4.1
France	−1.7
Germany	−19.1
Greece	+21.2
Ireland	+24
Italy	+3.9
Luxembourg	−45.1
Netherlands	+2.6
Portugal	+30.1
Spain	+33.7
Sweden	−1.9
United Kingdom	−12.6
EU 15 member states	−3.5
US	+12[a]

Sources: European Environment Agency, "EU Reaches CO_2 Stabilization Despite Upturn in Greenhouse Gas Emissions, Annex," Copenhagen, April 29, 2002; US Environmental Protection Agency, "Climate Action Report 2002. United States of America's Third National Communication Under the United Nations Framework Convention on Climate Change," p. 5.
[a] Data are for 1990–1999.

Looking to the Future: Trends in Greenhouse Gas Emissions in Europe and the US

Levels of greenhouse gas emissions in Europe and the US rose steadily for most of the twentieth century. The goal of the UNFCCC was to change this trend. The EU has done better than the US in this regard (see table 8.2). According to *Environmental Signals 2002*, a report issued by the European Environment Agency in May 2002, between 1990 and 2000 EU greenhouse gas emissions fell by 3.5 percent (CO_2 emissions alone dropped by 0.5 percent). The report acknowledged that about half of this cut was due to developments in Germany (the collapse of the East

German economy and the rapid shutdown of heavily polluting industries) and the UK (as a result of that country's switch from coal to natural
gas). However, between 1999 and 2000, emissions of CO_2 actually rose
by 0.5 percent, and other greenhouse gases increased by 0.3 percent
during this year. What these trends suggest is that the large and relatively
easy cuts have already been made and the EU will have to make substantial changes in its current energy use patterns if it is to achieve further
substantial cuts as required by the Kyoto Protocol.

The situation is not uniform across the EU. Germany, the largest EU
emitter, recorded a 19.1 percent decrease in its greenhouse gas emissions
between 1990 and 2000, putting it very close to the 21 percent reduction it is required to make as part of the EU bubble. Spain, on the other
hand, has seen a 33.7 percent growth in its greenhouse gas emissions.
Given that it is only allowed a 15 percent increase as part of the EU
bubble, this means that Spain will have to make substantial cuts in
emissions over the course of the next decade. The EU countries that
will have the hardest time meeting their commitments are Austria,
Belgium, Denmark, Greece, Ireland, Italy, the Netherlands, Portugal, and
Spain.[38]

Trends in the US are alarming. Total US greenhouse gas emissions in
1999 were 12 percent above 1990 levels. Eighty-two percent of total US
greenhouse gas emissions are from CO_2, and these emissions grew by 13
percent between 1990 and 1999.[39]

Conclusion

The divide between the EU and the US on climate change has substantially slowed the momentum that existed a decade ago to address climate
change internationally. This is troubling given that greenhouse gas emissions in the rest of the world are expected to rise substantially in coming
decades and the developing world looks to the US and the EU for signals.

Whether the EU will succeed in its efforts to keep the Kyoto Protocol
alive remains to be seen. If it succeeds, the EU may continue its slow
reduction in greenhouse gas emissions even if it fails to implement in full
the 8 percent reduction required under the protocol. If it fails, it remains
to be seen how the EU will act. It does not look as if the US has any

intention of returning to the Kyoto framework. Instead, it appears intent on waiting for a technological transformation toward a less carbon-intensive economy to occur. The serious question that must be raised is whether this will happen without sufficient incentives from the government.

In the meantime, efforts to establish a dialogue across the Atlantic on mitigation of climate change must continue. Perhaps where the EU and the US can cooperate is in the development of strategies for action in cooperation with developing countries. Also, it is possible to develop cooperative initiatives and an information exchange between the EU and the US at a more decentralized level—among cities, counties, and prefectures, US states and EU member countries, firms, and nongovernmental organizations.

Notes

1. Kyoto Protocol to the United Nations Framework Convention on Climate Change, December 10, 1997, UN Doc. FCCC/CP/197/L.7/Add. 1.

2. "EU Delivery Marks Important Achievement in Climate Treaty Process, EU Position on Climate Change Contrasts with Recent U.S. Decisions," US Newswire, May 31, 2002.

3. Eric Pianin, "U.S. Aims to Pull Out of Warming Treaty: 'No Interest' in Implementing Kyoto Pact," *Washington Post*, March 28, 2001, p. 1.

4. See Social Learning Group, William C. Clark, Jill Jaeger, Josee van Eijndhoven, and Nancy M. Dickson, eds., *Learning to Manage Global Environmental Risks*, Vol. 1. *A Comparative History of Social Responses to Climate Change, Ozone Depletion, and Acid Rain* (Cambridge: MIT Press, 2000).

5. European Community Treaty, Article 174(2). The European Commission subsequently produced a communiqué on the precautionary principle outlining its thoughts on when and how this principle should be used within the EU and internationally. Commission of the European Communities, Communication from the Commission on the Precautionary Principle, Brussels, February 2, 2000. It should be noted that the precautionary principle is also in the 1992 Rio Declaration that the United States signed.

6. For a more detailed discussion of the "no regrets" policy, see Paul G. Harris, ed., *Climate Change and American Foreign Policy* (New York: St. Martin's, 2000).

7. For a history of the negotiations at Rio, see J. A. Leonard and Irving M. Mintzer, *Negotiating Climate Change: The Inside Story* (Cambridge: Cambridge University Press, 1994).

8. See Albert Gore, *Earth in the Balance: Ecology and the Human Spirit* (New York: Plume, 1993).

9. To see how the tax was portrayed by conservative interests, see Jonathan H. Adler, "Clinton's Stealth BTU Tax," Competitive Enterprise Institute Washington, D.C., October 1, 1996.

10. Michael Grubb with Christiaan Vrolijk and Duncan Brack, *The Kyoto Protocol, A Guide and Assessment* (London: Royal Institute of International Affairs, 1999), p. 54.

11. Byrd–Hagel Resolution, 105th Cong., 1st Session, S. Res. 98, Report No. 105–54, July 25, 1997.

12. See Michele M. Betsill, "The United States and the Evolution of International Climate Change Norms," in Harris, ed., *Climate Change and American Foreign Policy*, pp. 215–216, 218–220.

13. See "Negotiators Discuss 'Differentiated' Emissions Cuts," CNN, December 2, 1997. Available at http://www.cnn.com/EARTH/9712/01/global.warming/.

14. This discussion relies heavily on Miranda A. Schreurs, "Competing Agendas and the Climate Change Negotiations: The United States, the European Union, and Japan," *Environmental Law Reporter* 31(10) (2001): 11218–11224. See also Miranda A. Schreurs, *Japan, Germany, and the United States* (Cambridge: Cambridge University Press, 2002).

15. CNN, World, March 31, 2001. http://www.cnn.com/2001/WORLD/Europe/italy/03/31/eu.Kyoto/.

16. "EU/UN Climate Change: MEPS Press Commission to Present Concrete Proposal," *European Report*, No. 2616, September 8, 2001.

17. *Washington Post*, November 11, 2001, p. A 2.

18. *Reliable, Affordable, and Environmentally Sound Energy for America's Future*, Report of the National Energy Policy Group (Washington, D.C.: White House, May 2001).

19. "In President's Words: 'A Leadership Role on the Issue of Climate Change,' " *New York Times*, June 12, 2001, p. 12.

20. "Economic Report of the President" (Washington, D.C.: Government Printing Office, 2002), p. 245.

21. See the White House press release on the hydrogen fuel initiative available at http://www.whitehouse.gov/news/releases/2003/01//20030128-25.html.

22. Andrew Revkin, "Climate Changing, U.S. Says in Report," *New York Times*, June 3, 2002, p. 1.

23. US Environmental Protection Agency, "Climate Action Report 2002. United States of America's Third National Communication Under the United Nations Framework Convention on Climate Change," p. 3. Available at http://www.epa.gov/globalwarming/publications/car/ch1.pdf.

24. Ibid.

25. Pew Center on Global Climate Change, "Climate Change Activities in the United States," June 2002. Available at http://www.pewclimate.org.

26. Quoted in Paul Brown, "US Dashes Hopes for Climate Deal," *Guardian* foreign pages, *Guardian,* p. 11.

27. Discussion with Dr. Ren Yong, State Environmental Protection Agency, Beijing, China, January 17, 2003.

28. Quoted in "In Ratifying Climate Pact, EU Asks U.S. to Reconsider," *Los Angeles Times,* June 1, 2002, p. 11.

29. Transport, Energy and Environment Section, Delegation of the European Commission to the US, September 2001, EU Law and Policy Overview, http://www.eurunion.org/legislat/climatechange.htm.

30. For a discussion of EU policy styles, see Albert Weale, Geoffrey Pridham, Michelle Cini, Dimitrios Konstadakopulos, Martin Porter, and Brendan Flynn, *Environmental Governance in Europe* (Oxford: Oxford University Press, 2000).

31. "Emissions Trading: EU Environment Commissioner Margot Wallström Welcomes Council Agreement as Landmark Decision for Combating Climate Change." Available at http://europa.eu.int/comm/environment/climat/emission.htm; "EU Parliament Launches Climate Emissions Trading," planet Ark, July 3, 2003. Available at http://www.planetark.org/dailynewsstory.cfm/newsid/21371/story.htm.

32. European Commission, "European Climate Change Program." Available at http://europa.eu.int/comm/environment/climat/eccp.htm.

33. Zogby International Omnibus Polling Results, "American Attitudes on Climate Change." Available at http://www.ucsusa.org/environment/zogby.html.

34. Associated Press state and local wires, November 10, 2001 and February 1, 2001.

35. *Los Angeles Times,* July 23, 2002, p. 1.

36. Gray Davis, "California Takes on Air Pollution," *Washington Post,* July 22, 2002, p. A 15.

37. Testimony of Eileen Claussen, president of Pew Center on Global Climate Change, before the Committee on Commerce, Science, and Transportation, US Senate, January 8, 2003. Available at http://www.pewclimate.org/media/testimony_01082003.cfm.

38. European Environment Agency, "EU Reaches CO_2 Stabilization Despite Upturn in Greenhouse Gas Emissions, Annex," Copenhagen, April 29, 2002.

39. US Environmental Protection Agency, "Climate Action Report 2002," p. 5.

9

Trade and the Environment in the Global Economy: Contrasting European and American Perspectives

David Vogel

This essay explores areas of agreement and disagreement between the US and the EU concerning the role of the World Trade Organization (WTO) in linking trade liberalization and environmental protection. It begins by tracing the background of the responses of the General Agreement on Tariffs and Trade (GATT) and the WTO to criticisms from environmentalists. After exploring the common interests of the EU and the US, it then explains the evolution of the American position with respect to trade and environment linkages over the past decade. The main part of the chapter examines the increasingly significant divergence between American and European perspectives and preferences on trade and environmental issues. On a number of critical issues, the US now favors the status quo. It believes that the current system of WTO trade and environmental rules allows it to challenge other countries' nontariff barriers (NTBS), but does not place its own rules in jeopardy. By contrast, the EU favors a renegotiation of WTO trade and environment provisions since they appear to make a number of its own standards vulnerable to challenges by the US and its other trading partners.

The Uruguay Round and the Creation of the Committee on Trade and Environment

In 1991 an international trade dispute settlement panel found that an American law banning imports of tuna caught in ways that harmed dolphins violated American obligations under the GATT, the predecessor organization to the WTO. The environmental community in both the US and the EU was outraged by the panel's decision. They urged that the GATT be changed so as to give governments wider latitude to maintain environmental regulations that restricted the imports of products produced in environmentally harmful ways. This highly controversial tuna-dolphin decision launched a decade-long, often heated debate over the compatibility of international trade rules (and their interpretation by dispute settlement panels) with environmental protection at the national, regional, and international levels.[1]

In marked contrast to the North American Free Trade Agreement (NAFTA), whose approval by Congress in November 1993 incorporated a number of provisions favored by environmental organizations, the Uruguay Round Agreement that concluded in 1994 addressed few of the principal concerns of the environmental community. However, responding to pressures from consumer activists, the United States did successfully demand a modification of the Standards Code. While an earlier draft had required that standards be "the least trade restrictive available," the final version imposed a less formidable hurdle. It stated that they may "not be more trade-restrictive than necessary to fulfill a legitimate objective, taking into account the risks nonfulfillment would create."[2] In addition, the Agreement on Subsidies and Countervailing Measures permitted governments to subsidize up to 20 percent of one-time capital investments to meet new environmental requirements, provided that the subsidies were "directly linked and proportionate" to environmental improvements. This provided a partial exemption for environmental subsidies from the WTO's broader restrictions on government subsidies of business.

Most important, at the initiative of the European Free Trade Association (EFTA), the WTO agreed to formally place the relationship between trade and environment on its own agenda. Following the tuna-dolphin decision, EFTA members had requested "a rule-based analytical discussion of the interrelationship between trade and environment . . . to ensure that the GATT system was well equipped to meet the challenges of environmental issues and to prevent disputes by . . . interpret(ing) or amend(ing) . . . certain provisions of the General Agreement."[3] Their request was strongly supported by both the United States and the EU.

US trade representative (USTR) Mickey Kantor expressed his support for "engag(ing) the GATT" with a "post-Uruguay Round work program on the environment."[4] For the EU, such a program was urgently needed in order to examine the relationship between WTO rules and multilateral environmental agreements (MEAs).

At the GATT's April 1994 ministerial meeting at which the Uruguay Round was formally ratified and the WTO established, an agreement was reached to undertake a systematic review of "trade policies and those trade-related aspects of environmental policies which may result in significant trade effects for its members."[5] The Committee on Trade and Environment (CTE) was formed to undertake this task. While the CTE has played a useful role in raising awareness and promoting discussion of trade and environment linkages, and has strengthened ties between the WTO and the secretariats responsible for administering international environmental treaties, it has been unable to agree on policy recommendations to submit to the WTO's membership. This is due to sharp differences among its members.

The principal points of conflict on trade and environment linkages within the WTO are between the EU and the US on one hand, and developing nations on the other.[6] The former favor a flexible interpretation of Article XX, which lists the grounds on which trade restrictions are permissible, as well as making the WTO dispute settlement process more transparent. Both positions are opposed by developing countries that face little or no pressure from domestic nongovernmental organizations (NGOs) to make the WTO more responsive to environmental concerns and that fear protectionist abuses of any new environmental provisions. The latter's trade policy preferences vis-à-vis the developed world are largely driven by domestic producers who want increased access to developed country markets—access which they see as threatened by rich country environmentalists who favor links between trade and environmental policies. Support for changing or clarifying WTO rules that govern environmental regulations that restrict trade has emerged primarily from the US and the EU. The positions of the EU and the US are complex. They both want a more open world economy, yet they also want to protect their own relatively strict environmental standards from being challenged as trade barriers.

Common EU and US Positions

As both major exporters and the political architects of the global trading system, the US and the EU favor a more open world economy, which in turn requires rules that restrict nontariff trade barriers. Indeed, both the Standards Code and the Agreement on Sanitary and Phytosanitary Measures (SPS) were included in the Uruguay Round Agreement largely at the insistence of the United States. Many American exporters felt they had been disadvantaged by the unfair application of technical, food and agricultural standards, and they wanted such standards to be subject to WTO scrutiny. For its part, the EU has had extensive experience in dealing with the role of regulations and standards as nontariff barriers (NTBs) in the context of its efforts to establish a single internal market. It has also favored rules that restrict discriminatory NTBs at the international level.

Yet both the US and the EU also have an extensive array of health, safety, and environmental regulations that they want to be able to protect from challenges through the WTO. Many of these regulations also command strong support from politically influential NGOs. The need to protect such regulations has, if anything, become more important in recent years. Owing to the increasing criticism of globalization in general and the role of the WTO in particular by activists and their supporters on both sides of the Atlantic, a successful legal challenge or even the threat of a successful legal challenge to a politically visible protective regulation would undermine public support for trade liberalization and the legal principles on which it is based.

Moreover, not all European and American producers benefit from liberal trade policies. In many cases, domestic producers want to maintain protective regulations that restrict imports or put them at a disadvantage. Alternatively, some environmental regulations impose a competitive disadvantage on domestic producers, which then gives the latter an interest in making their foreign competitors comply with them as well. This, for example, occurred in the case of American restrictions on chlorofluorocarbons (CFCs). Once their use was banned in the United States, major American producers supported an international agreement to phase out their worldwide use. The "export" of American or

European environmental standards is often also strongly supported by domestic NGOs because it both reduces business opposition to the imposition of stricter domestic regulatory requirements and serves to strengthen environmental standards in other countries. Health, safety, and environmental regulations backed by coalitions of NGOs and producers—so-called "Baptist–bootlegger" coalitions—are a common feature of trade politics in the US and Europe.[7]

Trade and Environment in American Politics

Congressional and Presidential Politics

While political support for reforming WTO rules to strike a "greener" balance between free trade and environmental protection is now much stronger in Europe than in the US, this was not always the case. The North American Free Trade Agreement approved by Congress in 1993 included, at the insistence of President Clinton, a Supplementary Agreement on the Environment (SAE) as well a set of environmental provisions in the trade agreement itself that was negotiated by the earlier Bush administration. Widely considered to be the "greenest" trade agreement ever negotiated, NAFTA appeared to represent a model for how to liberalize trade while at the same time safeguarding, even improving, environmental quality. Building upon its precedent, U.S. trade representative Mickey Kantor proposed to Congress in mid-1994 that the American legislation implementing the Uruguay Round WTO Agreement include, along with an extension of fast-track negotiating authority for the administration, a commitment to making "trade and the environment" one of seven "principal negotiating objectives" for the US in any future trade agreement.[8]

However, several congressional Republicans whose support had been critical to congressional approval of NAFTA strongly opposed this formulation. They had agreed to the SAE as a necessary price for the passage of NAFTA, but now the administration was proposing to elevate the status of the environment to a core provision in any future trade agreements negotiated by the US. To some of them, this went too far. They were particularly upset because the side agreement negotiated by Kantor had included provisions for trade sanctions in the event of noncompliance with some of its environmental provisions, which they regarded as a dangerous precedent. Accordingly, a number of congressional Republicans, along with important segments of the business community, insisted that fast-track legislation explicitly exclude any agreement on either labor or environmental standards. The Clinton administration backed down; when it finally submitted legislation authorizing the renewal of fast-track negotiations in the fall of 2000, environmental concerns were muted. But this in turn outraged many environmentalists and their congressional Democratic allies. Accordingly, when fast-track renewal finally came to a vote in the House of Representatives, it received virtually no support from Democratic representatives and was resoundingly defeated.

The failure of the American Congress to renew fast-track negotiating authority during the remainder of the Clinton administration had several causes, including the growing strength of protectionist forces within the Democratic Party and the reluctance of many congressional Republicans to hand President Clinton a political victory. Prominent among them was the impasse within the Congress over trade and environment links. Many Republicans, whose party controlled both houses of Congress after 1994, strongly opposed any such linkage, particularly if it provided trade sanctions for environmental nonperformance. They worried that environmental "safeguards" are really disguised forms of protectionism and that incorporating them would obviate the purpose of trade liberalization and represent a backdoor way to advance the green agenda. However, many congressional Democrats continued to insist on effective linkages, including provisions for sanctions.

The Bush administration, while publicly acknowledging that trade policies should also improve environmental quality, initially opposed any formal linkages between the two. USTR Robert Zoellick cautioned that, "while there are many ways to support international environmental . . . objectives, you have to be very careful to do so in a way that doesn't become a form of protectionism," adding that he shared the concern of developing countries that "this is a new way to slow their growth."[9] He also explicitly characterized the trade and environmental agenda as protectionist. However, in an attempt at compromise, the fast-track authorization narrowly approved by the House of Representatives in

December 2001 did state that one American trade negotiating objective would be to make trade and environmental objectives mutually supportive. When Congress finally approved the granting of Trade Promotion Authority in August 2002, 8 years after it had expired, congressional Republicans agreed to include a provision instructing American trade negotiators to regard labor and environmental goals as "principal negotiating objectives," although it did not bind the US to achieving any particular objectives.[10]

American International Initiatives
Sharp domestic political differences on trade and environment linkages have made it difficult for the US to take a consistent leadership position with respect to trade and environmental issues before the WTO. Many American proposals to the CTE have tended to emphasize procedural rather than substantive issues.

In a communication from the US on trade and sustainable development issued as part of the preparations for the 1999 Ministerial Conference in Singapore, the US proposed that the CTE conduct ongoing reviews of the links between the WTO's "negotiating agenda and the environment and public health."[11] These reviews "would identify and discuss issues, but not try to reach conclusions or negotiate these issues in the CTE itself." The US has also encouraged all WTO members to conduct reviews of the potential environmental effects of any trade proposals. Shortly before he left office, President Clinton issued Executive Order 13141 requiring written environmental reviews of major trade agreements. This order institutionalized a practice that had begun with the first Bush administration's review of the environmental impact of NAFTA, and President George W. Bush reaffirmed it in April 2001.

The United States has taken a leadership role within the WTO, especially at the ministerial meetings in Seattle, in attempting to promote increased transparency and openness. In a Declaration of Principles on Trade and Environment, the US noted that it "has been a staunch advocate for WTO reforms, including greater interaction and exchange of information with the public through the creation of consultative mechanisms," adding that "transparency and openness are vital to ensuring public understanding of and support for the WTO and all international institutions."[12]

The WTO has responded to a number of these suggestions. It has invited NGOs to participate in a number of conferences and seminars and has issued a steady stream of studies on the environmental impacts of trade liberalization. The dispute settlement process has also become more public, largely through the Internet, which now provides considerably more information on the progress of dispute settlement proceedings. In the shrimp-turtle case, the appellate panel did invite the views of experts in marine biology and it also permitted representations by NGOs, although these were formally required to be part of the American legal brief. These initiatives have been supported by the EU as well, although it has placed less priority on them than has the US.

The most important American policy initiative relating to trade and the environment has to do with the highly controversial area of subsidies, specifically in the areas of fisheries and agriculture. The US has long sought to restrict the EU's extensive agricultural subsidies, particularly its export subsidies, as well as subsidies for its fishing fleets. Both sets of subsidies adversely affect American producers, and American efforts to restrict them long predate the emergence of environmental concerns over international trade. However, with the growth of concern about the environmental impact of trade liberalization, the American position is now that these subsidies are environmentally harmful—a position that is supported by a number of WTO studies and reports.[13]

The US argues that by reducing such trade-distorting subsidies, "trade liberalization can promote competition and more efficient resource use, as well as contribute to higher standards of living and a cleaner environment."[14] In a related proposal, the United States supports what it describes as another "win-win" opportunity: the elimination of tariffs on environmental goods, such as pollution control technologies, and the liberalization of trade in environmental services. In short, for the Americans, the most constructive way to "green" the WTO is not to expand the grounds on which a nation can restrict trade to prevent environmental harms, but rather for the WTO to encourage governments to reduce their financial support for environmentally harmful economic activities.

EU–US Differences

Multilateral Environmental Agreements

From the very outset of trade and environmental discussions within the WTO, the EU urged the CTE to recommend that trade restrictions sanctioned by multilateral environmental agreements (MEAs) be protected from challenges through the WTO. (Of the approximately 200 multilateral environmental agreements, 20 contain trade provisions.) The EU has been concerned about the possibility that under current trade law, a country that belonged to the WTO but had not signed an international environmental agreement could legally challenge trade restrictions that were permitted or mandated by an MEA. This would not only "undermine international efforts to tackle environmental problems (but) it would also fuel the arguments of those opposed to the WTO."[15] While acknowledging that no trade measure taken pursuant to an MEA has yet been challenged in the WTO by a nonparty, the EU believes that "the legal ambiguity surrounding the possibilities of such a challenge causes uncertainty and doubt over the effectiveness and legal status of such measures and thus weakens MEAs." Accordingly, the EU wants the WTO to "clarify that . . . multilateral environmental agreements and associated trade measures are also respected by trade law."[16]

The American position is that no such clarification is necessary because "the WTO broadly accommodates trade measures in MEAs."[17] The US has expressed confidence that the WTO would not sustain a challenge to an MEA—a position that it believed to be confirmed by the appellate body ruling in the shrimp-turtle case. This case did not technically concern an MEA, since at issue was the US embargo on shrimp caught in ways that killed sea turtles. The most relevant MEA, the Convention on International Trade in Endangered Species (CITES), prohibits trade in sea turtles, not in shrimp. Nor does it provide for trade restrictions of related products as a means of enforcing its provisions. Nonetheless, the fact that the American trade restriction was intended to protect a species that was officially protected by an MEA was explicitly noted by the panel.

Underlying these transatlantic differences is the changing position of the US and the EU with respect to MEAs. Historically, MEAs have reflected a broad international consensus, one that has included both the US and the EU, with the former frequently playing a leadership role. But more recently, such agreements have reflected sharp differences between the two. An important example is the Montreal Protocol on Biodiversity.[18] The EU supported an international treaty that was consistent with its domestic restrictions on the planting, sale, and labeling of genetically modified foods and seeds. For its part, the US, as a major exporter of such crops, wanted to limit the basis on which trade in genetically modified foods and seeds could be restricted. The two parties specifically differed on the application of the precautionary principle to import bans and labeling requirements, whether the protocol should include bulk commodities intended for consumption (i.e., crops) or be limited to seeds, and the relationship between the protocol and WTO rules.

The result was a compromise; on the critical point of the relationship between the protocol and WTO, the former is deliberately ambiguous. However, if the US were to bring a claim before the WTO over an EU restriction on genetically modified organisms (GMOs), the Biosafety Protocol, which has been ratified by more than 130 countries, could be invoked by the EU as evidence of a strong international consensus. (The EU was unable to cite any such international consensus in its defense in the beef hormone case.) Whether this would enable the EU to prevail remains unclear, but it certainly would make its case stronger. In this context, it is not surprising that the EU urgently wants the WTO to "clarify" the legal relationship between MEAs and the WTO in a way that specifies the circumstances under which the former are subject to the latter. The US officially claims that no such clarification is needed because no nation has filed a challenge to a trade restriction sanctioned by an MEA. But clearly the US also wants to avoid having the WTO defer to an MEA that it does not support—a category that is steadily expanding.

Specifically, both the Basel Convention on the Export of Hazardous Wastes and the Kyoto Protocol have been ratified by the EU, but not by the US. Accordingly, the EU would like assurances that any trade restrictions that flowed from these agreements would withstand WTO scrutiny, a concern the US does not share.

Precautionary Principle

Within the EU, the precautionary principle has emerged as an important basis for the adoption of a wide range of risk-averse health, safety, and environmental policies, including restrictions on genetically modified foods and seeds (see chapter 1). It has been an explicit component of EU environmental policy since 1992 and is defined as one of the key principles of EU environmental law in both the Maastricht and Amsterdam Treaties. In order to better defend its regulations from possible legal challenges from the US and other WTO members, the EU wants the precautionary principle to be incorporated into international trade law. One way to accomplish this objective is to include this principle in as many international environmental agreements as possible and then to have these agreements accorded some kind of legal status by the WTO. For its part, the US wants to maintain the legal supremacy of the SPS Agreement, whose more demanding scientific standards for trade-restrictive regulatory policies enabled the US to prevail in its dispute over the EU's ban on beef hormones.

Not surprisingly, there were sharp differences between the EU and the US over whether the precautionary principle should be included in the Montreal Protocol on Biodiversity. As a compromise, Article 10 of the protocol incorporates the precautionary principle, though without explicitly mentioning it. A country is permitted to reject the importation of a "living modified organism for intentional introduction into the environment" where there is "lack of scientific certainty" regarding the extent of its potential adverse effects on either human health or biodiversity.[19] Most observers believe that this language reduces the amount of scientific evidence that would be needed to justify an important ban.

The EU and the US are also divided about the legal status of the precautionary principle in international trade law. During the Uruguay Round negotiations in the early 1990s, it was the United States which had insisted on changes in the SPS Agreement to make it easier for relatively risk-averse regulatory standards to pass the scrutiny of WTO dispute panels. This position reflected the relative stringency of many American health, safety, and environmental standards compared with these in the rest of the world, including the EU. But over the past decade, the EU has adopted a number of standards that are stricter or broader than their American counterparts. Accordingly, it is now the EU that is insisting that WTO rules be modified so that they can more easily defend their regulatory standards from trade challenges, including those from the US.

One such modification would be for the WTO to accord legal recognition to the precautionary principle—in effect harmonizing EU and WTO approaches to formation if regulatory policy in the face of scientific uncertainty. While the European Commission believes that measures based on the precautionary principle are a priori compatible with WTO rules, it nonetheless wishes to "clarify this relationship" and, in addition, "to promote the international acceptance of the precautionary principle." According to the EU, "this will help ensure that measures based on a legitimate resort to the precautionary principle, including those that are necessary to promote sustainable development, can be taken without the risk of trade disputes."[20] In this context, it is worth recalling that the EC did invoke the precautionary principle in the beef hormone case, only to have the WTO's Appellate Body decide that "the precautionary principle cannot override our finding . . . namely that the EU import ban . . . is not based on risk assessment" as required by the SPS Agreement.[21] Clearly, the EU would prefer that any trade dispute regarding genetically modified agriculture be decided on a different basis.

Once again, the US does not consider a change in WTO rules to be necessary. According to the Americans, not only is a "precautionary element . . . fully consistent with WTO rules, (but) it is an essential element of the US regulatory system." However, the US cautions that "precaution must be exercised as part of a science-based approach to regulation, not a substitute for such an approach."[22] While this is not inconsistent with the way the precautionary principle has been interpreted within the EU, the US remains concerned that as applied by the EU, there is a danger that the precautionary principle will become a "guise for protectionist measures." The US is satisfied with provisions of the SPS Agreement which permit a country to set high standards even when the scientific evidence on risk is uncertain, with the stipulation that such standards be regarded as provisional and thus subject to modification as more evidence becomes available. But the US is concerned that "explicitly embedding a precautionary principle in the SPS or TBT sections of the WTO framework would . . . allow countries to block imports on environmental or health grounds in the absence of any scientific evidence of significant risk."[23]

Process and Production Methods

Historically, the most important source of trade conflict between the US and its trading partners, including the EU, has stemmed from American efforts to unilaterally employ trade restrictions to impose its domestic environmental policies on other countries. This was the essence of the dispute in both the tuna-dolphin and the shrimp-turtle cases. But while the US lost both cases, including a second tuna-dolphin case, which was brought by the EU, the political significance of the two marine protection cases was substantially different.

In the shrimp-turtle case, the WTO's Appellate Body, in an opinion that sharply contrasted with the dispute panel decision in the tuna-dolphin cases, agreed that the US *could* limit imports on the basis of how a product was produced outside its borders in order to pursue legitimate environmental objectives—provided certain conditions were met. The WTO did not object to the goal of American policy, but rather the means the US had employed to achieve it. This meant that only minor changes were required to make US turtle protection regulations consistent with the WTO. Following these changes, the American regulations were subsequently upheld by another WTO dispute panel.

While many American environmentalists failed to appreciate the significance of the appellate body ruling, the US government has not. It regards the outcome of the shrimp-turtle case as a major political triumph. The WTO had effectively revised its legal interpretation of the rules governing one of the most persistent sources of trade conflict between the US and its trading partners.[24] The office of the USTR headlined a press release announcing that a second dispute panel had found America's slightly revised implementation of its sea turtle protection law to be fully consistent with the decision of the appellate panel, "U.S. Wins on WTO Case in Sea Turtle Conservation." Zoellick commented, "We have long maintained that the WTO Agreements recognize the legitimate environmental concerns of Members, and this report confirms our view.

I am pleased that the arguments we have made in this and other disputes have contributed to the body of cases illustrating the WTO's sensitivity to environmental concerns."[25]

The EU was also pleased with the outcome of this case since a number of environmental policies that its trading partners, including the US, have challenged have also revolved around the extraterritorial application of environmental regulations. However, the EU does not share America's satisfaction with the extent to which the shrimp-turtle dispute panel decision has "greened" the WTO. It wants WTO rules to be clarified in order to significantly broaden the basis upon which a country can regulate or restrict imports based on how they were produced outside its borders. According to the EU, "It is increasingly clear that how a good is made is important and can no longer be dismissed as a luxury or detail of concern only to developed countries."[26]

The EU's position on the appropriate status of environmentally related trade restrictions under the GATT/WTO has shifted markedly over the past decade. In 1991, the EU, along with virtually every other GATT member, applauded the dispute panel ruling against the US in the tuna-dolphin case for striking a much-needed blow against America's unilateral efforts to extend the scope of its environmental standards outside its borders. Now it is the EU that is in the forefront of urging the WTO to permit a wide range of environmentally related trade restrictions to protect the global environment—even in the absence of an international treaty. This change in the EU's position largely reflects its increasingly active leadership role in addressing international environmental issues—a role formerly filled by the US.

Ecolabels

A related point of contention between the EU and the US involves the legal status of environmental labels. Both the US and the EU support the use of ecolabels, both for the environmental impact of the product itself as well as for how it is produced. However, the use of ecolabels is much more common in Europe, where at both the national and European level they have become a major instrument of environmental policy. The US has periodically expressed concern about the EU's criteria for awarding ecolabels on the grounds that the European system has a "potential for discrimination against US firms whose production processes and methods differ from those used in the EU while having comparable environmental impacts."[27] In one of his annual reports to Congress, the USTR listed the EU's ecolabel scheme as a "topic of continuing concern," although the US has not filed a formal complaint.

Within the CTE, the relationship of ecolabels to the WTO has emerged as a major point of contention between the US and the EU. One key issue is their legal status. Specifically, does the Technical Barriers to Trade Agreement (TBT), which covers both technical regulations and standards, include ecolabels? The US claims that it does since the definition of both standards and technical regulations in the TBT explicitly includes labeling requirements as they apply to a product, process, or production method. This would make national, or in the case of the EU, regional, ecolabels subject to the same WTO discipline as any other technical standard, meaning they would be required to treat products from all WTO member countries equally and could not be prepared, adopted, or applied with the intention or effect of creating "unnecessary obstacles to trade." However, the US has not advocated a change in WTO rules; rather, it believes that the TBT is already sufficiently flexible to protect the use of ecolabels based on process and production methods and to subject them to WTO scrutiny.

The EU initially argued that the TBT does not cover ecolabels at all, a position that it based on the absence of specific references to environmental labels, as distinguished from "labeling requirements" in the TBT. However, as environmental concerns in Europe have grown, the EU's position has shifted. As in the case of the trade status of MEAs, the EU now wants the relationship between WTO rules and nonproduct-related process and production methods (NPRPPM, usually referred to as PPMs) to be "clarified." It particularly supports explicit recognition of the WTO compatibility of ecolabeling schemes based on a life-cycle approach. According to the European Commission, "EU consumers are increasingly concerned about a growing range of NPRPPM issues which they feel affect their everyday lives." Accordingly, "subject to ... important procedural safeguards, there should be scope within WTO rules to use such market based, non-discriminatory non-protectionist instruments as a means of achieving environmental objectives."[28]

WTO Dispute Settlement

Underlying the differences between the EU and US views toward modifying WTO rules governing regulatory standards that restrict trade is a divergence in their perceptions of the adequacy of WTO rules to protect legitimate health, safety, and environmental regulations and their interpretation by dispute panels. According to the US, "WTO rules recognize that there can be legitimate differences of view on scientific and technical issues in the development of health, safety and environmental measures. . . . WTO dispute settlement decisions in this area already reflect a considerable degree of deference to domestic regulatory authorities on health and safety matters."[29] The US has expressed confidence that "WTO panels will show . . . deference to U.S. regulators given the integrity, rigor, and open and participatory nature of the U.S. regulatory system."[30] Clearly this confidence was significantly reinforced by the ultimate outcome of the shrimp-turtle dispute.

However, the EU does not share this rather sanguine view of the WTO dispute settlement process, for the obvious reasons that a number of EU health and environmental regulations either have been or are likely to become vulnerable to challenges by Europe's partners, including the US. The most dramatic example, of course, is the EU's beef hormone ban, which prohibited the administration of growth hormones to cattle and the sale of any meat from cattle treated with these hormones. The overturning of this ban on American meat imports from cattle given growth hormones represented a highly visible challenge to a regulation that the EU and many of its citizens regarded as both important and necessary. Clearly in this case, the WTO dispute panel appeared to show inadequate "deference" to the EU regulatory process and the values and preferences of its citizens.

Even in the absence of formal dispute proceedings, WTO rules have made EU regulations vulnerable. For example, the EU was forced to modify its politically popular ban on the imports of furs from countries that permitted the use of leghold traps when it faced the likelihood of a successful legal challenge by the US and Canada. The EU has also found its efforts to develop forest certification schemes that would restrict imports of tropical timber undermined by questions about their consistency with WTO rules. The US has periodically raised questions about the WTO consistency of the EU's ecolabeling standard for paper products. More recently, US electronic producers, backed by the USTR, expressed concern about the trade implications of the EU's directive on Restrictions on Hazardous Substances in Electronic and Electronic Equipment. This directive, which was approved in 2002, requires phasing out the use of heavy metals in electronic products in order to protect landfills. Since similar restrictions have not been approved in the US, American exporters face the challenge of modifying the composition of their products in order to enjoy continued access to the European market.

Most important, the EU's restrictions on genetically modified foods and seeds remain an ongoing source of trade tension with the US.[31] As of mid-2003, the EU had not approved a new biotechnology crop for more than 4 years, owing in large measure to the inability of member states to agree on criteria for labeling and traceability. This has been very frustrating to both American government officials and much of the farm industry. The American view is that the EU's concerns about the safety of genetically modified agriculture have no scientific basis. Not only has the moratorium reduced American corn and soy exports to the EU by approximately $350 million per year, but European policies have encouraged other countries to adopt similar restrictions, thus reducing the market for American agricultural exports. After repeated delays, in May 2003 the US filed a formal complaint with the WTO challenging the EU's regulatory regime for GM foods and seeds.

The US experience has been quite different. American fuel economy standards were essentially found to be consistent with GATT in a case brought by the EU that was decided shortly before the Uruguay Round Agreement was submitted to Congress. While the first trade dispute adjudicated by the newly formed WTO did declare a US Environmental Protection Agency rule governing the composition of reformulated gasoline to be inconsistent with the WTO, the dispute had no substantive implications for American environmental standards. Indeed, the Clinton administration privately recognized that the US had imposed a trade barrier masquerading as an environmental regulation and was actually pleased with the outcome. American environmentalists sharply criticized the WTO ruling but were unable to generate much public interest in the dispute. As already noted, the appellate body in the shrimp-turtle case essentially endorsed American regulations aimed at protecting sea turtles outside its borders, in effect reversing much of the holding of the dispute panel in the tuna-dolphin case.

More broadly, with the exception of the second tuna-dolphin case, the US has never lost an environmentally related trade dispute with the EU (although the EU did formally support Venezuela in the reformulated gasoline case). Nor has it been forced to modify any of its environmental regulations because of fears that the EU might file a formal complaint with the GATT/WTO. Nor do any significant American health, safety, or environmental regulations now appear vulnerable to international trade legal challenges from any WTO member, including the EU. It is important to note that since 1994, every transatlantic environment-related trade dispute between the US and the EU has stemmed from American accusations that EU regulations were NTBs. For a politically influential segment of American producers, the most important health, safety, or environmental NTBs are now those imposed by the EU. (Fifteen years ago, the phrase "nontariff trade barrier" evoked Japan.) Alternatively, for Europeans, it is the US that represents the most important external threat to their ability to maintain their regulatory standards.

Subsidies

Just as the US has begun to challenge the EU's agricultural subsidies on the grounds that they are environmentally harmful, the EU's defense of them has increasingly rested on their environmental as well as social benefits. The EU contends that agriculture makes an essential contribution to the achievement of a number of important social goals beyond the production of food and fiber. The "multifunctional" roles of farming include the preservation and enhancement of the rural landscape, environmental protection, and the viability of rural areas.[32] In the case of subsidies for fisheries, the EU's position is more nuanced. While acknowledging that fisheries suffer from the tragedy of the commons, it argues that the focus within the CTE on subsidies, particularly those granted to their fleets by developed countries, and their possible encouragement of overcapacity and thus overfishing, is simplistic. Not only is there no clear definition as to what constitutes a subsidy, but in fact the vast majority of developed country support for fisheries has been devoted to general services such as infrastructure and research, which do not directly contribute to overcapacity.

Conclusion

Why has the EU identified trade and environment as one of three new areas in which it wants negotiations at the next international trade round, even though the WTO dispute settlement decisions have become increasingly responsive to environmental considerations? The *Economist* suggests that it has to do with different transatlantic legal traditions. "Anglo-Saxons may be happy with case law, but politicians in continental Europe, where laws are based on a civil code, like to write rules in advance."[33] Indeed, many of the differences between the EU and the US do have to do with means rather than ends. After all, both want the WTO to be relatively flexible in accommodating a range of environmentally related trade restrictions. The US, however, believes that such an accommodation is adequately taking place through the decisions of the appellate body, while the EU disagrees and wants it to be rulebased.

Yet The *Economist* is only partially correct, for there are also substantive disagreements. The EU *is* more vulnerable to having its protective regulations challenged through the WTO than is the US. And this in turn reflects the significant changes in regulatory politics that have taken place in Europe and the US over the past decade. Since 1990, the rate at which the US has strengthened or expanded the scope of its environmental standards has significantly declined. In the critical area of international environmental policy, the US no longer plays a leadership role. It has ratified neither the Basel Convention on Hazardous Wastes nor the Kyoto Protocol, and it only reluctantly signed the Biosafety Protocol. The Bush administration is highly unlikely to change this pattern.

By contrast, environmental policy in the EU has become increasingly vigorous over the past decade. Fifteen years ago it was unusual to find a European health, safety, or environmental standard that was stricter than its American counterpart. Now there are many. These include the

EU's bans on beef and dairy hormones, antibiotics in animal feed, and the use of leghold traps; its increasingly rigorous recycling requirements for products ranging from cars to computers to phones; its extensive eco-labeling schemes; and its wide-ranging restrictions on genetically modified foods and seeds. At the global level, it is Europe that has taken a leadership role in seeking to restrict trade in hazardous wastes, protect rain forests, maintain biodiversity, and reduce carbon emissions. In short, since the early 1990s, environmental issues have been much more politically salient in Europe than in the US.[34]

It is precisely those EU regulatory standards that are more stringent than their American counterparts that are most likely to be subject to trade disputes. (Note, however, that stringency should not be confused with effectiveness; more stringent regulations may or may not be more effective.) To be sure, domestic pressures in the US may inhibit the filing of another formal challenge to a politically popular EU health, safety, or environmental regulation. After all, the US does not want to provoke a further political backlash against globalization or further strain transatlantic relations. But for the Europeans, this is insufficient. In addition, according to an EU official, the American position on the Kyoto global climate change agreement has "reverberated into the politics of trade and environment and trade negotiations," making the EU less trustful of the American commitment to environmental protection and thus even more determined to have these issues addressed in the next trade round.[35]

These differences were largely papered over in the November 2001 declaration of the Fourth Ministerial Conference in Doha, Qatar, which officially launched the next WTO trade round. Thus the US agreed to negotiations aimed at clarifying how WTO rules apply to MEAs that contain trade provisions, and to explore if existing WTO rules stand in the way of ecolabeling policies. For its part, the EU agreed to clarify and improve WTO rules that apply to fisheries subsidies. Both supported negotiations aimed at reducing trade barriers on environmental goods and services and to expand cooperation between WTO officials and those governing MEAs. Nevertheless, the substantive differences between the EU and the US on trade and environment linkages are likely to reemerge as the Doha Round continues.

Notes

1. See, for example, David Vogel, *Trading Up: Consumer and Environmental Regulation in a Global Economy* (Cambridge, Mass.: Harvard University Press, 1995); Daniel Esty, *Greening the GATT* (Washington, D.C.: Institute for International Economics, 1994).

2. Quoted in Vogel, *Trading Up*, p. 136.

3. Ibid., p. 137.

4. Quoted in William Lash III, *The Clinton Trade Tango: One Step Forward, Two Steps Back*. Center for the Study of American Business (St. Louis, Mo.: Washington University, 1994) p. 7.

5. Quoted in Vogel, *Trading Up*, p. 137.

6. For an excellent analysis of the politics of the CTE, see Gregory Shaffer, "The World Trade Organization Under Challenge: Democracy and the Law and Politics of the WTO's Treatment of Trade and Environment Matters," *Harvard Environmental Law Review* 24(2) (2000): 563–592.

7. For the United States, see Elizabeth DeSombre, *Domestic Sources of International Environmental Policy* (Cambridge, Mass.: MIT Press, 2000).

8. Quoted in I. M. Destler and Peter J. Balint, *The New Politics of American Trade: Trade, Labor and the Environment* (Washington, D.C.: Institute for International Economics, 1999), p. 10.

9. "Calming the Waters: A Talk with the US Trade Rep," *Business Week*, July 23, 2001, p. 41.

10. "Promoting the Noble Cause of Commerce," *Economist*, August 3, 2002, p. 57.

11. Preparations for the 1999 Ministerial Conference, A Communication from the United States, July 30, 1999.

12. Office of USTR, Declaration of Principles on Trade and Environment, p. 3.

13. See, for example, Hakan Nordstrom and Scott Vaughan, *Trade and Environment*. Special Studies 4, World Trade Organization, 1999, Part II: A, E.

14. Office of USTR, Declaration of Principles on Trade and Environment, p. 2.

15. "The Non-Trade Impacts of Trade Policy—Asking Questions, Seeking Sustainable Development." Informal Discussion Paper Rev. 1, Directorate-General Trade, EU, January 8, 2001, p. 21.

16. Ibid.

17. Office of USTR, 2000 Annual Report, p. 81.

18. This section draws on Mark Pollack and Gregory Shaffer, "The Challenge of Food Safety in Transatlantic Relations," in Mark Pollack and Gregory Shaffer, eds., *Transatlantic Governance in the World Economy* (Lanham, Md.: Rowman & Littlefield, 2001), pp. 170–172.

19. Quoted in Jonathan Adler, "The Cartagena Protocol and Biological Diversity: Bio-Safe or Bio-Sorry?" *Georgetown International Law Review* 12 (2000), p. 771.

20. "The Non-Trade Impacts."

21. Quoted in James Cameron, "The Precautionary Principle," in Gary Sampson and W. Bradnee Chambers, eds., *Trade, Environment, and the Millennium* (New York: United Nations University Press, 1999), p. 259.

22. Office of USTR, Declaration of Principles, p. 5.

23. Michael Weinstein and Steve Charnovitz, "The Greening of the WTO," *Foreign Affairs* (November/December 2001): 154–155.

24. See, for example, Carrie Wofford, "A Greener Future at the WTO: The Refinement of WTO Jurisprudence on Environmental Exceptions to GATT," *Harvard International Law Review* 24(2) (2000): 564–592 .

25. Office of USTR press release, "U.S. Wins WTO Case on Sea Turtle Conservation," June 15, 2001.

26. "The Non-Trade Impacts," p. 21.

27. Quoted in David Vogel, *Barriers or Benefits? Regulation in Transatlantic Trade* (Washington, D.C.: Brookings Institution Press, 1997), p. 49.

28. "The Non-Trade Impacts," p. 20.

29. Office of USTR, Declaration of Principles, p. 5

30. Ibid.

31. Neil King, Jr., "Zoellick Slams EU for 'Immoral' View on Biotech Crops," *Wall Street Journal*, January 10, 2002, p. A8.

32. "The Non-Trade Impacts," p. 16.

33. "Playing Games with Prosperity," *Economist*, July 28, 2001, p. 26.

34. For a broader analysis of this development, see David Vogel, "The Hare and the Turtle: The New Politics of Consumer and Environment Regulation in Europe," *British Journal of Political Science* 33 (October 2003): 47–70.

35. Ibid.

10

International Development Assistance and Burden Sharing

Paul G. Harris

It is impossible to talk about effective environmental protection in contemporary international relations without also talking about development assistance and burden sharing. Financial resources are essential to aid the developing world in implementing environmentally sustainable development. In many cases, poor countries simply cannot deal with environmental problems on their own, and when they can, they may choose not to do so because that would divert resources from higher short- and medium-term priorities such as poverty reduction and economic development. Development assistance is also a matter of international equity and fairness. The developing countries feel that they deserve assistance from the developed countries. They believe that developed countries should—owing to their history of pollution, their control of technological resources, and their relative wealth—take on a greater share of the burdens associated with global environmental changes. In short, the developing world needs and demands resources at the North's disposal, and the poorer countries will be much less willing and able to take the necessary steps to protect the environment if more affluent countries do not first take on their fair share of global environmental burdens.

More and more developed countries, particularly in western Europe, are starting to share this view. The 1992 UN Conference on Environment and Development (UNCED), popularly known as the Earth Summit, and related international agreements resulted in many statements and pledges of additional assistance to poor countries to aid them in sustainable development. There have been more recent pledges by developed countries to take on additional burdens associated with global environmental changes and sustainable development.[1]

Whether the EU and the US will make a more constructive contribution to protecting the global environment depends, therefore, in large measure on the degree to which they assist the world's poorer countries. It also depends on their willingness to share fairly the burdens of environmental changes through, for example, reducing their emissions of carbon dioxide and other greenhouse gases (GHGs) contributing to climate change.

Recent events, such as transatlantic disagreements over how to implement the climate change regime, have highlighted sharp philosophical differences between America and Europe on these issues. The EU and its member countries are generally forthright in declaring support for increased aid for sustainable development. They also consistently assert their support for the notion of "common but differentiated responsibility," which requires developed countries to address international environmental problems before requiring the developing countries to do likewise. In contrast, the administration of George W. Bush came into office espousing vigorous support for national self-reliance in the developing world, and it firmly rejected demands that the US reduce its impact on the global environment. Under Bush, the US instead has tried to shift much of the blame for global pollution—and the responsibility for limiting future contributions to it—to the world's large developing countries.

On the surface, therefore, it seems that the EU and US are heading in completely different, possibly irreconcilable, directions on international environmental issues. However, the more one looks at their attitudes, policies, and actions on development assistance, the more EU and US actions—if less so their rhetoric—start to look similar. The greatest divergence may be in burden sharing. The Europeans have been more active in consciously trying to reduce their own impact on the Earth's environment. The upshot is that neither of these great powers is doing enough, given the magnitude of environmental changes under way, the profound difficulties these changes are causing for most of the world's developing countries, and the degree to which the US and the EU countries are themselves largely to blame for them.

To better understand the role of the US and the EU in international environmental protection, this chapter first considers their development assistance policies.[2] I then briefly describe EU and US rhetoric and policies with regard to international environmental burden sharing (what some might prefer to call international environmental equity, fairness, or justice). The Americans and Europeans moved closer in this regard during the 1990s. They agreed that they ought to take on a greater share of global environmental burdens and help the developing countries cope with environmental changes. However, the new Bush administration seems to have reversed US policy. The Europeans are clearly now more favorable toward the equity demands of the world's poorer countries. I then explore some variables shaping the policies described, with an eye to understanding ways of increasing US and EU development assistance and environmental burden sharing. Short-term prospects are not favorable, but environmental changes themselves may eventually push the entire Euro-Atlantic community to be more generous in promoting environmentally sustainable development at home *and* abroad.

International Development Assistance

For several decades following World War II, the US was the leading international aid donor, although on a per capita basis several other countries exceeded its generosity, and by the early 1990s Japan was providing more total aid. In recent decades, the northern and western European countries have been the most generous, with some giving two to ten times as much aid as the US when measured on a per capita basis. The development assistance policies of the US and EU countries mirror their domestic social welfare priorities.[3] Hence, the Europeans have generally been more generous than the Americans in helping the world's poor, much as they are more generous in helping their own poor. What is more, the foreign aid of the western European countries has been motivated by a greater degree of altruism and a desire to help those most in need than has US aid. Thus, compared with the US, more European aid goes to the world's poorest countries, more goes toward meeting basic human needs, and more is donated to multilateral development agencies.[4]

Over the past three decades the US and the EU have increased development assistance for environmental objectives. This increase does not necessarily fit the preferences of the world's poorer countries, however.

At the insistence of developing countries, the UNCED declarations and agreements called for *new and additional* funding for environmental purposes. This was codified in Agenda 21, the Earth Summit's exhaustive policy document describing sustainable development goals:

For developing countries, particularly the least developed countries, ODA [official development assistance] is a main source of external funding, and substantial new and additional funding for sustainable development and implementation of Agenda 21 will be required. Developed countries reaffirm their commitments to reach the accepted United Nations target of 0.7 per cent of GNP for ODA and, to the extent that they have not yet achieved that target, agree to augment their aid programmes in order to reach that target as soon as possible and to ensure prompt and effective implementation of Agenda 21.[5]

The amount of new money called for at the Earth Summit would (conveniently, perhaps) raise aggregate ODA to the UN target of 0.7 percent of gross national product (GNP). However, actual funding has never come more than about halfway toward that target. The US, following decades of policy, refused to be committed to the 0.7 percent ODA target level, and EC countries were divided.

In the end, the developed countries refused to be committed at the Earth Summit to providing new development funds for sustainable development above and beyond existing aid. Indeed, overall development aid from rich to poor countries fell during the 1990s. Donor countries suffered "aid fatigue" as economies slowed at home and poverty increased in developing countries despite decades of ODA. Some analysts like to point to sometimes very large increases in direct foreign investment as an alternative for ODA. But very little of this aid reaches the poorest countries, and most of it is directed at a few developing countries. The countries that rely the most on outside aid received the least of it. *All* of sub-Saharan Africa received less than 1 percent of investment flowing to developing countries in 2000.[6]

During the UNCED process, there was—and there remains—concern among developing countries (1) that foreign aid would be diverted from economic development per se to sustainable development, which may not always be what they want and (2) that aid would be directed at global environmental objectives (e.g., climate change, international water pollution, stratospheric ozone depletion), rather than local and national environmental objectives that most concern developing countries (e.g., urban pollution, sanitation, water scarcities). The share of ODA going to environmental and resource conservation does seem to have been increasing.[7] However, because overall aid flows have dropped, aggregate funding falls far short of the needs foreseen a decade ago. According to the UN Development Program, "Since the Earth Summit, official international financing for sustainable development has remained well below the level necessary to implement Agenda 21."[8]

As developing countries feared, funds have indeed been directed at global environmental problems. In one sense this can be viewed as progress because some money is being spent on important environmental problems like climate change and ozone depletion, as demonstrated by the Montreal Protocol (ozone) Multilateral Fund and the Global Environment Facility (which funds the incremental costs of domestic actions that produce global environmental benefits). But from the perspective of the developing countries, these funds exist mostly to promote the goals of rich countries, not goals that may be of highest priority for the world's poor. What is more, even these international funds have received inadequate support from donor countries. The developing countries' fears were exacerbated in the 1990s by concerns that money would be diverted away from their needs to allow increased funding to the former communist states of Europe and the Former Soviet Union (FSU). To a substantial degree these fears have been realized.

The United States and Foreign Aid

The US government's Agency for International Development (USAID) acknowledges that as a percentage of GNP, the US provides less aid than all other major industrialized countries.[9] The US gives only 0.1 percent of its GNP in ODA, which is less than one-third that of the EU countries combined, and very much less than some EU member countries, notably Denmark at over 1 percent of its GNP.[10] Nevertheless, Americans overwhelmingly support foreign aid in principle, although they think much of it is lost through waste and corruption, and they believe that the US government gives forty times as much aid as it actually does (!).[11]

Only about one-tenth of the one-half of 1 percent of the US federal budget spent on foreign aid—an amount equal to about $600 million each year—is spent on environmental programs, a figure that is lower than some other donor countries (e.g., it is about one-third of what Germany spends on environmental programs).[12] This aid is spread over more than seventy countries.[13] Congress has usually been more willing to provide aid to developing countries for sustainable development than has the White House. Congress has directed that US foreign aid be spent to help "developing countries to protect and manage their environment and natural resources. Special efforts shall be made to maintain and where possible to restore the land, vegetation, water, wildlife, and other resources upon which depend economic growth and human well-being, especially for the poor."[14] However, aggregate US funding for sustainable development has suffered because overall foreign aid has dropped since the cold war.

Following UNCED, there was much talk of increased foreign aid for sustainable development, but the actual amounts were quite small. For example, in 1997 the Clinton administration announced a $1 billion 5-year Developing Country Climate Change Initiative to help poorer countries limit their GHG emissions or increase their carbon sinks. This built on the congressionally mandated Global Warming Initiative and followed several years of funding by USAID for projects to promote energy efficiency and sustainable development practices. The US under Clinton also provided some support to developing countries through its Country Studies Program and the US Initiative on Joint Implementation. However, much of this spending came from funds already included in USAID's budget projections, and hence only about one-fourth of Clinton's proposed initiative would have been new money. What is more, most of the aid was directed at a few countries. As one analyst pointed out at the time, "Despite the [Clinton] Administration's characterization of its Developing Country Climate Change Initiative as something new, it appears to add little substance to what USAID is already doing as a result of the Global Warming Initiative begun at congressional impetus in 1990."[15]

When it came into office, the Bush administration was less sympathetic to development assistance than was the Clinton administration, and spending for sustainable development seems unlikely to see substantial increases. Bush administration proposals for foreign aid spending in fiscal year 2002 called for a small increase in overall aid, with an increasing reliance on the private sector and private nongovernmental actors in the development field (a move started under Clinton).[16] Administration officials highlighted some environmental programs, but aid for them was, characteristically, to be diverted from the AID budget, and in any case was quite small. The major focus was on encouraging private direct investment.

A potential change in US policy was announced by President Bush on the eve of the March 2002 UN Conference on Financing Development. He proposed increasing US foreign aid by 50 percent—to $15 billion annually—by 2006,[17] although there was skepticism about this proposal even within the administration, and disagreement within the White House about its details.[18] The president's proposal focused on good governance and private sector initiatives, suggesting that the money would not go to the most needy countries. Observers noted that the increases were to be phased in gradually, were subject to congressional approval, would not take effect until after Bush's first term (when he might not control the White House), and fell far short of need.[19] Even if the increase becomes reality, the US will remain far behind the EU and most of its members in provision of aid as a percentage of GNP. Indeed, at the same financing conference, the EU proposed even more new aid than the US did.

The European Union's Development Assistance Policies

Compared with the US, the EU countries are generous in their official development assistance. Of total world ODA, over half comes from the EU and its member countries, and they donate two-thirds of all grant aid.[20] While few countries will make the UN target of 0.7 percent of GNP a binding one, EU countries generally view it as a firm goal, and several have achieved or exceeded it. The EU countries combined give 0.33 percent of their GNPs in ODA, and some member countries give considerably more (e.g., Denmark, 1.06 percent; Sweden, 0.81 percent; Norway, 0.80 percent).[21] Indeed, the only industrialized countries reaching the UN target by the late 1990s were west European.[22] What is more, the EU and its member countries tend to give much more assistance, as a percentage of GNP, to the world's poorest countries. About one-fifth of member countries' foreign aid is channeled through the EU. The

Maastricht Treaty requires the EU (i.e., the European Commission) to use foreign aid to reduce poverty through sustainable and social development. However, programs for the *least* developed countries have actually decreased (despite increases in EU organization aid generally).[23] This may be because, by the mid-1990s, the focus of much EU development policy had shifted away from the poor countries of Africa, the Caribbean, Central America, and Asia, to the countries of eastern Europe, the Mediterranean region, and South America.[24] Political developments have made cooperation with central and eastern Europe a priority.[25] Indeed, many pressing environmental problems are found to the East, not the least being suspect nuclear power facilities and existing sources of air and water pollution.

Like the US, many EU member countries are involved in pilot projects in developing countries for "joint implementation" of GHG goals. But the Europeans seem to be much more serious, and many of their projects are more advanced. The EU has generally given great rhetorical support to providing additional assistance to poorer countries, including funds for sustainable development. Especially in the past decade, the EU has devoted increasing financial and technical assistance to help developing countries address environmental problems and raise their environmental standards,[26] although it has had only limited success in pushing for increased funding from member countries for sustainable development in poor countries. Having said this, it is nearly impossible to determine precisely how much assistance the EU provides for environmentally sustainable development.[27] However, by way of example, of about 3.9 billion European Currency Units (ECUs) disbursed to developing countries in 1994, approximately 815 million was spent on environmental priority areas.[28] During the 1990s, about 10 percent of aid to African, Caribbean, and Pacific countries under the Lomé Convention, as well as the same amount of aid to Asian and Latin American countries, was allocated to environmental projects or to meeting environmental needs, with similar portions of aid to central and eastern European countries, and a smaller percentage of aid to Mediterranean countries going to environmental projects.[29] Small island states, which are particularly vulnerable to climate change, have received some special attention, particularly in the context of the Lomé Convention.[30]

As the poorer countries feared at UNCED, money from the EU and its members has often been diverted to Eastern Europe and the FSU countries, and money for sustainable development has often been diverted from existing development aid. Currently, by the European Commission's own admission, "Community funding for environmental purposes [in developing countries] remains modest compared to other EU aid," and there is some emphasis on helping developing countries "respond to *global* environmental issues and to implement the major UN environmental conventions on climate, biodiversity and desertification"—just as developing countries have feared.[31]

Sharing the Burdens of Environmental Change

The developing world has been arguing for decades that the developed world ought to bear a much greater share of the burdens associated with addressing adverse changes to the world's natural environment. Developing countries believe that the world's wealthy countries are inordinately responsible for environmental changes because they have polluted the Earth so much since the Industrial Revolution, because their personal consumption and per capita emissions of pollutants are so high relative to the rest of the world, and because they have financial resources—in large part a consequence of not paying the full price for natural resources and environmental services—to address the problems of environmental change. The developing countries have acknowledged that they also bear some responsibility, but they argue that their increasing pollution of the environment—unlike the wealthy countries' pollution—is largely a consequence of their struggle to end poverty and achieve modest economic development.

Thus the developed and developing countries have generally agreed that they share *common but differentiated responsibility* for dealing with global environmental problems. This means that they will share the burdens, but that the developed countries should do more to limit their own impacts on the natural environment before expecting the poorer countries to do likewise. The EU countries have come to accept this notion in principle, and they have started to act upon it. Similarly, the US under Clinton eventually agreed, and prominent Clinton administration officials argued that the US ought to share much more of global environmental burdens simply because it polluted so much. However, the advent of the new Bush administration has dealt this position a setback, with the Europeans now clearly more favorable toward the equity demands of the world's poorer countries.

The United States and Global Environmental Burdens

The first Bush administration was generally opposed to any language in UNCED agreements or treaties that might legally establish US responsibility for past pollution or its contemporary global consequences. While the Bush administration did sign on to such language in the Rio Declaration and agreements, it did not embrace the notion that the developed countries—and especially the US—should bear the brunt of efforts to protect the global environment in the near and medium term. To emphasize his opposition to reducing US consumption, shortly before the Earth Summit, President Bush said that "the American way of life is not negotiable," making it nearly impossible for US delegates to UNCED to make concessions with regard to consumption patterns.[32] Similarly, in the case of the early climate change negotiations, the US was the only industrialized country refusing to negotiate targets and timetables for controlling GHG emissions. President Bush's attendance at the Earth Summit was made contingent on the Climate Change Convention not requiring specific targets and timetables for reduction in emissions of GHGs. In the end, the US efforts in this respect were successful. Emissions limitations would be voluntary. According to Philip Shabecoff, "Adamant opposition to the inclusion of binding targets and timetables for limiting the emission of carbon dioxide was the *sine qua non* of the US negotiating posture" at the climate negotiations before the Earth Summit.[33]

In contrast, the Clinton administration was more forthcoming with regard to US and Northern responsibility for global pollution and environmental change, at least in its rhetoric. For example, in June 1993, Vice President Al Gore told the UN Commission on Sustainable Development that the US and other developed countries

have a disproportionate impact on the global environment. We have less than a quarter of the world's population, but we use three-quarters of the world's raw materials and create three-quarters of all solid waste. One way to put it is this:

A child born in the United States will have 30 times more impact on the earth's environment during his or her lifetime than a child born in India. The affluent of the world have a responsibility to deal with their disproportionate impact.[34]

Indeed, on the issue of common but differentiated responsibility and consumption patterns, there was a substantial shift between the Bush and Clinton administrations.

On climate change, the Clinton administration committed the US to reducing its GHG emissions to 1990 levels by the year 2000.[35] Policies to achieve this objective were largely voluntary in nature, and they ultimately failed to achieve this goal—much as other developed countries fell short of the goal. However, in what was uniformly interpreted as a U-turn in US policy, in 1996 the Clinton administration said that it would support a legally binding instrument or protocol to the climate change convention with specific targets and timetables for reductions of GHG emissions. This was codified in the 1997 Kyoto Protocol, in which the US agreed to a "binding" requirement to reduce its GHG emissions by 7 percent before the end of 2012. At subsequent climate change conferences, the US tried to maximize its freedom in using "flexible mechanisms," such as carbon trading and agricultural sinks, to ease its implementation of the Kyoto Protocol. While this made negotiations difficult, the Clinton administration never said that the US was *not responsible* for its inordinate share of GHG emissions. Arguably, the Bush administration has reversed this policy.

In mid-2001 President Bush repeated almost word-for-word what President Clinton said during his tenure: "We want to work cooperatively with these [developing] countries in their efforts to reduce greenhouse emissions and maintain economic growth."[36] But a month later he proposed *reducing* foreign aid to help developing countries lower their GHG emissions (cutting one program by $41 million from the previous year's $165 million), instead hoping that industry would do more to reduce GHG pollution in developing countries.[37] What is more, at least in the case of climate change, the US has recently tried to shift the burden to developing countries. President Bush said, "I oppose the Kyoto Protocol because it exempts 80 percent of the world, including major population centers such as China and India."[38] This is despite these countries' per capita carbon dioxide emissions being tiny compared with those in the US (which are at least 10 times those of China and 25 times those of India)[39] and their aggregate contributions to the problem being much smaller than that of the US. (China and India's combined twentieth-century carbon dioxide emissions were 9 percent of the global total, compared with the US's 30 percent. According to some statistics, China *reduced* its GHG emissions in the late 1990s, despite its growing economy.[40]) Given the historical responsibility of the US for enormous amounts of emissions, President Bush's argument has largely fallen on deaf ears abroad.

Thus, the US, while agreeing to the notion of common but differentiated responsibility under Clinton, has done relatively little to act on it, and recently it has backed away from the principle.[41] Instead, President Bush has sought to blame poor countries, particularly China, for greenhouse gas pollution and a failure to agree to firm commitments to reduce GHG emissions—despite strong evidence that China and other poorer countries are restricting their emissions more than the US. Having said this, many Americans, their municipalities, their states, and even some large corporate polluters have realized their disproportionate contribution to pollution of the atmosphere, and consequently they are starting to take action to reduce their emissions of GHGs. In this respect they are closer to Europeans and the EU on climate change than is the federal government.

Environmental Burden Sharing and the European Union

The EU and its member countries have demonstrated a willingness in recent years to take on a greater share of the burdens associated with global environmental changes. This is especially evident in their rhetoric regarding climate change, although it would be wrong to say that they have always embraced the notion or that they have done as much as many would argue they ought to do. But they are at least moving well ahead of the US toward greater burden sharing. The EU has also expressed a willingness to take on a greater share of global environmental governance by increasing funds to the UN Environment Program and a proposed World Environment Organization.[42] In its preparations for the 2002 Rio+10 World Summit on Sustainable Development (WSSD), the European Commission suggested to the Council of Ministers and the European Parliament four "strategic objectives," including "increased global equity," and it has recommended that the EU work to ensure that financial and technical assistance is "channeled so that progress made is not counteracted by negative impacts on environmental resources or equity."[43]

The commission has itself acknowledged that since UNCED, "*The world has not become more equitable*. . . . A relatively small percentage of the world's people and nations still use most of the world's economic and natural resources."[44] The commission has argued for the Rio+10 summit to promote an "atmosphere of partnership" between the developed and developing countries by integrating protection of the environment and eradication of poverty: "One way is to reassure developing countries that the developed countries' global environmental concerns do not take precedence over the economic goals. . . . Furthermore, there needs to be a clear sense of equity in the preparations for the outcome of the [2002] Summit, making reality of the notion of common but differentiated responsibility."[45] At least rhetorically, therefore, the EU seems much more willing to embrace international environmental equity and burden sharing than it was 10 years earlier in the run-up to the Earth Summit, and much more so than both of the Bush administrations.

With respect to climate change, the EU has expressed dismay at the US's insistence that developing countries take on new commitments for reducing GHG emissions. An EU minister is reported to have said that "The EU believes that it is not realistic to ask the developing countries to reduce or limit their emissions if we cannot show that we, as the biggest emitters, have done something ourselves."[46] The EU's environment commissioner has stated emphatically that:

action has to be based on the principle of *common but differentiated responsibility*, as enshrined in the Climate Convention. We can not expect the developing countries to do something that many industrialized countries in the world, with all their research and technological capabilities, have not been able to do. The EU does not think this is realistic or fair, taking into account historical responsibility, current capability and actual per capita emissions.[47]

Instead, the stated EU policy, and that of member governments, is to reduce EU-wide GHG emissions and to assist developing countries through aid and technical know-how to promote sustainable development.[48] Toward that end, at the WSSD and the 2002 New Delhi conference of the parties to the climate convention, EU diplomats tried to push for greater commitments on the part of developed countries to undertake GHG reductions beyond those in the Kyoto Protocol. The US strongly resisted these efforts, instead forming a new alliance with large developing countries to move the debate away from *mitigating* global warming (by cutting GHG emissions) to providing aid to developing countries to help them *adapt* to the inevitable impacts of climate change.[49]

The EU has been more forthcoming than the US with regard to recognizing its responsibilities and rhetorically supporting the equity demands of developing countries. However, it would be an overstatement to say that it has been very forthcoming in acting upon these sentiments. Actions by the EU countries to reduce environmental impacts were greater than those of the US over the past decade, and the EU certainly has been more supportive of developing country demands in international environmental negotiations. Admittedly, many EU member countries, like the US, continue to increase their emissions of global pollutants. However, the Europeans are now committed to taking firmer (first) steps toward sharing the burdens of climate change—much more so than the US—and some EU countries, such as Germany, have started taking very serious steps to fulfill their GHG reduction obligations under the Kyoto Protocol.

Variables Shaping American and European Policies

There are many variables shaping the sometimes-convergent, often-divergent policies of the US and the EU and its members with respect to international development assistance and environmental burden sharing. The following discussion is intended to briefly highlight only some of these variables.

Differing Conceptions of Economics, Foreign Aid, and Science

There are differences in philosophy regarding economics, foreign aid, and science in the EU and the US. Their publics seem to have somewhat different perspectives on development assistance and burden sharing. Europeans are more willing to bear increased costs to take on their fair share of global environmental burdens, possibly because they are more accustomed to higher taxes intended to shape their behaviors. As suggested earlier, foreign aid mirrors domestic welfare spending. Hence, the foreign aid of the EU countries resembles their relatively generous domestic welfare systems. For various reasons, Europeans have developed stronger feelings of obligation to those in need, and they more often view government policies and programs as effective ways to address domestic and international poverty and human suffering. The Europeans believe that "there's a moral argument for helping people in poor countries,"[50] and public opinion is generally very favorable toward development cooperation with poor countries.[51]

In contrast, while the US has a welfare system that is generous relative to most of the world, compared to Europe it is quite stingy with government aid, both domestically and internationally. Americans are more hostile to foreign aid—not because they don't want to help the world's poor, but because they wrongly believe that it is a huge portion of the government's spending. US politicians have chosen to exploit public ignorance of foreign aid spending. Furthermore, Americans and the US government are more favorable toward private sector solutions to collective problems, and they often view government intervention as more of a problem than a solution. For the US, foreign aid has a largely prudential purpose: promoting US national interests. This is as true of environmental aid as other forms of aid. Thus, for example, Congress has declared that it is "in the economic and security interests of the US to provide leadership both in thoroughly reassessing policies relating to natural resources and the environment, and in cooperating extensively with developing countries in order to achieve environmentally sound development."[52] The EU countries also often want benefits from their environmental aid, but they also see indirect benefits, such as a reputation for being international leaders in sustainable development.

While the US and its European counterparts now generally share a preference for free market economics, the details of implementing this preference vary and affect their respective environmental foreign policies. The US prefers that developing countries adopt policies to attract private foreign investment instead of foreign aid; it directs more of its aid to countries that have the greatest economic potential; and more of its aid is tied to the purchase of US products. Indeed, the US obsession with the free market often extends to its environment-related international policies. Hence, insofar as it agrees to take on a greater share of the world's environmental burdens, it seeks voluntary measures (which rarely have much impact) and free market solutions, such as emissions trading. In contrast, the EU and its member countries are more willing to provide grant aid to some of the poorest countries, and domestically and often internationally they accept that "command-and-control" measures can be effective in bringing about efficiencies that lead to fewer polluting emissions. While both sides shy away from mandatory regulation of domestic industries, the EU countries are more willing to accept such regulation.

What is more, Americans and particularly their present political leadership, seem more skeptical than the Europeans of scientific findings related to the environment that might have significant economic consequences or will require lifestyle changes. While both sides accept the precautionary approach, it is weaker in the US. The US government—if less so Americans themselves—explains away US droughts and wildfires as unfortunate but normal, whereas the EU and Europeans more easily make connections (even if they cannot be proved) between their floods and emissions of GHGs. Something closer to concrete proof of harm is often required before the US government is willing to act, especially with regard to *international* environmental issues; this is particularly true of the George W. Bush administration. The Europeans seem more willing to leave science to scientists, even more so in this issue area. As in other issue areas (e.g., genetically modified organisms and hormone-treated beef[53]), the Europeans have a stronger desire for proof that potentially dangerous substances are *safe*, whereas the Americans are more comfortable with activities and substances not proved *un*safe. Furthermore, in the United States, environmental science becomes politicized and used to promote particularistic interests more easily than it does in Europe (although it happens there as well).

Political Pluralism

There are varying levels of pluralism among EU countries and between the EU and the US. Civil society actors—nongovernmental organizations, business interests, and citizens, for example—have access to politics in EU countries, and through the European Parliament and their governments influence the EU organization and collective policies on development assistance and burden sharing. Civil society actors also sometimes have a profound influence in shaping the policies of member governments, and hence EU-wide policies. But this process is arguably even more profound in the US, where interest groups have myriad access points to the policy process. Business interests and political ideologues, often those who resist changes in US policies on foreign aid and the changes in American lifestyles necessary to reduce global pollution, are able to influence members of Congress and the White House. This does not mean that US policy is predetermined—the public and opposing groups could use the pluralistic policy process to push for new policies—but it does mean that US foreign policy is resistant to changes that threaten powerful interests, barring shocking events.

Americans recognize the need to combat global environmental problems, and they are willing to change their behaviors toward this end. However, if the cost of doing so is even modest, their willingness falters. This ease with which they change their attitudes on burden sharing is exploited by politicians and business interests hoping to limit US action on climate change and other international environmental problems.

European policies are often more consciously cooperative efforts among various societal interests to achieve common goals. As a consequence, the EU and its members may be better equipped to transform policy over time toward addressing questions of environmental assistance and burden sharing, whereas the US will resist change until pressure—from the public, from the international community, or from the environment itself—reaches a level sufficient to overcome entrenched interests and their representatives in the policy process. Environmental skeptics, whether genuinely questioning the need for burden sharing or acting in support of industrial interests that might be harmed by it, have relatively easy access to the policy process and can thwart change. Indeed, business interests that do not support international environmental burden sharing still have the upper hand in the US, whereas in Europe business sees the need for gradual change and is more likely to see future benefit in it. The policies of the US may, however, eventually converge with those of the EU as US businesses realize that greater regulation of the present activities is inevitable in the long term, and as businesses that will benefit from greater environmental action put pressure on the policy process.

"United" States versus the European "Union"

Policies on sharing the burdens of climate change illustrate some important differences between the US and the EU. For example, one reason for the EU's willingness to reduce its GHG emissions more than the US can be found in an aspect of the EU that usually complicates policy coordination: the lack of a more powerful central government able to represent all member countries at once. Arriving at the GHG reductions agreed upon among EU countries required major negotiation among them. This makes agreement difficult. But once agreements are reached, it then becomes a matter of EU policy to implement them. Thus, according to Sbragia and Damro, "the very characteristic which makes the EU so problematic for traditional global negotiations—an uncertain, or mixed, international identity—becomes a strength when it comes to ratification and implementation of an agreement."[54] In contrast, the US must continue internal negotiations on international agreements, and in the case of climate change, the Congress and the current administration are unwilling to seriously consider ratifying the reductions agreed upon at Kyoto by the Clinton administration.[55]

What is more, the US is always reluctant to allow international rules to dictate its behavior, especially in sensitive areas in which it has not already decided to act. This contrasts with the EU situation, where "In psychological terms, the kinds of global restrictions being discussed in relation to climate change do not, therefore, represent a loss of sovereignty in the way they do for the United States. The member states have already 'pooled' so much sovereignty in the field of environmental protection that the issue of sovereignty understood as unilateral decision-making is far less salient than in the United States."[56] What is more, unlike the US, which has viewed the burdens of climate change as something to be avoided, the EU "chose to use the issue to demonstrate its competency and identity as an international actor, to make its mark on the international scene."[57] This is combined with the "inviolable requirement that the EU, a nascent society, take a unified approach" when asserting its willingness to share the burdens of climate change.[58]

Conclusion

The foreign aid policies of the US and the EU share some common characteristics. They evolved from World War II; they were often directed at achieving economic benefits for the donors; and during the 1990s aid fell. There are also differences. EU countries have been much more generous with their aid on a per capita basis; more of the EU's aid has gone to the poorest countries; and more is given to multilateral agencies. Environmentalists may be heartened by the increasing percentages of development aid devoted to environmental goals. However, much of that increase is not directed toward the environmental projects sought by poor countries, but instead is going to projects with global benefits, and in any case, the money comes from a shrinking pool of ODA.

Where EU and US policies diverge most noticeably—and now seem to be diverging quite radically—is on the question of burden sharing. Much as his father refused at the Earth Summit to apologize for Americans' profligate lifestyles, President George W. Bush has declared his opposition to requiring Americans to take on their fair share of the burdens associated with global environmental change. Current slowdowns in both the US and world economies mean that this policy is unlikely to change in the near term, and as a consequence perhaps the EU will grow resistant to environmental burden sharing. However, it seems highly unlikely that the EU or its members countries will do what the new Bush administration has done, namely, blame poor countries for global environmental problems and demand that they clean up before the developed countries do more. It is this fundamental difference in attitudes—one that blames the poor for collective problems, one that seeks to share the burden and help the poor who suffer from those problems—that is perhaps the most striking (and worrying) divergence between US and EU policies.

Acknowledgments

The author wishes to thank the editors, contributors, and anonymous readers for their comments on earlier drafts of this chapter.

Notes

1. See Paul G. Harris, *International Equity and Global Environmental Politics: Power and Principles in US Foreign Policy* (Aldershot, UK: Ashgate, 2001), pp. 44–88.

2. Here I largely repeat arguments found in Paul G. Harris, "Sharing the Burdens of Environmental Change," *Journal of Environment and Development* 11(4) (December 2002):380–401.

3. Alain Noel and Jean-Philippe Therien, "From Domestic to International Justice: The Welfare State and Foreign Aid," *International Organization* 49(3) (summer 1995):523–553.

4. David H. Lumsdaine, *Moral Vision in International Politics: The Foreign Aid Regime, 1949–1989* (Princeton, N. H.: Princeton University Press, 1993).

5. Agenda 21, para. 33.13, reproduced in *Report of the United Nations Conference on Environment and Development,* Vol. I: *Resolutions Adopted by the Conference* (New York: United Nations Press, 1993).

6. UNCTAD, "World Investment Report 2001, available at http://www.unctad.org/wir/contents/wir01content.en.htm.

7. Gareth Porter, Janet Welsh Brown, and Pamela S. Chasek, *Global Politics of the Environment*, 3rd ed. (Boulder, Col.: Westview Press, 2000), p. 158.

8. United Nations Environment Program, *Global Environment Outlook* (London: UNEP/Earthscan, 1999), pp. 211–212.

9. USAID, "USAID and the Environment: What We Spend," May 15, 2001, http://www.usaid.gov/environment/whatwespend.html, p. 1.

10 Organization for Economic Cooperation and Development (OECD), "Net Official Development Assistance Flows in 2000," http://www.oecd.org/dac/htm/dacstats.htm.

11. USAID, "USAID and the Environment: What We Spend," May 15, 2001, http://www.usaid.gov/environment/whatwespend.html, p. 1.

12. Ibid., pp. 1–2.

13. USAID, "USAID and the Environment: Who We Are," http://www.usaid.gov/environment/whoweare.html, p. 1. See the environment section of USAID's strategic plan, "Strategic Goal Five: Protect the Environment," February 2000, http://www.usaid.gov/environment/index.html.

14. US Foreign Assistance Act, Part I, Section 117, "Environment and Natural Resources," http://www.usaid.gov/environment/faa_section_117.htm, Sec.117.\71\(b).

15. Curt Tarnoff, "Global Climate Change: The Role of US Foreign Assistance," *Congressional Research Service Report for Congress*, November 21, 1997 (available at http://countingcalifornia.cd/ib.org/crs/ascii/97-1015).

16. USAID, "Summary of USAID Fiscal Year 2002 Budget," August 6, 2001, http:www.usaid.gov/pubs/cbj2002/request.html, pp. 4–5.

17. George W. Bush, "President Proposes $5 Billion Plan to Help Developing Countries," Remarks by the President on Global Development, Inter-American Development Bank, March 14, 2002, http://www.whitehouse.gov/news/releases/2002/03/20020314-7.html.

18. Joseph Kahn, "As Panacea, Globalization Comes Down to Earth," *International Herald Tribune*, March 22, 2002, p. 4. Skepticism continued into 2003, with prospects that only ten to fifteen countries would meet the political and market-opening reforms necessary to receive the aid. Agence France-Press, "US Offers Extra Aid as Reward for 'Sound Policies,'" *South China Morning Post*, February 5, 2003, p. 9.

19. Joseph Kahn, "In Mexico, A Global Focus on Plight of Poor," *International Herald Tribune*, March 21, 2002, pp. 1, 4.

20. European Commission, "Ten Years after Rio: Preparing for the World Summit on Sustainable Development in 2002," Communication from the Commission to the Council and European Parliament, Brussels, February 6, 2001, photocopy, p. 13.

21. "Net Official Development Assistance Flows in 2000," http://www.oecd.org/dac/htm/dacstats.htm.

22. OECD, "Aid and Private Flows in 1997," OECD news release, June 18, 1998, cited in Porter, Brown, and Chasek, *Global Politics of the Environment*, p. 158.

23. Mirjam Van Reissen, *The North-South Policy of the European Union* (Utrecht the Netherlands: International Books, 1999), p. 38.

24. Marjorie Lister, "Europe's New Development Policy," in Marjorie Lister, ed., *European Union Development Policy* (London: Macmillan, 1998), p. 17.

25. Directorate-General XI (Environment, Nuclear Safety and Civil Protection) [EU], *Towards Sustainability: The European Commission's Progress Report and Action Plan on the Fifth Programme of Policy and Action in Relation to the Environment and Sustainable Development* (Luxembourg: Office of Official Publications of the European Communities, 1997).

26. Joseph A. McMahon, *The Development Cooperation Policy of the EC* (The Hague: Kluwer Law International, 1998), pp. 249–250.

27. Even the EU's environment directorate cannot determine the extent of environmental aid. See Directorate-General XI, *Towards Sustainability*, p. 133.

28. Ibid.

29. Ibid., pp. 133–137.

30. During 1996–2000 the EU gave twenty-six small island states over 1 billion euros, although it is impossible to say how much of this was devoted to strictly environmental concerns. "UN: Globalization, Erosion of Trade Preferences Undermine Small Island States, Commission Told," M2 Newswire, April 26, 1999, p. 7, http://www.umi.com.

31. European Commission, "Integration of Environmental Policy: Integrating Sustainable Development into Community Cooperation Policy," n.d., p. 2, http://europa.eu.int/scadplus/leg/en/lvb/ 128114.htm.

32. Philip Shabecoff, A New Name for Peace: International Environmentalism, Sustainable Development, and Democracy (Hanover, N.H.: University Press of New England, 1996), p. 153.

33. Ibid., p. 152.

34. Albert Gore, "US Support for Global Commitment to Sustainable Development," Address to the Commission on Sustainable Development, United Nations, New York City, June 14, 1993, US Department of State Dispatch, pp. 4, 24 (June 14, 1993).

35. See Paul G. Harris, ed., Climate Change and American Foreign Policy (New York: St. Martin's, 2000).

36. John Heilprin/Associated Press, "Bush Aims to Cut Aid on Global Warming," South China Morning Post, July 8, 2001, p. 6.

37. Ibid.

38. Letter from President George Bush to Senators Hagel, Helms, Craig, and Roberts, March 13, 2001 (cited in Kevin A. Baumert and Nancy Kete, "The US, Developing Countries, and Climate Protection: Leadership or Stalemate?," World Resources Institute Climate Issue Brief, June 2001, p. 1).

39. Carbon Dioxide Information Analysis Center, Trends: A Compendium of Data on Global Change (Oak Ridge, Tenn.: Oak Ridge National Laboratory, 2001), http://cdiac.esd.ornl.gov/ trends/emis/meth_reg.htm.

40. Energy Information Administration, International Energy Annual 1999 (Washington, D.C.: US Energy Information Administration, 2001), http://www.eia.doe.gov/emeu/iea/carbon.html. These reductions are disputed, but they at least suggest that China is already doing more than the US to control its carbon dioxide emissions.

41. See Harris, International Equity and Global Environmental Politics.

42. "Main Results of the Environment Council 18–19 December 2000," EU memorandum, Brussels, December 21, 2000, http://europa.eu.int.

43. European Commission, "Ten Years after Rio," pp. 2–3.

44. Ibid., p. 9 (original emphasis).

45. Ibid., pp. 13–14.

46. Jayanta Bhattacharya, "Kyoto Protocol: Rich Nations Must Make Sacrifices," *The Statesman*, June 10, 2001, p. 2, http://global.umi.com.

47. European Commission, "EU Reaction to the Speech of US President Bush on Climate Change," Brussels, June 12, 2001, photocopy (original emphasis).

48. See European Commission, "Commission Reacts to US Statements on Kyoto Protocol," Brussels, March 29, 2001, p. 1, http://europa.eu.int.

49. I say more about this US policy shift in Paul G. Harris, "Fairness, Responsibility, and Climate Change," *Ethics and International Affairs* 17: 1 (2003): 149–156.

50. Christopher Patten, EU external affairs commissioner, interviewed in "Correspondent," BBC World television, June 19, 2001.

51. Lister, *European Union Development Policy*, p. 18.

52. US Foreign Assistance Act, Part I, Section 117, "Environment and Natural Resources," reproduced at http://www.usaid.gov/environment/faa_section_117.htm, Sec. 117.\71\(a).

53. I am grateful to an anonymous reader for reminding me of this point.

54. A. M. Sbragia and C. Damro, "The Changing Role of the European Union in International Environmental Politics: Institution Building and the Politics of Climate Change," *Environment and Planning C: Government and Policy* 17 (1999):67.

55. See Robert L. Paarlberg, "Lapsed Leadership: US International Environmental Policy Since Rio" in Norman J. Vig and Regina S. Axelrod, eds., *The Global Environment: Institutions, Law, and Policy* (Washington: CQ Press, 1999), pp. 236–255.

56. Sbragia and Damro, "The Changing Role of the European Union," pp. 63–64.

57. Ibid., p. 66.

58. David G. Victor, "The Regulation of Greenhouse Gases: Does Fairness Matter?" in Ferenc L. Toth, ed., *Fair Weather: Equity Concerns in Climate Change* (London: Earthscan, 1999), p. 197.

11

Sustainable Development: Comparative Understandings and Responses

Susan Baker and John McCormick

The term "sustainable development" has achieved a notable prominence in discussions about environmental policy since the mid-1980s. Following its central role in the conclusions of the Brundtland Commission (1984–1987), it appeared with increasing frequency in academic studies and in government reports, initially in relation to the environmental problems of less developed countries, but subsequently in relation also to those of industrialized countries. So well used did it become that within a matter of years there were concerns that it had become a cliché.[1] This is curious, because despite the extensive literature on the subject, and the debate about the role that the concept plays—or should play—in public policy, there is almost no agreement on what "sustainable development" means. There have been lengthy debates about its political, economic, and social dimensions, but absent from those debates has been much discussion about how the concept can be turned into practical policy change, or how it can be measured. Some even doubt that it is a practical goal, charging that sustainable development is an oxymoron.[2]

This chapter examines the impact of the idea of sustainable development on environmental policies in the United States and the European Union. We do not try to add to the discussion about the meaning of sustainable development, but instead set out to show how the concept of sustainable development evolved, and to examine its impact upon thinking on both sides of the Atlantic. We conclude that while the principles of sustainable development can be traced back to nineteenth-century ideas about nature conservation and forest management on both sides of the Atlantic, the concept has more recently achieved greater prominence in policy discussions in the European Union. At the same time, however, we argue that while there has been much discussion about the need to incorporate the principles of sustainable development into policy on both sides of the Atlantic, there has so far been little practical change in policy.

After examining the record in the European Union and the United States, the chapter ends with a discussion about the difficulties inherent in incorporating sustainable development principles into policy. These include questions about the appropriate level at which the principles should be implemented, debates about the extent of environmental problems and thus about the unsustainability of existing policies and norms, and questions about the extent to which sustainability can be achieved in two regions of the world whose economies are so heavily driven by consumption.

Understanding Sustainable Development

As a concept, sustainable development is nothing new. It has become conventional in the literature on sustainable development to suggest that it was first introduced into the environmental debate in 1987, with the final report of the World Commission on Environment and Development (the Brundtland Commission).[3] The occasional author might point out that it was used as early as 1980 by selected governments, nongovernmental organizations (NGOs), and by specialized agencies in the United Nations. Very few note that it is in fact much older, and that—far from introducing a new idea—the Brundtland Commission was merely reviving a concept that had been a part of discussions about how to respond to environmental problems for the better part of a century.

Most discussions about sustainable development use as their benchmark the definition provided by the Brundtland Commission: development that "meets the needs of the present without compromising the ability of future generations to meet their own needs."[4] There has been much analysis in the literature about just what this means.[5] The common conclusion is that there is no simple answer, nor are there even straightforward answers to the meanings of the words "sustainable" or "development." Discussions about sustainable development often include such terms as "vague" and "ambiguous."

Modern discussions about sustainable development trace their roots back to the debate about conservation that accompanied plans for the creation of the United Nations and its specialized agencies after World War II. The rational management of resources was a central element, for example, in the work of the UN Economic and Social Council (ECOSOC), and of the UN Food and Agriculture Organization (FAO), whose constitution listed "the conservation of natural resources" as one of the means of achieving the organization's goal of global food security. In 1949, the UN hosted an international conference at Lake Success, New York, which explored the theme of the "continuous development and widespread application of the techniques of resource conservation and utilization."[6]

In the late 1960s the issue of conservation—by then more commonly known as sustainable development—was revisited during preparations for the 1972 UN Conference on the Human Environment, held in Stockholm. While much of the political debate about environmental issues in the 1960s had been driven by the priorities of industrialized countries (notably their concerns about pollution), preparations for Stockholm were heavily influenced by the priorities of developing countries, which were concerned that environmental regulation might retard their economic growth. The list of principles agreed upon at Stockholm included the arguments that natural resources should be conserved, the earth's capacity to produce renewable resources should be maintained, development and environmental concerns should be combined, and poorer states should be given every incentive possible to promote rational environmental management.[7]

Meetings and studies after Stockholm sponsored by the UN subjected these ideas to further analysis. A 1973 FAO-sponsored report argued that the definition of conservation as "the rational use of the earth's resources to achieve the highest quality of living for mankind" could be extended to define the goals of economic development.[8] The term "sustainable development" was used in discussions at a 1974 UN-sponsored conference in Cocoyoc, Mexico, which emphasized the importance of implementing policies aimed at satisfying the basic needs of the world's poor while ensuring adequate conservation of resources and protection of the environment. At the same time, the newly created United Nations

Environment Program (UNEP) was talking about the role of "ecodevelopment," defined as development that gave attention "to the adequate and rational use of natural resources, and to applications of technological styles."[9]

In the late 1970s a number of international environmental nongovernmental organizations—notably the International Union for Conservation of Nature and Natural Resources (IUCN)—worked with several UN specialized agencies to develop the World Conservation Strategy, published in 1980. Designed to strengthen efforts to protect nature and natural resources, the strategy noted the importance of advancing "the achievement of sustainable development through the conservation of living resources." Conservation was defined as "the management of human use of the biosphere so that it may yield the greatest sustainable benefit to present generations while maintaining its potential to meet the needs and aspirations of future generations."[10] The Brundtland Commission definition of sustainable development was to sound very similar.

In 1983, the UN General Assembly passed a resolution calling for the creation of a commission charged with investigating the relationship between development and the environment. The following year the World Commission on Environment and Development was created under the chairmanship of former Norwegian Prime Minister Gro Harlem Brundtland. Its final report was published in 1987, and argued that environment and development were inextricably linked, that the goals of policy-making institutions were too focused on increasing output rather than sustaining environmental resource capital, and that the environmental dimensions of policy should be considered at the same time as economic, energy, agricultural, and other dimensions.[11] Sustainable development, which until that time had been the subject of discussion only by a relatively small circle of environmental activists and NGOs, together with selected UN specialized agencies, was brought to the attention of a broader constituency.

Its new role in the environmental debate was reflected in the title of the next big global summit on the environment, the UN Conference on Environment and Development, held in Rio de Janeiro in June 1992. One of the outcomes of Rio was the publication of Agenda 21, a lengthy plan of action for implementing the conclusions of the conference, in which sustainable development featured prominently. Rio also resulted in the creation in December 1992 of the Commission on Sustainable Development (CSD), a UN body whose task was to periodically bring together government representatives and to monitor progress on the implementation of the principles in Agenda 21, and on the development of national sustainability plans, which were to be completed by 2002.

The Earth Council was subsequently established with headquarters in San Jose, Costa Rica, to promote and monitor the work of national councils for sustainable development (NCSDs), and to sponsor contacts between these national councils and the CSD. By 2002, the list of NCSDs and of national sustainable development strategies was still modest—no more than two dozen countries had founded councils or drafted strategies, the majority of them developing countries.[12] The CSD was meanwhile organizing another high-visibility international conference, the World Summit on Sustainable Development, held in Johannesburg, South Africa, in August–September 2002. One of its tasks was to review progress on the implementation of Agenda 21 and on the development of national sustainable development strategies.

The European Response

European Union initiatives in the field of environmental protection began after 1972, when the member states agreed to take collective action on the environment. There was no "constitutional" basis for this, because there was no explicit reference to the environment in the 1957 Treaty of Rome, which founded the European Economic Community. It was only with the changes made by the 1987 Single European Act (SEA) that a clear base for environmental policy was provided in Community law.

Although the SEA did not use the phrase "sustainable development," it defined one of the Community's environmental objectives as the "prudent and rational utilization of natural resources." The SEA stated that environmental policy should be guided by the principles of prevention and of sectoral policy integration (that is, environmental protection requirements should be a component of other policies). Further amendments made by the 1992 Maastricht Treaty on European Union called explicitly for "sustainable and non-inflationary growth respecting the environment."

Semantic confusion caused by the term "sustainable growth" was addressed by the 1997 Treaty of Amsterdam, which called for "balanced and sustainable development of economic activities," and adopted sectoral policy integration as a key means to achieve sustainable development. More important, the Treaty of Amsterdam made sustainable development one of the objectives of the Community; thus it is now applicable to the general activities of the EU, not just its activities in the sphere of the environment. As a result of the amendments brought by the SEA, Maastricht, and Amsterdam, it has been argued that "there is probably no single government or other association of States with such a strong 'constitutional' commitment to sustainable development."[13]

The EU's commitment to the promotion of sustainable development is also driven by a sense of moral obligation. As the European Commission has argued:

As Europeans and as part of some of the wealthiest societies in the world, we are very conscious of our role and responsibilities. . . . Along with other developed countries, we are major contributors to global environmental problems such as greenhouse gas emissions and we consume a major, and some would argue an unfair, share of the planet's renewable and non-renewable resources.[14]

For the commission, playing a leadership role in international efforts to promote sustainable development is one way of meeting its moral obligations.[15] The aim is to have sustainable development accepted as a guiding principle or norm of international politics.[16] Getting its own house in order by taking steps to change Europe's unsustainable patterns of production and consumption is seen as an important first step in bringing that leadership role to fruition.[17]

Environmental policy in the EU is largely framed by environmental action programs (EAPs), of which there have been six to date. Although the first four EAPs did not have sustainable development as their explicit focus, they were nevertheless influential:

• The First EAP (1973–1976) acknowledged that economic growth was not an end in itself, but a means of obtaining a more environmentally sustainable and equitable form of social development.[18]

• The Second EAP (1977–1981) referred to the physical limits on material growth stemming from natural resource limitations, stressed that the "harmonious development of economic activities and continuous and balanced expansion" of the Community could not be achieved without environmental protection, and affirmed that "economic growth should not be viewed solely in its quantitative aspects."[19]

• The Third EAP (1982–1986) emphasized the importance of environmental policy as an element in the Community's industrial strategy, not least because it could stimulate technological innovation. Environmental protection has since been seen as having the potential to enhance the competitiveness of the EU's economy.[20] The limits to growth argument was rejected in favor of a belief in continued economic growth based on environmental protection. As the commission later stated: "the main message is that we need to change growth, not limit growth."[21]

• The Fourth EAP (1987–1992) emphasized that "ecological modernization" could offer competitive advantage to European industry. Ecologically modernized industry treats the environment not as a free resource, but as a factor of production that has to be priced. In the short term, this results in ecoefficient businesses (those that use fewer natural resource inputs for a given level of economic outputs or value added). In the long term, ecological modernization can protect the resource base upon which further economic development depends, and environmental protection can stimulate technological innovation, which can open up or expand markets. This thinking was critically important in facilitating the acceptance of sustainable development as a norm of EU policy.

A change of emphasis came with the Fifth EAP (1993–2000), which made the first explicit policy commitment to the promotion of sustainable development, defined as "continued economic and social development without detriment to the environment and natural resources, on the quality of which continued human activity and further development depend."[22] It also called for the use of a wide range of policy instruments, including fiscal and voluntary measures to improve implementation.

Among the many reviews of the Fifth EAP, the commission's own *Global Assessment* found that while it "set out an ambitious vision" for sustainable development, progress had been limited. Sectoral policy integration remained weak and "shared responsibility" (involving different levels of government) still needed to take more widespread hold. More seriously, it found that there had been no reversal in economic and social trends harmful to the environment, particularly in relation to transport, consumer goods, and tourism. It concluded, "unless more fundamental changes are made, the prospects of promoting sustainable development remain poor."[23]

The Sixth EAP (2001–2010) attempts to address some of these shortcomings. It highlights climate change, overuse of natural resources, loss of biodiversity, and accumulation of persistent toxic chemicals in the environment as central issues. It makes sectoral policy integration one of five key "thematic strategies," alongside more effective policy implementation, enhanced citizen and business engagement, and developing a more environmentally conscious attitude to land use.

Concerns about the lack of progress also led to the launch in 1998 of the Cardiff Process.[24] Named after the location of the June 1998 European Council meeting, its aim is to promote sustainable development through a focus on sector-specific integration strategies, identification of sustainability indicators, and construction of monitoring mechanisms. It is evidence of recognition that environmental law is not enough to promote sustainable development, especially when developments in areas that create environmental pressures, such as transport, energy, or agriculture, often outweigh the benefits of new regulations.

Unfortunately, there has been a great deal of unevenness in the response of the EU councils to the Cardiff Process, with the Agriculture Directorate-General (DG) having made the most progress, and the Internal Market DG among the least.[25] Moreover, member states with less progressive environmental policies still continue to halt progress toward sustainable development.

The difficulties the EU faces in trying to realize its commitment to sectoral policy integration can be seen more clearly by examining some key policy areas:

• EU transport policy has long been a key source of environmental stress, particularly with the large-scale infrastructure projects that have come as part of the European single market program. The integration of environmental considerations into these projects has been slow; "sustainable transport" remains poorly conceptualized in policy terms; and long-term targets have not been developed.[26] A lack of commitment in many member states also remains a problem; this is especially true in peripheral areas and in east and central European states that have applied for EU membership, where road building represents a strategic response to the competitive challenges posed by the completion of the European single market.

• The energy sector has made some progress in integrating environmental considerations into policy. The need to meet the obligations of international agreements on global warming has resulted in institutional capacity building; there is an EU-level Energy Framework Program (1998–2002), and there are programs to promote renewable energy (Altener) and energy efficiency (SAVE). However, concerns about security of supply and ensuring that energy prices do not threaten European competitiveness take priority over environmental considerations. Furthermore, the primary responsibility for energy policy remains with the member states.

• The goals of agricultural policy have long been contrary to the goals of sustainable development, mainly because the EU's Common Agricultural Policy (CAP) has promoted intensive agriculture, resulting in the pollution of the aquatic environment, altered natural habitats, and reduced biodiversity. Fortunately, recent reforms of CAP provided an opening through which environmental considerations could begin to be integrated into agricultural policy.[27] For example, direct agricultural aid payments to farmers are now linked to environmental criteria, and payments are available for farmers who, on a voluntary and contractual basis, protect the environment and maintain the countryside. Environmental considerations have also been attached to the EU's structural funds (which support economic and social development, particularly in peripheral and rural areas), although they still cause environmental problems by focusing on the funding of infrastructure development.

However, the promotion of sustainable agriculture remains hampered by six main problems:

• Member states have much discretion over how to meet their obligation to undertake environmental measures in the agricultural sector, and the level of commitment varies from one state to another.

• The agricultural sector is plagued by poor compliance with environmental legislation.

• Eastern enlargement threatens to cause new problems because it will bring states with less progressive environmental policies into the EU, several of which have given agricultural modernization priority over environmental protection.

• While the EU hopes that its biodiversity strategy will promote sustainable agriculture,[28] there are continuing problems with the implementation of key pieces of the law on biodiversity.

• In the industrial sector, many large European companies have already reaped the rewards of ecoefficiency,[29] and an integrated product policy is developing, while the commission increasingly makes use of the principle of shared responsibility in its dealings with industry. However, there are serious weaknesses in efforts to promote sustainable development, including those within the strategy on policy integration. Meanwhile, small, medium, and domestically oriented businesses have made little or no commitment to the promotion of sustainable development.[30]

• The EU's Common Fisheries Policy (CFP) was designed to resolve conflicts among member states over territorial fishing rights and to prevent overfishing. Annual quotas have been imposed on the take of Atlantic and North Sea fish, and there are regulations on fishing areas and equipment, including limits on the mesh size of fishing nets and on the size of fish caught. New ways in which a balance can be found between fishing activity and stock levels are being explored,[31] and new efforts are being made to integrate wider nature conservation objectives into the CFP.[32] However, most CFP reforms have been driven by the fishing crisis caused by overexploitation. Much still needs to be done to support the conservation and sustainable use of commercial stocks and marine ecosystems, especially since efforts to date have not halted the decline in fish stocks.

As this review shows, when analysis moves from exploration of the European Union's constitutional and declaratory commitments to its implementation efforts, especially at the sectoral level, a different, and altogether more pessimistic picture emerges of the promotion of sustainable development in the EU. The planned expansion of the EU eastward could see this capability–expectation gap,[33] a gap between policy outcome and declaratory intent, grow ever wider.

Enlargement through the addition of eastern states, the first phase of which is expected to take place in May 2004, will see several former communist-bloc countries become member states of the EU. Unfortunately, official Community discussions about the institutional and policy reforms that the EU needs to make in preparation for its unprecedented expansion have so far failed to address the thorny issue of how enlargement can be reconciled with the commitment to sustainable development. That an enlarged Europe will be a greener Europe is far from certain. With enlargement, Europe's environmental future will be in the hands of a European Union that has accepted into its fold several environmentally laggard states that are bent on economic modernization and development. A predilection to sacrifice environmental protection to the goal of economic development already marks the countries of the region.[34]

The rush to development—in part to meet the membership criteria laid down by the EU and, ironically, also partly funded by the Union—has already seen massive infrastructure development, rising consumerism, and a push to modernize agriculture. These threaten the rich biodiversity of the region and are giving rise to growing problems of consumer waste. With enlargement, the further danger is that through the exercise of their new voting rights in the council and driven by the centrality of their economic goals, the new member states will exert a downward pressure on European environmental policy. Here, we can justifiably be fearful that in the new European Union collectively, the implementation deficit that has long plagued the EU's commitment to the promotion of sustainable development will become all the more marked and all the more challenging to confront.

However, this is not to deny the importantce of the EU's declaratory and legal commitments to the promotion of sustainable development. On the contrary, declaratory political statements are important because they oil the wheel of European integration politics. They help consolidate the integration process by providing a basis upon which Europe is articulating the values that will shape its shared, environmental future.[35]

More generally, ideas and values act as a vehicle though which the European Union, *as a group*, is defining its identity. As well as helping in the construction of the identity of the EU, values and norms help in legitimizing the integration project.[36] This need for legitimization stems from the fact that the EU is both an emerging and a hybrid entity, which is neither a state nor a nation, but a unique combination of supranational and international forms of governance. As such, the EU is forced to pay a great deal of attention to the identification and articulation of shared values and legitimizing principles. Such values and principles help in the mobilization of support for the integration project.[37] In the early days of the European Community, this mobilization was driven by elites. However, in the post-Maastricht Europe, mobilization of a wider support base is seen as increasingly important.

The idea of sustainable development has many of the key elements needed for it to act as a legitimizing, mobilizing value for the EU integration process. It conforms to deep-seated European social constructs. It is undeniable that the European integration project was founded on economic values, especially belief in the achievement of economic prosperity through the construction of a single, European, free market. However, it is also the case that the integration project has roots in a deep-seated belief in the ethos of collective societal responsibility for the welfare of the community as a whole.[38] This has allowed Europeans to see environmental protection as part of the protection of the common good. The promotion of sustainable development resonates with this belief and, more important, it provides a framework for the reconciliation of ecological, economic, and social goals. By combining the social and economic underpinnings of the integration project, the commitment to sustainable development allows the European integration project to be seen as part of the construction of a new European society, one that is based upon and develops shared, European values grounded in the idea of social responsibility.

Beyond the borders of the EU, the commitment to the promotion of sustainable development can also help to shape the EU's identity by singling it out as different from other actors. European Union norms, including its constitutional principles, such as the principle of sustainable development, act as constitutive factors determining the international identity of the EU. This allows the EU to act as a normative power (as opposed to a military power) in international politics.[39] This marks a major difference with the United States.

Thus, while we may be sceptical about the implementation of the principle of sustainable development in the EU, the declaratory and constitutional commitments made by the EU to this principle are of deep and, we hope, lasting significance. The diffusion of this norm, not least through the conditionality clauses laid down for new applicant states in east and central Europe, offers one hope: that enlargement may diffuse the norm of sustainable development eastward. Here we would be foolish to ignore the power of ideas in the development of politics. To deny the importance of the EU's commitment to sustainable development is to see the European integration project only in terms of its structural, procedural, and material components. European integration is also a project based upon, but simultaneously rearticulating, shared European values. We can thus see the commitment to sustainable development as the Community's contribution to what Weiler calls "the flow of European intellectual history."[40] Until the time comes to write the history of twenty-first-century European ideas, we must wait to see how the EU has shaped the environmental politics and ethics of the new Europe. In waiting, and in knowing that the European integration process is also built upon the construction of new environmental values, we remain somewhat optimistic about the environmental future of the new Europe.

The American Response

The idea that exploitation should be more carefully managed began to enter policy thinking in the United States in the late nineteenth century, as the new western territories were being opened up to white settlement. The forester Gifford Pinchot, who was instrumental in the creation in 1905 of the US Forest Service, argued that environmental management should be driven by conservation, which he defined as "wise use" or the planned development of resources.[41] Building on a tradition of progressive, scientific agriculture that dated back to the eighteenth century,[42] Pinchot argued that conservation should be based on three principles: development (using natural resources for the needs of the present generation), the prevention of waste, and the management of natural resources for the many rather than the few.[43]

Conservationist ideas found support in President Theodore Roosevelt, who sought the counsel of Pinchot and made resource management a principle of federal policy. At the core of that policy was a belief in efficiency and the "scientific" management of natural resources. For example, sustained-yield forest management—by which the cutting of trees was balanced with growth to ensure a continuous supply of wood—was promoted by the US Forest Service. Meanwhile, water management was promoted by a short-lived Inland Waterways Commission, created in 1907. In the 1930s, President Franklin D. Roosevelt argued that professional resource management and efficiency should be a cornerstone of economic recovery following the Great Depression, and such objectives were at the core of several New Deal programs. For example, the Tennessee Valley Authority became the apogee of the belief that natural resources should be sustainably exploited for multiple purposes, and the Soil Conservation Service was created in 1935 to help farmers fight soil erosion.

Elements of the sustainable development rationale can also be found in approaches to the management of public land: nearly 650 million acres of land (covering about 29 percent of the land area of the United States) that are owned and managed by the federal government. The use of that land has been driven since 1960 by the Multiple Use-Sustained Yield Act and other legislation that requires that public land be used for different purposes and in a sustainable manner. Multiple use is defined in Section 531 of the act as the management of natural resources so that they "will best meet the needs of the American people . . . and harmonious and coordinated management of the various resources . . . without impairment of the productivity of the land." Sustained yield is defined in the same section as "the achievement and maintenance in perpetuity of a high-level annual or regular periodic output of the various renewable resources of the national forests without impairment of the productivity of the land."

While particular environmental problems have been influenced by the principle of sustainable development, this principle has yet to be adopted as part of a generalized policy on the environment in the US. Even since Rio, sustainable development is a term that has been used only on the margins of the debates about environmental management. One of the first attempts to bring it into the mainstream was made by the National Commission on the Environment, a group that included former administrators of the US Environmental Protection Agency, and that issued a report in 1992 that argued that "US leadership should be based on the concept of sustainable development. By the close of the twentieth century, economic development and environmental protection must come together in a new synthesis . . . sustainable development can and should constitute a central guiding principle for national environmental and economic policymaking."[44]

The following year the Council on Environmental Quality, a presidential advisory body, called for the establishment of a President's Council on Sustainable Development (PCSD). This was created in 1993 by the Clinton administration, and was charged with working to forge agreement among government (including seven cabinet-level government departments), business, industry, labor, NGOs, and private citizens, and to develop a sustainable development strategy based on that agreement. A preliminary report was duly published,[45] which included policy goals in ten areas, including health, nature, population, and education, and a set of recommendations for changes in government, business, and individual behavior. Multiple meetings were held, and the final report of the council[46] was published as the council was wound up in June 1999.

This final report is long on rhetoric and short on substance. It concluded "a sustainable United States will have a growing economy that provides equitable opportunities for satisfying livelihoods and a safe, healthy, high quality of life for current and future generations." The "national goals towards sustainable development" included a healthy environment, economic prosperity, justice for all, the conservation of nature, sustainable communities, civic engagement, and a leadership role for the US in the development and implementation of global sustainable development policies. These were all laudable goals, to be sure, but they have been among the goals of government for generations. Besides, who would *not* be in favor of "a healthy US economy that grows sufficiently to create meaningful jobs, reduce poverty, and provide the opportunity for a high quality of life for all?"[47] The problem lies in finding the means to achieving the goal, and then making the necessary practical changes.

In its 1999–2000 report on progress in the development of national sustainable development strategies, the Earth Council noted that while the PCSD had a diverse and high-level membership, participation in its meetings had declined in its closing years, attendance by the seven cabinet secretaries was sporadic, the implementation of the council's recommendations was not well tracked, and attempts to encourage Congress to implement the recommendations had been difficult. "Sustainable development is not yet a mainstream idea in the United States," the Earth Council concluded. "A . . . critical issue is the difficulty in overcoming established patterns of activity, and . . . creating change. Without a crisis, it is difficult to motivate people to accept a new way of doing things."[48]

Symbolic of the difficulties was the approach adopted by two presidents to one of the more specific recommendations of the PCSD that dealt with climate change. The council suggested that climate protection policies should be "fundamentally linked" to any national agenda for economic growth, outlined an incentive-based program designed to reduce greenhouse gas emissions, and suggested policies for the development and deployment of climate-friendly technologies. Yet little was done under the Clinton administration to address the problem of climate change; road traffic and gas consumption grew in tandem in the United States; and the Bush administration—within weeks of taking office—withdrew the US from any agreement to meet the requirements of the Kyoto Protocol to the 1992 UN climate change convention.

While action at the federal level suggests that there is only marginal political interest in promoting sustainable development, there may be indications of more tangible changes at the local level. Indeed, one of the key recommendations of the PCSD was for the development of urban and rural strategies for sustainable communities. One of the proposals made by the PCSD was for the creation of regional councils, which were subsequently founded in the Pacific Northwest and in the San Francisco Bay area. However, its conclusion[49] that "sustainable and livable community concepts *have become mainstream*, and communities *all across the country* are implementing innovative initiatives and projects" [emphasis added] is patently an exaggeration.

Mazmanian and Kraft argue that there has been a shift in the United States away from an era of environmental regulation to one in which a more diversified set of environmental polices have been adopted, based on the goals of achieving sustainability and making decisions increasingly at the local and regional levels.[50] They offer case studies of how local communities have experimented with innovative approaches such as market incentives, but it is questionable whether the cases are reflective of a broad-based trend in the United States toward the use of sustainable development principles. The authors admit that the cases are indicative only of what "could be" a transition toward sustainable communities, whose effects are difficult to predict.[51]

The US Department of Energy maintains a Center of Excellence for Sustainable Development, which provides information "on how your community can adopt sustainable development as a strategy for well-being."[52] The center maintains a list of "sustainable development" projects around the country, which include community recycling programs, attempts to encourage residential communities to become more sustainable, and the development of sustainability programs by whole cities and states. However, the reports of these projects are replete with terms such as "vision," "objectives," "heightening awareness," and "aspire"—there are few examples of sustainability projects with a track record of measurable results.[53]

Attempts to make sustainable development part of policy thinking in the United States are undermined by several substantial handicaps, related to the attitudes of government and corporations, and to individual consumer taste. Regulation and "big government" is anathema to US corporations, and the idea that limits should be placed on exploiting resources wins little corporate or public sympathy. The United States is a high-consumption society, in which there has long been an assumption that the exploitation of land and natural resources is central to the establishment of economic and personal independence. In a country where there is still considerable space, considerable untapped resources (including forests, coal, and minerals), and where vehicle ownership is regarded almost as a basic human right, the idea that the present generation should curb its wants because of concern about future generations is a hard one to sell.

Conclusions

There is little doubt that the European Union is committed to the promotion of sustainable development. From earlier concern about resource management and ensuring that economic development resulted in an "improvement in the quality of life," EU environmental policy has become ever more deeply engaged in the promotion of sustainable development. This engagement is boosted by involvement in international environmental management regimes and by EU treaty obligations. The promotion of sustainable development is now an objective of the EU and, as such, it has become embedded in the EU integration process. Resonating as it does with deeper European social constructs, sustainable development now acts as a legitimizing and mobilizing value for the EU integration process. In addition, sustainable development has become a norm of EU policy, especially at the international level.

For the United States, sustainable development has long been an element in selected arenas of environmental policy, most notably in the management of national forests and public lands. Within the past 10 years, the United States has formed a blue ribbon commission to make recommendations on how sustainable development could be integrated into policy. It reached the same conclusions as the EU about the value of sustainable development, although the objectives it outlined were not formally adopted by the federal government. Sustainable development has not become an objective of national government in the same way as it has become an objective of collective European governance.

However, Mazmanian and Kraft note that "the ultimate test of sustainability will not be in its rhetoric but in real-world applications."[54] In reading the literature on sustainable development, and in studying the policies of the European Union and its member states, and of the United States, it is difficult not to come to the conclusion that while there have been many words written and spoken in support of the general principle of sustainable development, and a strong legal, moral, and political commitment on the part of the EU, significant policy results are hard to find.

Take the case of the European Union. While economic growth has contributed to an improved quality of life, growing consumption has increased the use of natural resources and increased pressure on the environment. EU environmental policy has had some success in combating the effects of these pressures, for example, in relation to cleaner fuel or reducing and preventing industrial discharges. However, policy has not been able to keep pace with the increasing pressures from road transport, energy production and use, tourism, the production and consumption of consumer goods, and intensive agriculture. Growth in these areas has simply outweighed the improvements attained by better technology and stricter environmental controls.

This growth continues to threaten both the biodiversity and the health of Europe. The decoupling of growth from resource consumption, polluting emissions, and waste generation has not been achieved, and only faltering steps have as yet been made toward reducing material consumption. The policy priorities of the EU, including the commitment to intensive agriculture, ensuring cohesion, completion of the single market, and preparation for eastern enlargement have been major contributory factors to the intensification of pressure on Europe's environment. As a result, the EU integration process continues to result in the encouragement, stimulation, and funding of obstacles to sustainable development.

Much the same can be said for the United States, only more so. The overall quality of life for most Americans has improved dramatically since World War II, and part of the improvement has come from a tightening of regulations on air and water quality. However, Americans face the same pressures as Europeans from an expansion in road transport, energy consumption, the production and consumption of consumer goods, and intensive agriculture. More worryingly, the concept of sustainable development is not understood or discussed in the hallways of federal or state government to the extent that it is among national leaders and bureaucrats in the EU. On both sides of the Atlantic, much remains to be done to translate declaratory and legal obligations into concrete output that actually protects the environment. Regardless of the varying degrees to which Europeans and the Americans have explored the nature of sustainable development, its practical implementation faces a number of substantial hurdles.

First, in order for planners and policymakers to understand whether development is sustainable, the political, economic, social, and scientific dimensions of that development must be understood, and built into the actions taken by governments. However, there is little political agreement on the existence of many of the most critical environmental problems. A good case in point is the greater reluctance of American policymakers, relative to their European counterparts, to accept that human behavior is having a detrimental impact on the Earth's climate. Even where there is such agreement, there are differences of opinion about the most effective responses to such problems. Similarly, the economics of environmental management are often poorly understood, most environmental problems being too complex to lend themselves to conventional cost-benefit analysis. The social dimension raises troubling questions about how to promote environmental justice (ensuring that the poor or minorities are not disproportionately exposed to environmental risk) and how to ensure that the imposition of new patterns of production and consumption will not lead to a loss of jobs or the flight of polluting industries to regions with looser regulations. Finally, the science of many environmental problems is either not fully understood, or is denied by policymakers or industrialists opposed to change.

Second, there are questions about the appropriate level at which the principles of sustainable development should be implemented. The Europeans have made it a goal of macrolevel regional integration, but have so far held member states to broad principles. There is a strong case to be made for national or regional policies, and indeed several EU member states have their own national plans and policies. However, when it comes to practical implementation of those policies, the US model of focusing on the local level—counties, cities, and neighborhoods—may offer the best chances of success. However, microlevel policy initiatives of this kind demand changes in the lifestyle of individuals. How is this to be achieved in societies where personal success is measured in large part by consumption?

Third, there are substantial institutional obstacles to the application of sustainable development. No national government has yet succeeded in establishing a network of environmental agencies that is effective, in part because of the doubts about how best to distribute policy responsibilities. There is an environmental dimension to the work of all government departments, and sustainable development cannot be applied effectively unless all those departments are working to the same plan. In order for this to happen, there needs to be a level of coordination—or at least a consistency of purpose—which to date has been lacking in most national environmental policy structures. Different national governments have defined different sets of priorities, and have created institutions that are often at odds with one another. The EU's commitment to sectoral policy integration may offer a solution to this problem.

Fourth, there are questions regarding the seriousness of environmental problems. There are some who suggest that environmentalists may have overstated their case, an intellectual thread that can be traced through from critiques of *The Limits to Growth* in the 1970s to the work of Julian Simon and Herman Kahn in the early 1980s (with their criticisms of the *Global 2000* report)[55] to the arguments made by Bjorn Lomborg.[56] If—as Lomborg argues—stocks of natural resources are able to meet demand, if agricultural production is keeping up with population growth, if threats to biodiversity have been overstated, and if the air is cleaner than at any time since the industrial revolution began, then the arguments in favor of sustainable development are moot. Or, at the very least, we are already living a sustainable lifestyle.

Finally, all the talk about sustainable development sounds hollow when it is applied to two regions of the world—western Europe and North America—where much development is apparently unsustainable. The data—if they are to be believed—suggest that while there have been achievements in some areas, notably the reduction of many kinds of air and water pollution, and rapid agreement on the actions needed to remove threats to the ozone layer, many other problems are becoming worse. There is more vehicle traffic on the roads, more consumption of fossil fuels, too little being invested in renewable sources of energy, continued urban sprawl, threats to biodiversity, ongoing problems with the production of waste (much of it hazardous or toxic), and few signs of significant responses to the problem of climate change.

In short, while sustainable development has become central to the environmental lexicon on both sides of the Atlantic (more so in the European Union than in the United States), it remains a declaratory commitment, an aspiration, or—in the case of the EU—a legal obligation. It has been applied to limited policy areas, but many difficulties remain in moving from declaratory politics to the implementation of policy, and to the kinds of changes in behavior that will really make a difference. In making that difference, however, we should never underestimate the power of ideas in the shaping of politics. When viewed in terms of substantive policy, the EU and US share a common default on implementation. However, when we view the US and the EU in terms of values and principles, a view that enables us to see European integration as a project based on the construction and rearticulation of shared values, a transatlantic divergence begins to emerge. Sustainable development has become a legitimizing and mobilizing value in the construction of the new Europe. As such, we expect that the future will see environmental values take a stronger grip on policy in Europe than in the US. Thus the current divergence on principles and values may well be translated into substantive transatlantic policy differences.

Notes

1. Johan Holmberg and Richard Sandbrook, "Sustainable Development: What Is to be Done?" in Johan Holmberg, ed., *Making Development Sustainable: Redefining Institutions, Policy, and Economics* (Washington, D.C.: Island Press, 1992), p. 2.

2. William Ophuls, "Unsustainable Liberty, Sustainable Freedom," in Denis Pirages, ed., *Building Sustainable Societies: A Blueprint for a Post-Industrial World* (Armonk, N.Y.: M. E. Sharpe, 1996), p. 34.

3. Desmond McNeill, "The Concept of Sustainable Development," in Keekok Lee, Alan Holland, and Desmond McNeill, eds., *Global Sustainable Development in the 21st Century* (Edinburgh: Edinburgh University Press, 2000), pp. 10–30.

4. World Commission on Environment and Development, *Our Common Future* (Oxford: Oxford University Press, 1987), p. 8.

5. See, for example, Michael Redclift, *Sustainable Development: Exploring the Contradictions* (London: Methuen, 1987); Ian Moffatt, *Sustainable Development: Principles, Analysis and Policies* (New York: Parthenon, 1996); Susan Baker, Maria Kousis, Dick Richardson and Stephen Young, "Introduction: The Theory and Practice of Sustainable Development in EU Perspective," in Baker, Kousis, Richardson and Young, eds., *The Politics of Sustainable Development: Theory, Policy and Practice Within the European Union* (London: Routledge,

1997), pp. 1–40; Ute Collier, "Sustainability, Subsidiarity and Deregulation: New Directions in EU Environmental Policy," *Environmental Politics* 6 (1997): 1–23; Dick Richardson, "The Politics of Sustainable Development," in Baker et al., *The Politics of Sustainable Development*, pp. 43–60; Daniel A. Mazmanian and Michael E. Kraft, eds., *Toward Sustainable Communities: Transition and Transformations in Environmental Policy* (Cambridge, Mass.: MIT Press, 1999), chap. 1; Lamont C. Hempel, "Conceptual and Analytical Challenges in Building Sustainable Communities," in Mazmanian and Kraft, *Toward Sustainable Communities*, pp. 43–74; and McNeill, "The Concept of Sustainable Development."

6. ECOSOC resolution cited by Robert Boardman, *International Organization and the Conservation of Nature* (Bloomington, Ind.: Indiana University Press, 1981), p. 39.

7. See John McCormick, *The Global Environmental Movement*, 2nd ed. (London: Wiley, 1995), chap. 5, for details.

8. Raymond F. Dasmann, J. P. Milton, and P. H. Freeman, *Ecological Principles for Economic Development* (Chichester, UK: Wiley, 1973), p. 12.

9. UN Environment Programme, *The Proposed Programme* (Nairobi: UNEP, 1975), p. 3.

10. IUCN/UNEP/World Wildlife Fund, *World Conservation Strategy* (Gland, Switzerland: IUCN, 1980), p. 1.

11. World Commission on Environment and Development, *Our Common Future*.

12. Earth Council website (2002): www.ecouncil.ac.cr.

13. European Consultative Forum on the Environment and Sustainable Development, *EU Sustainable Development: A Test Case for Good Governance* (Luxembourg: Office for Official Publications of the European Communities, 2001), p. 5.

14. Commission of the European Communities, *Environment 2010: Our Future, Our Choice*, COM (2001), 31, final, 11.

15. Commission of the European Communities, *A Sustainable Europe for a Better World: A European Union Strategy for Sustainable Development*, COM (2001), 264, final, 2.

16. Susan Baker, "The European Union: Integration, Competition, Growth—and Sustainability," in William M. Lafferty and James Meadowcroft, eds., *Implementing Sustainable Development: Strategies and Initiatives in High Consumption Societies* (Oxford: Oxford University Press, 2000), pp. 303–336.

17. Commission of the European Communities, *A Sustainable Europe for a Better World*, p. 4.

18. Andrew Jordan, "Editorial Introduction: The Construction of a Multilevel Environmental Governance System," *Environment and Planning* 17 (1999): 1–18.

19. Commission of the European Communities, "Second Environmental Action Programme, 1977–1981," *Official Journal of the European Communities*, C 139, June 13, 1977, p. 8.

20. Commission of the European Communities, "Third Environmental Action Programme," *Official Journal of the European Communities*, C 46, February 17, 1983, p. 5.

21. Commission of the European Communities, *Communication from the Commission to the Council and European Parliament: Ten Years after Rio: Preparing for the World Summit on Sustainable Development in 2002*, COM (2001) 53 final, 16.

22. Commission of the European Communities, "Towards Sustainability: A European Community Programme of Policy and Action in Relation to the Environment and Sustainable Development, 1993–2000," *Official Journal of the European Communities*, No. C 138/5, May 17, 1993, p. 12.

23. European Communities, *Global Assessment: Europe's Environment: What Directions for the Future?* (Luxembourg: Office for Official Publications of the European Communities, 2000), p. 9.

24. European Environment Agency, *Environment in the European Union at the Turn of the Century* (Copenhagen: EEA, 1999).

25. R. Andreas Kraemer, *Results of the "Cardiff-Processes": Assessing the State of Development and Charting the Way Ahead*. Report to the German Federal Environmental Agency and the German Federal Ministry for the Environment, Nature Conservation and Nuclear Safety. Research Report N0. 299 19 120 (UFOPLAN) (Berlin: Ecologic Institute for International and European Environmental Policy, n.d.).

26. Commission of the European Communities, "From Cardiff to Helsinki and Beyond: Report to the European Council on Integrating Environmental Concerns and Sustainable Development into Community Policies," Working Paper, SEC (1999), 1941, final, 8.

27. Commission of the European Communities, *Directions Towards Sustainable Agriculture*, COM (1999) 22.

28. Commission of the European Communities, *Report to the UN Commission on Sustainable Development: Eighth Session* (Brussels: Commission of the European Communities, Directorate-General Environment, April 2000).

29. Commission of the European Communities, *Progress Report from the Commission on the Implementation of the European Community Programme of Policy and Action in Relation the Environment and Sustainable Development*, *"Towards Sustainability,"* Commission of the European Communities, COM (95) 624 final, 3.

30. Ibid.

31. Commission of the European Communities, *Communication on Fisheries Management and Nature Conservation in the Marine Environment*, COM (1999) 363.

32. Commission of the European Communities, *Report of the Council (Fisheries) to the European Council (Santa Maria de Feira) on Integrating Environmental Issues and Sustainable Development into the Common Fisheries Policy June 2000*, 9386/00.

33. C. Hill, "The Capability-Expectations Gap, or Conceptualising Europe's International Role," *Journal of Common Market Studies* 31(3) (1993): 305–328.

34. Susan Baker, "Environmental Politics and Transition," in F. W. Carter and D. Turnock, eds., *Environmental Problems of East Central Europe*, 2nd ed. (London: Routledge, 2002), pp. 22–39.

35. Susan Baker, "The Values Underlying EU Global Environmental Policy," in S. Lucarelli and I. Manners, eds., *Values, Images and Principles in EU Global Action* (London: Routledge, forthcoming 2003).

36. J. Weiler, "Ideals and Idolatry in the European Construct," in B. McSweeney, ed., *Moral Issues in International Affairs: Problems of European Integration* (London: Macmillan, 1998), pp. 55–88; I. Manners, "Normative Power in Europe: A Contradiction in Terms?" *Journal of Common Market Studies* 40(2) (2002): 235–258.

37. Weiler, "Ideals and Idolatry in the European Construct," p. 60.

38. Ibid., p. 62.

39. Manners, "Normative Power in Europe."

40. Weiler, "Ideals and Idolatry in the European Construct," p. 61.

41. Roderick Nash, *Wilderness and the American Mind* (New Haven, Conn.: Yale University Press, 1973), p. 129.

42. Donald Worster, *Nature's Economy* (San Francisco: Sierra Club Books, 1977).

43. Gifford Pinchot, *The Fight for Conservation* (New York: Doubleday Page, 1910).

44. National Commission on the Environment, *Choosing a Sustainable Future* (Washington, D.C.: World Wildlife Fund, 1992), p. 42.

45. President's Council on Sustainable Development, *Sustainable America: A New Consensus for Prosperity, Opportunity, and a Healthy Environment for the Future* (Washington, D.C.: Government Printing Office, 1996).

46. President's Council on Sustainable Development, *Towards a Sustainable America: Advancing Prosperity, Opportunity, and a Healthy Environment for the 21st Century* (Washington, D.C.: Government Printing Office, 1999).

47. Ibid., p. 1.

48. Earth Council, *NCSD Report 1999–2000* (San Jose, Costa Rica: Earth Council, 2000), pp. 102–103.

49. President's Council on Sustainable Development, *Towards a Sustainable America*, p. 5.

50. Mazmanian and Kraft, *Toward Sustainable Communities*.

51. Ibid., p. 285.

52. Department of Energy website (2002): www.energy.gov.

53. But see also Kent E. Portney, *Taking Sustainable Cities Seriously: Economic Development, the Environment and Quality of Life in American Cities* (Cambridge, Mass.: MIT Press, 2003).

54. Mazmanian and Kraft, *Toward Sustainable Communities*, p. 26.

55. See McCormick, *The Global Environmental Movement*, pp. 98–106, 228–229.

56. Bjorn Lomborg, *The Skeptical Environmentalist* (Cambridge: Cambridge University Press, 2001).

IV
Transnational Networks and Dialogue

12

Emerging Transnational Policy Networks: The European Environmental Advisory Councils

Richard Macrory and Ingeborg Niestroy[1]

Various forms of official advisory bodies have long been part of the process by which environmental policy is developed by national governments. Traditionally these have taken the form of specialized committees that provide expert advice on particular topics, such as hazardous substances, biotechnology, or air pollution, and many such bodies continue to exist, both at national and European Community level. However, the identification of environmental policy as a distinct and new dynamic in the 1970s and 1980s saw a number of governments in Europe establishing advisory councils with a far broader environmental policy mandate. Many retained an expert basis, although increasingly one has also seen the establishment of bodies with a more representative structure of stakeholders. The pattern has now been adopted in many western European countries, and increasingly so in central and eastern Europe. In more recent years the growth of sustainable development as an underlying policy concern of government has also seen the emergence of similar bodies concerned exclusively with sustainable development, or the extension of existing environmental advisory bodies to encompass the issue.

In contrast, while the US federal and state agencies have a number of advisory committees, these appear to be more specialized by subject area and to offer mainly scientific and technical rather than policy advice.[2] A more comparable body, the President's Commission for Sustainable Development (PCSD), was wound up in 1999 after several years of work and appears to have had little influence (see chapter 11).[3] The Council for Environmental Quality (CEQ) established under the National Environmental Policy Act of 1969, is supposed to provide environmental policy advice to the president,[4] but it is located within the Executive Office of the President[5] and in that respect would not appear to be as independently structured as the equivalent bodies that have emerged within Europe, and which are considered in this chapter.

The role and influence of such bodies on national governments will differ according to their structure and the political traditions of individual countries. They rarely substitute for the influence and perspectives of nongovernmental organizations (NGOs) or other interest groups, but their official standing within the broad structure of government can provide a distinctive and important source of independent policy influence. Within the European context, many areas of environmental policy are now handled by governments at a supranational level, both as a matter of legal division of powers within the European Union, and as a matter of functional necessity because of close physical, economic, and cultural connections across European boundaries. In a mirror image of this process, a number of these national official environmental policy bodies in 1993 began informal cooperation with each other, eventually leading to the establishment of the European Environmental Advisory Councils (EEAC), a loose federation that now consists of around thirty such bodies across both western and central and eastern European countries.[6]

This chapter considers the nature of these bodies,[7] and the emergence of the EEAC as a distinctive form of a transnational policy community. It becomes apparent that the diversity of institutional types of councils cooperating in this network, and also their individual different activities, make it difficult to classify it as either a "policy network" or as an "epistemic community" as these terms are sometimes defined;[8] it seems to fall in between. Nevertheless, the overall characteristic of the individual councils is one that is based on expertise and knowledge (and in a wider sense than simply scientific knowledge), and the network is centered on the consensual sharing of that knowledge and experience rather than the collective representation of interests. In that sense the network does share characteristics of other epistemic communities, although perhaps it is distinct from more familiar forms with respect to dealing with broad subject areas, a much bigger range of disciplines, and a broader understanding of "knowledge" (although many council members individually belong to research communities). As we will show, it differs from interest groups that influence European policy-making in that its preferred goal is "giving advice" rather than "influencing" as such. Policy advice and recommendations are based on rigorous analysis and it is this factor rather than representation of interests that gives its work weight among policymakers. As to shared beliefs and principles, another defining characteristic of epistemic communities, member councils all have environmental concerns and a recognition of the importance of sustainable development as an important overall policy goal; but these are of course incredibly broad and hardly in themselves the type of shared beliefs that often characterize epistemic communities. In recent years, though, EEAC councils have been working closely together to develop common perspectives both on how the environment should be handled and how to influence the concept of sustainable development. Councils, by their very nature, will always retain an independent outlook and resist being constrained or bound by preset principles. Nevertheless, the increasing amount of cooperative work that has been undertaken in recent years is laying the basis for a more systematic set of shared beliefs.

The EEAC has not been officially sanctioned by national governments or Community institutions, but developed from the dynamics of European cooperation and policy-making. In doing so, the network has had to both respect the independence of individual councils, and be wary of jeopardizing their relationships with their own national governments. At the same time, it has had to be sensitive to what are sometimes very different national perspectives on the future of European environmental policy.

The Nature and Types of Advisory Councils for Environmental Policy and Sustainable Development in Europe

Establishment of Advisory Councils
The earliest initiatives for advisory councils in the late 1960s and early 1970s took place in Sweden (Environmental Council, 1968), the United Kingdom (Royal Commission for Environmental Pollution, 1970) and Germany (Council of Environmental Advisors, 1971). Denmark and the

Netherlands also have a long-standing tradition for governmental advisory councils. This first phase originated from the emergence of the environment as a distinct policy field following the Stockholm conference in 1972, and the need for governments to have access to a source of independent advice and information concerning the environment. The official nature of the bodies provided them with a distinctive authority, and the more cynically minded could see that for governments faced with difficult new challenges, institutional initiatives provided an impression of action without the need to undertake a substantive policy change. This first round was followed by the establishment of similar bodies in other European countries such as Austria, Belgium, and Finland, and a little later in Denmark, Ireland, and Spain, together with a considerable amount of restructuring in other countries (see table 12.1). Similar bodies were later created in some central and eastern European (CEE) countries, namely, in Poland, Slovenia, Hungary, and Lithuania, and in all but one case (Polish State Council for Nature Protection) it was, as in most western European countries, an initiative of the government (Hungary, Lithuania, Poland) or parliament (Slovenia).

The major political influence for the second wave of advisory councils was the United Nations Conference on Environment and Development (UNCED) in Rio in 1992. Following the recommendations of Agenda 21 (in chapters 8 and 37), many countries have created national mechanisms to follow up the Rio agreements, including processes to formalize the participation of relevant stakeholders in sustainable development. Often these two functions are combined with the creation of a body in the form of a national commission for sustainable development (NCSD)[9] that is engaged in creating national sustainable development strategies; in line with Agenda 21 principles, these bodies tend to be made up of stakeholders rather than experts, a distinction explored further later. Examples within Europe include:[10]

· Belgian Federal Council for Sustainable Development (FRDO–CFDD), in 1993–1997

· Finnish National Commission for Sustainable Development (FNCSD), in 1993

· Estonian Commission on Sustainable Development (ECSD), in 1996

- French Commission on Sustainable Development (CFDD), in 1995–1999
- Irish National Sustainable Development Partnership (Comhar), in 1998
- Slovakian Council of the Government for Sustainable Development, in 1999
- UK Sustainable Development Commission, in 2000[11]
- Lithuanian National Council for Sustainable Development (LNCSD), in 2000
- Czech Advisory Council for Sustainable Development, in 2000
- German Council for Sustainable Development (RNE), in 2001

The model of a separate stakeholder commission or council concerned with sustainable development has not, however, been followed in all countries. In Germany, for example, the Advisory Council for Global Change (WBGU), also founded in 1992 in the context of Rio, is an expert body concerned with the global dimensions of environmental policy. The Portuguese National Council on Environment and Sustainable Development (1998) has advisory tasks for both the environment and sustainable development, as does the new Greek National Council for Physical Planning and Sustainable Development, established in 2001. In some countries, existing environmental advisory bodies were restructured to include sustainable development within their scope, as in the case of the Walloon Environmental Council in 1994.

Establishment of councils is no guarantee of their permanence or success. A new government in Denmark dissolved the Danish Nature Council in 2002, and the French Sustainable Development Commission was terminated in 2003 for similar reasons, although in this case a successor was established. The Spanish Environmental Advisory Council stopped convening in 1998 owing to lack of support from the environment minister and difficulties in relating to civil society representatives. An attempt in 2002 to reestablish it failed. In other cases, notably in central and eastern Europe but also, for example, in Greece, the councils have not met on a regular basis (Estonia) or have met only a few times (Czech and Slovak Republic, Lithuania, Latvia). Usually the existence of an independent secretariat indicates both sufficient political will and budgetary support for a council to function well.

Table 12.1
European Environmental Advisory Councils and Predecessors

Country	Council Name (acronym)	Year Founded*
EU Member States		
Austria (A)	Clean Air Commission of the Academy of Sciences (KRL)	1962
	Austrian Association for Agricultural Research (OeVAF)	1989
	Austrian Council for Sustainable Development (OERNE)	1997, 2002
Belgium (B)	Walloon Environmental Council for Sustainable Development (CWEDD)	1985, 1994
	Environment Council of the Region Brussels (CERBC)	1990
	Environment and Nature Council of Flanders (MiNa-Raad)	1991
	Federal Council for Sustainable Development (FRDO-CFDD)	1993, 1997
Denmark (DK)	Danish Nature Council (DNC)	1998** [2002]***
Finland (FIN)	Finnish Council for Natural Resources (FCNR)	1999**
	Finnish National Commission on Sustainable Development (FNCSD)	1993
France (F)	Commission on Sustainable Development (CFDD)	1995, 1999 [2002]
Germany (D)	Council for Land Stewardship (DRL)	1962
	Council of Environmental Advisors (SRU)	1971
	Advisory Council on Global Change (WBGU)	1992
	Council for Sustainable Development (RNE)	2001
Greece (HE)	Council for Physical Planning and Sustainable Development (CPPSD)	2001 [2002]
Ireland (IRE)	Heritage Council (HC)	1995
	National Sustainable Development Partnership (Comhar)	1998
Netherlands (NL)	Scientific Council on Government Policy (WRR)	1972, 1976
	Advisory Council for Research on Spatial Planning, Nature and Environment (RMNO)	1981
	Council for the Rural Area (RLG)	1996**
	Council for Housing, Spatial Planning and the Environment (VROM-raad)	1996**

Table 12.1
(continued)

Country	Council Name (acronym)	Year Founded*
Portugal (P)	National Council on Environment and Sustainable Development (CNADS)	1998
Spain (E)	Environmental Advisory Council (CAMA)	1994 [1998]
	Advisory Council for the Sustainable Development of Catalonia (CADS)	1998
Sweden (S)	Environmental Advisory Council (MVB)	1968
United Kingdom (UK)	Royal Commission on Environmental Pollution (RCEP)	1970
	English Nature (EN)	1990**
	Countryside Council for Wales (CCW)	1991**
	Scottish Natural Heritage (SNH)	1991**
	Joint Nature Conservation Committee (JNCC)	1991**
	UK Sustainable Development Commission (UK SDC)	2000

EU Candidate States and Others

Country	Council Name (acronym)	Year Founded*
Croatia (HR)	Council for Spatial Planning and Environmental Protection	2001
Czech Republic (CZ)	Advisory Council for Sustainable Development	2000 [2001]***
Estonia (EE)	Commission for Sustainable Development	1996
Hungary (HUN)	Hungarian National Council on the Environment (OKT)	1996
Lithuania (LT)	Advisory Council to the Ministry of Environment	1990 [2000]***
	National Council for Sustainable Development (LNCSD)	2000 [2002]***
Poland (PL)	State Environmental Council of Poland (PROS)	1993**
	State Council for Nature Protection (PROP/SCNP)	1991**
Slovak Republic (SR)	Council of the Government for Sustainable Development (RV TUR)	1999 [2001]***
Slovenia (SLO)	Council for Environmental Protection (CEPRS)	1990

* If more than one year given, later years indicate dates of reorganization.
** Indicates that earlier predecessors existed in 1980s or before.
*** Bracketed dates indicate year council was terminated or became inactive.

The Diversity of Structures

Figure 12.1 shows an attempt to represent the range of different bodies and their key characteristics. It is clear that even though the motivation for establishing these types of councils may have arisen from the shared international political agenda of the environment and sustainable development, the actual structure of such bodies, their relationship with their governments, and their overall reputation and influence differ considerably, and are under constant development.[12] The "classic" institutional type is a government decision establishing an independent council of experts with an advisory task and their own budget. There are examples of bodies, however, that did not originate from a government initiative but were essentially composed of scientists in environmentally related fields who gathered together of their own accord and gained an advisory function through their reputation and activities (e.g., the German Council for Land Stewardship). Others were originally primarily concerned with research issues but have adopted some policy advisory functions (e.g., the Austrian Clean Air Commission). The United Kingdom has a strong administrative tradition of establishing "nondepartmental public bodies," agencies of government operating somewhat at an arm's length from departmental ministries and with specific operational responsibilities such as the licensing of industrial installations. Theoretically such bodies are charged with implementing existing government policy, but in practice they often exercise considerable influence on its development, and in some cases, such as English Nature, the legislation establishing them gives an express power to provide independent advice on policy.

Many of the environmental councils have an extremely broad authorization that allows them to advise on not only environmental issues per se—such as biodiversity, water, waste management, or chemicals—but also on the environmental implications of a wide range of sectoral policies such as agriculture, fisheries, transport, energy, or fiscal structures. Others, as their names suggest, have a rather narrower area of responsibility; examples include those concerned with nature, rural areas, heritage, and marine issues. What is equally clear is that despite the more recent creation of specific sustainable development councils, many of the environmental councils have now found themselves engaged in the

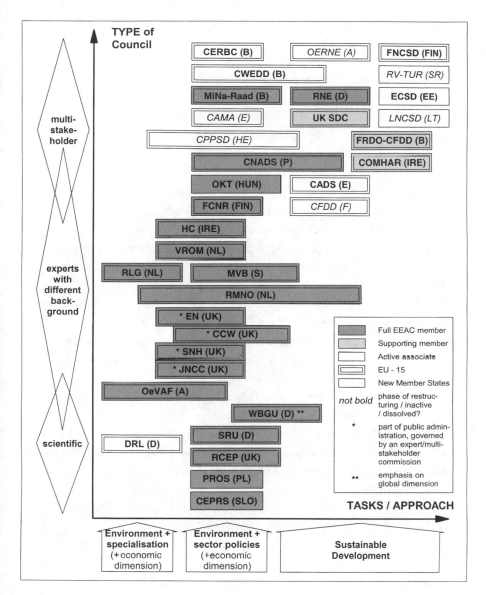

Figure 12.1
Classification of advisory councils: institutional differences and missions. Some councils listed in table 12.1 are not included here as they do not appear to be active currently.

language of sustainable development, albeit from an environmental perspective.

There is a similar range of differences in the relationships of councils to their government or parliamentary structures. Many advise ministries of the environment or their equivalent, or are at least organizationally related to such a ministry, which may provide secretariat support or financial supervision. Even the UK Royal Commission on Environmental Pollution, which jealously guards its role as strictly advising the Queen and is therefore not a departmental advisory body as such, has a special administrative relationship with the environment ministry. Even with sustainable development councils, whose responsibilities and council members explicitly cover the three "pillars" of sustainable development, the secretariat often lies with the ministry of environment (Austrian Council for Sustainable Development, French Commission on Sustainable Development, Finnish National Commission on Sustainable Development, UK Sustainable Development Commission, Estonian Commission for Sustainable Development), although in France there has been considerable discussion about the most appropriate governmental links for the sustainable development council. In Belgium (Federal Council for Sustainable Development–CFDD), Germany (Council for Sustainable Development), and Ireland (National Sustainable Development Partnership) though, there is an independent secretariat, and advice is directed to the chancellor or the government as a whole. The Portuguese National Council on Environment and Sustainable Development has an independent secretariat and so far directs it advice to the ministry of environment.

Scientific-Expert versus Multistakeholder Councils

In terms of organizational structure, the most explicit distinction is between scientific-expert and stakeholder bodies.[13] Most of the environmental councils have been established as expert bodies,[14] generally exclusively or at least dominated by scientific expertise, with members appointed by the government.[15] As such their functions could be described as (1) gathering and analyzing information; (2) raising awareness and minimizing risks through "early warning"; (3) critical analysis of planned or existing policies; (4) informing and enlightening individual politicians, the government, and the public; and (5) rationalizing debate, and thereby assisting in mediating and defusing conflicts.[16] The independent authority that such bodies can provide clearly represents an important legitimization and underpinning of governmental action, and many governments continue to insist that environmental policy must be based on "sound science." The unquestioned authority of such expert advice which prevailed in the 1970s and 1980s is, however, no longer so prevalent, with governments and the public increasingly aware of the uncertainties inherent in the scientific process, especially in the environmental field.[17]

Most of the environmental councils have retained their expert and independent qualities, but a number of these, such as the UK Royal Commission on Environmental Pollution, the Dutch Advisory Council for Research on Spatial Planning, Nature and Environment (RMNO) and the Swedish Environment Council, have become engaged in public discourse on the appropriate role of scientific and expert advice in the policy process. There are examples, though, of environmental councils that are explicitly established as stakeholder bodies, and these include the Environment and Nature Council of Flanders and the Hungarian National Council on the Environment. When it comes to sustainable development councils, the stakeholder model predominates, with membership from different societal interests, typically including trade unions, employer organizations, farm groups, the church, and environmental and/or development NGOs. One main institutional variable is the degree of direct governmental or parliamentary involvement. At the "highest" level one sees the Finnish example of representation of all ministries, with the prime minister as chair of this council,[18] while other councils have some degree of government representation, typically with observer status (e.g., Belgian FRDO-CFDD, Irish Comhar, Portuguese CNADS), and still others have no government representation or participation at all.[19]

The provision of independent advice remains the core activity of both the environmental and sustainable development councils, even though members of the latter at least reflect different interest groups. They are appointed to ensure that their membership reflects such societal groups, but are asked not to act as representatives or delegates of such groups in terms of negotiation and discussion. The role of governments in such councils remains a delicate issue. The degree to which the councils respond to requests from the government or initiate their own subjects for study varies, as will their mechanisms for providing advice, although this will typically take the form of published reports. In addition to this, and in contrast to the environmental bodies, the sustainable development councils are often involved to some degree in the development of the national sustainable development strategy,[20] and some are charged with communicating to the public issues concerning sustainable development.[21]

Figure 12.1 shows a systematic overview of the different types of councils and their areas of responsibility.

Networking the EEAC Councils

Early and Systematic Development of the Network

The origins of the current network can be traced to the early 1990s when the UK Royal Commission and the German Council of Environmental Advisors met informally to exchange information and experiences and to explore common themes, especially those that involved a European Community dimension. Being confronted with the increasing importance of EU environmental policy for national policy-making, advisory councils from Belgium, Germany, the Netherlands, and the UK in 1993 then took the initiative for a more systematic exchange of information on issues of common interest and concern, largely based on an annual meeting focused on a single topic. At the annual meeting in 1996, significant steps were taken toward institutionalizing a network. A subscription fee was introduced to finance a secretariat post ("Focal Point") and a common website. In addition, a small steering committee was elected to provide overall direction and strategy and to assist in representing the network during the year. The development of the underlying structure of the network, however, has not followed a conventional organizational path, and has had to be especially sensitive to the independent nature of the councils, the diversity of their structures and authority, and the dangers of jeopardizing their relationships with national governments. These requirements are reflected in a number of principles that have guided the development of the network to date. It has not been formed into a separate legal entity, and rather than being heavily rule based, it operates largely as a loose federation working by consensus and discussion.[22] The organization of the network deliberately avoids a top-down and hierarchical structure, but is designed to ensure a wide distribution of responsibility and a large degree of rotation.

Objectives and Distinctiveness of the EEAC Network
The core objectives for the network are

- to enrich the advice the individual councils give to their national and regional governments;
- to profit from the experiences and work of councils in other countries;
- to assist in anticipating strategic issues at a European level; and
- where appropriate, to exert an influence on policy developments at EU level by acting cooperatively.

This last function has, hardly surprisingly, proved to be the most sensitive area for development. The councils recognize that European policy-making continues to exert an enormous influence on national capacities to handle environmental challenges, yet in formal terms the responsibility of the councils is generally to advise on national and regional policy, albeit increasingly with a European dimension. Their constitutional role is generally to advise their national governments, not the Community institutions, and although they may possess the independence to do so, acting as a distinct lobby group in Brussels is likely to jeopardize their relationships and authority with their own national governments. On a more practical level, the rich variety of councils—by their very nature, independent-thinking bodies that do not readily sign on to common statements—means that a common European policy position is likely to be the exception rather than the rule. Dialogue with Brussels, however, remains significant. Regular meetings with senior commission officials are held, not with a view to lobbying on particular policies, but to ensure that Brussels is fully aware of the nature of the network and particular activities taking place within it. However, the process works two ways. One of the key features of the EEAC councils, which distinguishes them from many other interest groups operating at a European level, is that their official status provides them with direct lines of influence within their national governments. By establishing good lines of communication in Brussels, the network can assist in ensuring that councils have authoritative information on European dimensions for policy development, and can improve their capacity to advise their own governments on these aspects.

It is nevertheless clear that different councils have different perspectives on the importance of activities in Brussels and the role of the network. Sometimes this reflects general national approaches toward Europe, with some arguing that greater European integration is inevitable and that Brussels will increasingly represent the heart of environmental policy-making. Others present a far more Euro-skeptical approach. For the network, this means steering a delicate path that recognizes the significance of the European level of policy-making, but respects the broad variety of views within the different councils. As a result of a major review of the network's activities and objectives in 2000 it was agreed that:

• There must be a balance of networking and influencing.

• Networking remains the core business, which reflects both practical and political realities. Without networking, the councils could not develop any common approaches or advice at a European level, and a common position on a particular proposed policy at EU level in any event demands a considerable degree of consensus building within the network. Given their primary functions of advice at national and regional levels, the capacity of the councils for direct engagement at a European level is limited.

Membership and Regional Expansion

In the past 3 years, the network has rapidly deepened in terms of intensity of collaboration and has widened in both geographic scope and numbers. This expansion has raised a number of challenges if the distinctive nature of the network is to be preserved.

Regional Expansion   The early development of the network was dominated by councils from northern Europe, and two particular initiatives were made to encourage a broader regional input, especially from southern Europe[23] and central and eastern Europe.[24] The establishment of an independent council and its structure and responsibility remains a decision for national or regional governments, but in those countries where such a body is lacking, the network has been able to demonstrate the value of independent advisory bodies as an element of contemporary good governance, and to provide information on the range of different models that can be adopted.

Characteristics of Membership   Early on the EEAC established itself as a network of independent but official environmental advisory bodies. As such it is to be distinguished from equivalent cooperation or networks of government departments or national enforcement agencies, and from environmental nongovernmental organizations.[25] However, as membership has widened, the network has found itself obliged to develop in rather more detail its criteria for independence in this context. Currently the view is that this requires (1) institutional independence from the organs of the state (purely interministerial or parliamentary bodies therefore have to be excluded); and (2) policy independence, that is, the capacity to think independently, even though officially funded. It is, however, nearly impossible to prescribe in precisely drafted rules the conditions for independence, and to a large degree it has been a matter of judgment in each case. The task, though, has been made somewhat easier by the distinction between core subscribing members who must satisfy these conditions, and hence exercise decision-making in the network, and associated councils where rather less strict criteria may apply.

Sustainable Development Councils   A more recent challenge for the network has been the emergence of distinct sustainable development councils who wish to participate as full members. Such bodies are likely to satisfy the conditions of independence,[26] and in some cases will be heavily engaged in the environmental dimensions of sustainable development. Many existing environmental councils find themselves engaged in the language of sustainable development, and some existing councils have been given explicit authority to encompass both the environment and sustainable development. The EEAC could not be blind to these developments, but equally wished to preserve the environmental perspective that its members brought to the debate. As it turned out, quite fortuitously at the same time the network was developing a collective paper, "Greening Sustainable Development Strategies," as an input to the European policy process on this subject (see later discussion). This exercise required the existing councils to consider collectively and in some depth their analysis of sustainable development. Following endorsement of the paper in principle by nearly all existing member councils, the steering committee stated that endorsement of the approach in the document represented a reasonable condition for full membership for all, including sustainable development councils.[27] Some such councils became full members on these grounds (French CFDD, German RNE, Irish Comhar), while others felt unable to endorse the emphasis given to environmental sustainability but remain associated councils, cooperating and engaging in dialogue with the network. A recent initiative of sustainable development councils and others working in this field has led to the establishment of a working group within the EEAC framework, which follows up on sustainable development strategies and their implementation.

The Core Structure

A Focal Point and a steering committee form the key elements of the network. The secretariat Focal Point acts as a neutral information exchange point both between the constituent members and with external interests. It supports the steering committee and the working groups, and provides a switchboard for contacts and information management. The Focal Point is "hosted" by a member council,[28] which provides office space and administrative support, and is responsible for financial management. The location of the Focal Point within an existing member council rather than as an independent office provides a certain reassurance to councils that its affairs will be managed professionally and with sensitivity.[29]

The main function of the steering committee is to assist in developing the operational and strategic direction of the network. Although to a certain extent it represents the external face of the network, it does not have the authority to bind members on policy positions, and deliberately possesses very little in the way of formally delegated powers. Its membership is approved by councils at the annual conference, but this is largely determined by the functional responsibilities taken on by the councils.[30] Both the Focal Point and the steering committee report to the annual plenary session of all councils, which is the main decision-making body of the network.

Work Style

Annual conferences and working groups are the two main mechanisms for cooperation within the network. The annual conference, hosted by one of the member councils, provides the key platform for discussing both national and European policy issues. Working groups have tended to emerge on the lines of a bottom-up approach, with a number of interested councils working together on a particular theme. Recent examples include working groups on governance, agriculture, marine environment, energy, and sustainable development. Some working groups may simply act as a means for discussing ideas on a certain topic, but others may be producing a common statement to influence a particular EU policy agenda, which the councils of the group can endorse in their individual capacity. Or they may elaborate a position that can later be supported by more councils or the EEAC network as a whole, a process epitomized in the case of sustainable development, discussed in the next section.[31]

Sesimbra to Gothenburg: Influencing the EU Sustainable Development Strategy

Developing a collectively agreed-upon view on a policy issue is clearly a challenging task for the EEAC, given the independent nature of the councils and the diversity of their interests. Yet it is these very qualities that mean that a common perspective can be highly influential, and one not easily ignored by policymakers. In 2001 the EEAC published a report, "Greening Sustainable Development Strategies," which was supported in principle by the majority of the councils, and which had a significant impact on the direction and content of the emerging European Community strategy on sustainable development. It was the first time that so many councils had collectively supported such a lengthy analysis, and the process by which it emerged and the steps taken to ensure that it reached the heart of European decision-making provide important insights into the network as an emerging source of European policy influence.

The 1997 Amsterdam Treaty amendments first inserted an explicit reference to sustainable development within the European Treaties,[32] but it was not until the Helsinki summit of December 1999 that heads of government decided that a formal Community strategy on the subject should be developed. The European Commission was instructed to draw up a draft strategy for consideration at the Gothenburg summit of the Swedish presidency in June 2001, in part with a view to the need to present a European strategy at the Rio + 10 conference in Johannesburg the following year. The EEAC recognized the policy significance of the initiative and set up a new working group to provide a possible input to the development of the EU strategy for sustainable development ("S.D. strategy").[33]

Within the European Commission, responsibility for developing a sustainable development strategy had to be taken, not by the Directorate-General (DG) Environment, but significantly, by the president of the commission, reflecting the overarching nature of sustainable development. The EEAC working group held early meetings in the summer of 2000 with both the Forward Studies Unit (FSU) within the president's office, which was at that time responsible for drafting the strategy for the so-called "Prodi group,"[34] and the Environmental Economics Unit within DG Environment, which was responsible for the contribution of this DG to the strategy. What became apparent was that there remained great uncertainty as to the nature of the sustainable development strategy, and, rather more disturbingly from the EEAC perspective, early signals suggested that the environmental dimension of sustainable development was being downplayed in favor of more conventional economic development perspectives.

The EEAC working group initially collected ideas from the work of individual EEAC councils and from positions emerging in their countries, together with inputs from other working groups. With the Prodi group still making fairly heavy weather of its own work, it became apparent that the output of the EEAC working group was assuming greater political importance than perhaps originally envisaged, and that it could be extremely timely in ensuring that the environmental significance of sustainable development was fully taken on board. By December 2000, the EEAC working group had completed a first draft of the statement, which was sent for comment to all EEAC councils, and by January a final version was ready for publication at a conference in Stockholm organized by the Swedish Environment Council. "Greening Sustainable Development Strategies" is a lengthy document, and it was recognized that it would be nearly impossible and certainly impracticable given the time scales to secure a line-by-line agreement by all councils; instead, what was sought was a general support of the orientation of the analysis, without necessarily implying endorsement of all the proposals. This approach resulted in the document securing support from some twenty-three councils in fifteen countries, including those from central and eastern Europe,[35] with further councils subsequently adding their endorsement.

The Stockholm conference, which was also part of the official program of the Swedish EU presidency, was attended by more than a hundred participants, including representatives from the commission, member states' governments, environmental NGOs, and businesses. Influencing policy processes effectively requires both a sound message and well-judged timing. As for timing, the statement was published at a critical moment when the commission had yet to complete its own formulation but was now under pressure to produce a document for the Gothenburg summit in 4 months' time. The statement itself combined a strongly worded analysis of the environmental implications of current trends in Europe, a set of principles that should underpin a sustainable development strategy, and a list of specific recommendations encompassing both policy and process within the Community. The underlying message was that in the longer term, sustainable development is not simply a question of balancing competing interests, but that a healthy environment is fundamental to economic development and human welfare. Fundamentally, a new concept of what is meant by development was required. The statement noted that in March 2000 the heads of government had endorsed the idea of an annual high-level review of the progress of the Community, set against a selected number of indicators, the so-called "Lisbon process." Only economic and social indicators had initially been included, and, although the Lisbon process was little known or understood at the time in the wider political community, it was apparent from discussions with DG Environment that it was emerging as an extremely important high-level policy dynamic within the Community. "Greening Sustainable Development Strategies" therefore strongly supported the idea that environmental indicators had to be included in the Lisbon assessment, but noted that simply adding a third set of "balancing" indicators to the Lisbon process was not sufficient—true sustainable development implied the need to reexamine and modify existing economic and social indicators in the light of environmental implications. This is a message, not surprisingly, that has not yet been reflected in official government statements.[36]

Councils were asked to take the key messages of "Greening Sustainable Development Strategies" to their own national governments to influence member state positions at the Gothenburg summit itself. It became clear that much of the analysis of this document did eventually find its way into the commission's own draft strategy. Core ideas concerning greening in the Lisbon process and the importance of indicators were endorsed by the heads of government at Gothenburg, although most of the quantitative targets proposed by the commission were not accepted by the heads of government, and the overarching character of the sustainable development strategy was either not understood or not acknowledged.[37] Many interest groups were lobbying to influence the process, but the independent yet official nature of the EEAC councils undoubtedly gave their perspective particular authority. The endorsement of the statement by a number of councils from CEE accession countries was especially significant and undermined any simplistic assumption that accession states were solely interested in seeing a Community pursuing conventional patterns of growth and economic development. The statement represents the most intense effort to date of the EEAC to collectively give advice and exert an influence on European policy-making.

Conclusions and Challenges for the Future

The EEAC has developed in the course of a few years from a set of informal discussions between a small number of advisory councils to what is now an extensive network that can act as a distinctive influence on environmental policy development, both at Community level and within individual countries. The network's structure had been deliberately based on principles that avoid hierarchies and overconcentrated powers, with decision-making largely carried out by a process of consensus building rather than formal voting. Those involved in developing the network have to be continually sensitive to the delicate positions that the councils have within their national systems, and to the very differing perspectives on Europe that many councils hold. The success of "Greening Sustainable Development Strategies" indicates how influential the councils can be when they act collectively, and it has allowed the councils to make a significant contribution to the debate on sustainable development without losing the environmental perspective that underlies the approach and expertise of the councils. The process has also raised the profile of the EEAC within European circles, and one of the key challenges for the EEAC in the future will be how to capitalize on that success without threatening the key strengths of the network. Commonly endorsed statements of the sort seen in "Greening Sustainable Development Strategies" are likely to remain rather rare, and the EEAC is never likely to develop into the more conventional type of European interest group seen in Brussels circles. It will need to keep abreast of key developments at the European level, but the process of influencing outcomes, both at national and European levels, involves a rather richer set of processes than those adopted by more typical European interest groups.

For the individual councils, the challenge will be to ensure that the significance of European policy and the importance of both linkages and occasional collective action with other European councils is acknowledged and does not get squeezed out by other more purely national priorities.

The EEAC will continue to promote the value of independent, official advisory bodies as an element of good governance, within European member states, other European countries, and in other regions of the world.[38] At the end of the day its greatest strength will rest on the willingness and capacity of the councils to network with each other, sharing experiences and perspectives, learning from each other, and recognizing those areas where collectively agreed-upon viewpoints that transcend national perspectives are vital for the future well-being of Europe's environment.

Notes

1. The views expressed in this chapter are personal to the authors and do not necessarily reflect the views of the EEAC or any of its councils.

2. On EPA committees, see, e.g., Sheila Jasanoff, *The Fifth Branch: Science Advisers as Policymakers* (Cambridge, Mass.: Harvard University Press, 1994).

3. See also http://clinton2.nara.gov/PCSD/.

4. See especially sec. 204 (4) National Environmental Policy Act (42 USC 4342): "to develop and recommend to the President national policies to foster and promote the improvement of environmental quality to meet the conservation, social, economic, health, and other requirements and goals of the nation."

5. National Environmental Policy Act, sec. 202.

6. For more information, see www.eeac-network.org.

7. In doing so, we do not attempt to evaluate the authority or influence of individual councils within their own countries, a task well outside our capacity, but where systematic research tends to be lacking. On the UK RCEP, see Susan Owens and Tim Rayner, "'When Knowledge Matters': The Role and Influence of the Royal Commission on Environmental Pollution," *Journal of Environmental Policy and Planning* 1 (1999): 7–24.

8. Peter M. Haas, "Epistemic Communities and International Policy Coordination," *International Organization* 46 (1992): 1–35.

9. According to the NCSD network, NCSDs have been established in more than seventy countries since 1992 (www.ncsdnetwork.org/background.htm). For the definition of NCSDs according to the Earth Council, see the same website.

10. Some of which have not really started to work, or convened only once, or their activity remains unclear or nonexistent (Czech Republic, Latvia, Lithuania, Slovak Republic).

11. Subsuming the UK Round Table on Sustainable Development and the British Government Panel on Sustainable Development (www.sd-commission.gov.uk/commission/02.htm).

12. Table 12.1 includes some councils that have been abolished or that do not meet on a regular basis, as explained earlier.

13. "Scientific" generally means that all or most of the council members are academics, with disciplinary fields that often include law, economics, and philosophy as well as the physical sciences.

14. Only the Flemish MiNa-Raad and the Hungarian NEC are multistakeholder councils. The Danish Nature Council, which was wound up in 2002, had a multistakeholder component with a board of representatives that advised the council on themes and its reports.

15. There are also council members who are appointed because of a professional expertise outside of academia. A distinctive case is the Finnish FCNR, where

many of the council members are members of parliament. The Council of State appoints them, taking into account the size of political parties in the parliament.

16. See G. I. Timm, *Die wissenschaftliche Beratung der Umweltpolitik. Der Rat von Sachverständigen für Umweltfragen* (Wiesbaden: Deutscher Universitätsverlag, 1989).

17. The experience of the bovine spongiform encephalitis (BSE) crisis (mad cow disease) has been particularly important in this context, and undoubtedly shaped European public concerns over genetically modified organisms.

18. This is also the case in Estonia, Lithuania, and the Slovak Republic.

19. The French CFDD, German RNE, and the UK Sustainable Development Commission.

20. The Belgian FRDO-CFDD is a powerful example.

21. Examples include the Belgian FRDO-CFDD, French CFDD, German RNE, and UK SDC. An interesting example of overlapping responsibilities has taken place in the Netherlands, where there is no specific NCSD but ten advisory councils came together in a conference to provide advice on the national sustainable development strategy (www.vromraad.nl/adviezen/f10330.htm). After further discussion in 2003 the Netherlands decided not to establish a new council.

22. Over the past few years it was nevertheless realized that it would be desirable to agree on set of rules as a basis for operation. Such a "codification" process started in 2002 and led to the endorsement of a "Framework for EEAC" in the same year.

23. In 1998 with councils and experts from Portugal, Spain, and Italy, which helped lead to the establishment of the Portuguese CNADS. A similar event did not occur in Italy though, and the Spanish CAMA stopped meeting approximately that year for (political) reasons that remain unclear. An expected reestablishment in 2002 has not so far not happened. Regarding Italy, the network is starting another attempt in 2003, when it holds its annual conference at the European University Institute in Florence, during the Italian EU presidency.

24. In 1999 in Hungary and 2000 in Slovenia. In 2002 EEAC deliberately elected a member from CEE countries, the president of the Slovenian Council for Environmental Protection, as new chairman of the steering committee.

25. Compare the European Network for the Implementation and Enforcement of Environmental Law (IMPEL) (http://europa.eu.int/comm/environment/impel/), European Environment Information and Observation Network (EIONET) (http://europa.eu.int/comm/environment/impel/), and European Environmental Bureau (www.eeb.org).

26. Nevertheless it remains unclear how having a minister or prime minister chair a council influences its independence.

27. The key message of this document later became one of the "principles" of the network, endorsed with the "Framework for EEAC."

28. English Nature (1997–1999), the German SRU (1999–2002), and after May 2002, the Dutch RMNO.

29. In 2001 it was nevertheless decided to install a "hotdesk" in Brussels.

30. Such as organizing an annual conference or hosting the Focal Point.

31. Subsequent examples are a statement elaborated by the agriculture working group, "A sustainable agricultural policy for Europe," which was finally endorsed by seventeen EEAC councils and presented at the annual conference in 2002; and a statement of the governance working group on "Environmental governance" to be presented at the annual conference in 2003.

32. Within both the Treaty on European Union (Article 2) and the Treaty establishing the European Community (Article 6); but only the latter refers to the environmental integration requirement.

33. Eight EEAC councils participated in this group, which was chaired by the Danish Nature Council.

34. The existing group of commissioners on "growth, competitiveness, employment" met for an inaugural discussion on sustainable development (the added responsibility) in January 2000.

35. Of the remaining councils, only one felt unable to support the underlying message, while the rest had internal, administrative reasons for not being in a position to sign at the launch date.

36. The process of the European Convention reveals that meanwhile even parts of the *aquis communautaire* are jeopardized, mainly a proper reference to sustainable development as an objective for the Community, and the environmental integration requirement (Article 6 TEC) with its wording that reflects the need to integrate environmental considerations in order to move toward sustainable development.

37. Reflected in the Presidency Conclusions of the Göteborg European Council, June 15 and 16, 2001.

38. A number of the EEAC member councils were represented at the World Summit on Sustainable Development 2002 in Johannesburg, and used the occasion to promote the value of such independent advisory councils. Obstacles to establishing and maintaining such bodies can already be observed across Europe, mainly, though not exclusively, in southern, central, and eastern European countries.

13

The Transatlantic Environmental Dialogue

Carl Lankowski[1]

On a warm and sunny May 2, 1999 in the top floor conference room of a modest hotel a few blocks from Brussels' Grande Place, seventy-one participants representing nongovernmental organizations (NGOs) concerned with the environment—twenty-three Americans and forty-eight Europeans—met to launch a novel experiment in transatlantic cooperation. The first conference of the Transatlantic Environment Dialogue (TAED) lasted 3 days and unfolded in two parts: consultations among the NGOs followed by an exchange of views between NGOs and representatives of EU and US officials in the soulless Centre Borschette. The aim of the exercise was never absolutely precise, but it was used as a forum to develop and communicate a set of common priorities for action and directed at governments. Its philosophy rested on the acknowledgment by governments on both sides of the Atlantic that there was no way around some sort of direct participation by environmentalists in setting the parameters for new trade efforts. It provided environmental NGOs with an additional channel of communication and a possible mechanism to advance their agendas. For the more visionary among the participants, it was about experimenting with a potential prototype of transnational governance in the age of globalization.

In all, the TAED organized three conferences: May 1999 (Brussels), October 1999 (Washington), and May 2000 (Brussels). The TAED process started with discreet discussions in Washington between some government officials and NGO representatives in 1997 and proceeded quickly to organizing a steering committee to plan the first conference. The size and format of the conferences were virtually constant. The NGO voices represented at them varied but little. TAED activities were suspended on November 7, 2000.

This chapter examines the origins, dynamics, results, and demise of the TAED as an experiment in transatlantic civil society dialogue. In particular, it focuses on the different perspectives of US and European participants and attempts to explain some of the reasons for these differences in terms of the structures and interests of American and European NGOs and their relationship to governance.[2]

TAED Origins: A Confluence of Three Processes

The TAED was born as a confluence of three processes. The first concerns transformation of the international agenda following the end of the cold war and the disintegration of the Soviet Union. Five years after the fall of the Berlin Wall, the Clinton administration was pressed to find a new way to articulate US–European relations. Economics partially eclipsed political–military relations. The biggest player aside from the US was not any individual European state, but the European Union, a point driven home by the EU's central role in the just completed Uruguay Round negotiations of the General Agreement on Tariffs and Trade (GATT) establishing the World Trade Organization (WTO). To accomplish almost anything in the realm of economics at the global level, the EU and the US needed to reach agreement. This was also true for a broad range of issues with political salience that were not economic in the narrow sense, such as stabilizing central and eastern Europe, coordinating development aid and monitoring the international financial institutions, fighting disease, and dealing with hot spots, such as the Balkans.

President Clinton's January 1994 visit to Brussels marks the commitment of his administration to building a multidimensional, continent-to-continent partnership with the Europeans. A concept for realizing this aim emerged at the December 1995 US–EU semiannual summit, held in Madrid. The two sides launched the New Transatlantic Agenda (NTA), listing dozens of initiatives under four rubrics. This development constituted a breakthrough in transatlantic relations. In the first place, the NTA elevated the European Union to a position of central importance in that relationship, highlighting Washington's acknowledgment of compelling advances in European integration and presaging a new orientation in its bilateral relationships with Europe's national capitals. Second, the NTA was animated by the insight that transnational forces with global reach presented challenges that required closer policy coordination with a view toward negotiations in multilateral forums. Recognition of the significant role played by nonstate actors both globally and in the transatlantic arena, as sources of policy ideas and partners in solutions, as well as potential agents of legitimization for transnational exercises of authority, constituted a third innovative element of the NTA.

Of particular significance for the genealogy of the TAED is the "people-to-people" dimension of the NTA, which appeared as "building bridges across the Atlantic" in the joint communiqué. Many bridges of a nongovernmental nature had already spanned the Atlantic, so the communiqué actually was about acknowledging the rise of transnational actors as significant elements in the formulation and conduct of "foreign" policy. This constituted the second process relevant to the origin of the TAED.

Not surprisingly, "private" nongovernmental organizations—business enterprises, pioneers of current-day globalization—were the first entities to seize the opportunity to identify common goals and present them to governments. Several dozen chief executive officers of leading corporations with significant positions in transatlantic trade and investment met in Madrid at the Transatlantic Business Dialogue (TABD) in the run-up to the December 1995 US–EU summit. They presented the summiteers with an agenda for attaining the transatlantic marketplace referred to in the NTA. In doing so, they established the precedent for other civil society dialogues. A TABD infrastructure of working groups under a rotating Euro-American co-chairmanship rapidly emerged. Business and government gained an attractive additional channel of communication for organizing and managing the world's densest trade and investment area bilaterally and vis-à-vis significant others, particularly the World Trade Organization (WTO). Perhaps its signal achievement was the negotiation of a mutual recognition agreement (MRA) worth about $50 billion. That agreement, geared to removing technical constraints on trade, is notable for the regulatory cooperation between the US and EU, a field in which the issue of sovereignty is especially pronounced.

A second class of NGOs—those purporting to represent public interests, particularly those connected to global processes and institutions such as migration and refugees, health, and the environment—had evolved rapidly in the 1980s. Transnational cooperation among environmental NGOs was desirable because the problems to be solved transcended national boundaries. This cooperation was mode possible by new communication technologies, English as an accepted *lingua franca*, and multilateral meetings held mostly under the organizational umbrella of the United Nations. Transnational cooperation between NGOs and parliamentary green parties (from opposition benches) in Europe concerning development aid projects go back to the 1980s. Beyond a doubt, the signal event in NGO presence was the June 1992 Earth Summit (UN Conference on Environment and Development, or UNCED) in Rio de Janeiro, out of which came Agenda 21, the Framework Convention on Climate Change, and the Convention on Biological Diversity. The North–South process generated by UNCED was global and it clearly implied the need for coordination of public policy among sovereign states to meet the challenge of reconciling tensions between trade and investment and environmental protection.

The same tensions were carried over to the trade arena in the ongoing GATT Uruguay Round, concluded at Marrakech in 1994. Imbued with an ethos of limits, for many environmental activists Marrakech was a "betrayal" of Rio, while for the United States, the two conferences represented action in different spheres that could be mutually supporting.

Originally, the concept behind the NTA's people-to-people dimension was more or less direct contact between active citizens engaged in projects of interest to their counterparts on the other side of the Atlantic— i.e., a transatlantic community of sharing, learning, and, wherever appropriate, emulation. This vision suffused a May 1997 meeting in Washington cosponsored by the US State Department and the Dutch EU Council presidency. Attended by more than 300 American and European representatives of nongovernmental organizations who ranged from educators to foundation executives, news media jockeys to vocational education officials, there was no particular environmental focus, and environmental NGOs were scarce at that event. However, two of the meeting's four discussion groups—those on "civil society" and "partners in a global economy"—clearly implied a role for NGOs—and by extension, environmental NGOs. The early products of this process featured innovations in information exchange, including a digital library project and an attempt to organize a mega-website for citizen-to-citizen interaction, launched as TIES (Transatlantic Information Exchange Service) at the Washington US–EU summit in December 1997. TIES offered an environment section as one of its initial foci.

TAED and the Trade Agenda

Despite this background, the TAED did not, in fact, come into existence primarily as the multifaceted process envisioned by those espousing the concept of a transatlantic learning community. Rather, the critical factor and third process at work in the origins of the TAED came from government in the form of a desire to communicate with groups that resisted the administration's trade agenda.

New trade issues became contentious for a variety of reasons, two of which deserve mention here. First, the US discovered the same set of problems the Europeans had encountered since the 1960s in creating markets between sovereign jurisdictions: barriers to trade and investment often took the form of legitimate exercises of authority in furtherance of public goods. A new type of trade authority and a new negotiating style were required to cope with this reality. When harmonization of legislation was no option, mutual recognition agreements were considered. Given the level of detail required to specify the domain of such agreements, they are difficult to technically define and manage. In addition, the sheer complexity of national regulatory regimes reflects the unique economic, social, political, and cultural histories of the participating jurisdictions. That being the case, matters that are important in one jurisdiction may not be important in another, leading to a problem of establishing consensus on what should be included and what excluded from the negotiating agenda.

Moreover, MRAs require a high level of trust between jurisdictions with respect to functional equivalencies in the level and scope of protection, a matter that touches upon the second dimension of contention. A new class of regulations associated with newly defined societal interests—concerning in particular environmental and consumer affairs—gained a lively and well-informed as well as potentially broad-based political clientele. Its NGO representatives defended consumer and environmental protection embodied in these regulations and sought to extend it. They were skeptical of revisions that accompanied trade liberalization efforts in the post-Uruguay setting. The broader point is that governments needed to generate public support for initiatives that skirted the margins of their legal authority. In light of the complexities involved, it is unlikely that such support would arise spontaneously. It had to be organized.

Political arithmetic at the national level also may have played a role in the Clinton administration's initiative. After the November 1994 federal elections, the Republicans controlled both houses of Congress. This outcome reinforced a verity of governance in Washington. Even more than before the elections, Democratic President Clinton had no choice but to govern from the center. The characteristic sharing of power across institutions in the American system of government is constitutionally specified in international economic affairs. Congress is also centrally involved in the entire panoply of domestic legislation associated with health, safety, and environment regulations that comprise an important part of the terrain of the post-Uruguay trade agenda. In 1994, the Clinton administration successfully carried the North American Free Trade Agreement (NAFTA), but with many more Republican than Democratic votes in the Senate. In that context, it was imperative that any new international trade agreement be supported by constituencies normally closer to the Democrats, prominent among them organized labor and the environmental lobby. In the politically polarized setting in the last half of the 1990s, President Clinton's attempts to get fast-track authority proved futile. All the more reason, then, for involving labor and environment NGOs in the policy development process as a critical dimension of the administration's strategy to advance trade with public support.

In the winter of 1998 the German Marshall Fund provided a small grant to the National Wildlife Federation (NWF), one of the largest US conservation groups, to explore the possibility of organizing a transatlantic dialogue on environmental policy issues from the American side. A US Information Agency grant of $100,000 was offered to facilitate the process, to be matched by a similar grant to the European Environmental Bureau (EEB) from the European Commission. The NWF initiative was met with skepticism by the mostly Washington-based NGOs, who feared a public relations exercise that would bring little change in government positions. Nine organizations—five from Europe, four from the United States—agreed, in effect, to sponsor the experiment by accepting an invitation to serve on an interim steering committee charged with defining the enterprise and expanding the circle of participants.

On the EU side, funding from the European Commission flowed to the EEB as the umbrella group in the best position to organize the participation of European NGOs. It embraced the dialogue idea with more enthusiasm than the American counterparts, hiring a professional environmental activist to organize it. Soon, other EEB full-time staff began to integrate the TAED into their work schedules. Eventually, an additional individual was hired to attend to the TAED's communication strategy in a collaborative venture with TIES-Environment.

TAED Organization and Process

Toward the end of 1998, planning began in earnest. The economic side of the NTA morphed into a program for Transatlantic Economic Partnership (TEP) that took as its main inspiration the agenda of TABD. On the other hand, a "red-green" government had just been installed in Germany, which was to assume the role of EU Council President in January 1999. The steering committee presented TAED-I as the NGO response to the challenge posed by the transatlantic agenda advanced to that point by business-government collaboration. At least some of the organizers hoped to be able to use the TAED as an additional channel to organize pressure on governments, especially in situations in which the NGO sector could work with one government against the other—e.g., in the trade area, to move toward more or less consensual NGO positions on agricultural subsidies (Common Agricultural Policy) or labeling of genetically modified organisms (GMOs) by the EU or US government.

After constituting itself, the TAED NGO plenum gave itself a limited mandate. It defined rights to participation broadly, extending an invitation to any nongovernmental environmental membership-based group. Henceforth elected by the plenum, the steering committee was composed of roughly equal numbers of US and European representatives. It was given a relatively narrow mandate to coordinate rather than initiate.

If the steering committee defined its mission in terms of resisting the spirit of the TEP, it also wanted to rescue the TAED from being entirely reactive to the TEP agenda. Working with the EEB-based TAED secretariat, it sought to provide the basis for adumbrating an alternative, ecological vision of global affairs through its choice of working groups. Five working groups were planned:

- climate change
- biodiversity
- multilateral trade and environment
- agriculture
- industry

Most of the NGO phase of the May 1999 TAED unfolded in the working groups, which produced statements that were conveyed to EU and government representatives during the rather formalistic second phase at the Centre Borschette. Those statements were, by and large, cautionary with respect to trade liberalization and supportive of EU concepts, especially in the area of agricultural biotechnology and farming more generally, and resisted further market encroachment (multifunctionality of agriculture for the latter, GMO bans or labeling for the former). There were hortatory statements about consumption preferences (gasoline) in the climate change statement. In return, the governments and EU pledged commitment to the process.

Developmental Dynamics

The path taken by the TAED revolved around several factors, of which the following played an important part:

Differences in the Environmental Issue Agenda

The European agenda, at least by comparison with the US, is embedded in a broad development strategy that enjoys significant support across the political spectrum. The EU level of governance plays a specific and creative role in fashioning regional responses to perceived environmental challenges. Concepts articulated in organic documents such as the Single European Act, the (Maastricht) Treaty on European Union, and the Amsterdam Treaty provide guidelines for policy development. They include the notions of policy integration and the precautionary principle, and more broadly, the sustainability language that found its way into the treaties.

In contrast, American NGOs and the Europeans encountered the American reality of a conflicted relationship between the president and the Congress in the US government. In America's pluralist society, the shifting configurations of interests produce different outcomes in different issue areas. There is no development plan or general vision assigning a specific role to environmental desiderata.

NGO Organizational Asymmetry

Continuing integration in many policy domains normalizes the pooling of sovereignty in Europe. European NGOs are becoming more adept at and at ease with operating in that kind of setting than their American counterparts. As a result, the Europeans have begun to create an organizational infrastructure designed to handle interjurisdictional and/or transnational coordination. The European Environmental Bureau is the clearest example. This organization played a central role in managing the TAED as a whole. It provided a natural locus for organizing the enterprise, and its leadership willingly took on the task, devoting significant resources and guidance.

The EU is a source of a more or less coherent environmental program. In its role as an EC-level interlocutor for environmental interests, the EEB is encouraged to react in the aggregate, holistically, as it were, to the entire panoply of environment issues. In contrast to the Europeans, the Americans do not possess an EEB-like mechanism; the American NGO sector is fragmented, each entity establishing its own priorities, lobbying effort, and funding sources.

One result of this organizational asymmetry was the piecemeal and defensive posture of American participation in the TAED. EU precepts served as de facto default positions:

- In agriculture, ambiguities in the EU "multifunctionality" concept went unexplored.
- On the biotechnology side of agriculture, the precautionary principle went unchallenged.
- Little attention was given to the political situation in the United States regarding climate change policy.

Participation Gap

In terms of membership and resource base, there was an asymmetry between the American and European contingents. The Europeans included representatives of the World Wide Fund for Nature (WWF), the EEB (representing well over a hundred groups), the German League for Nature and Environment (Germany's largest environmental NGO), Friends of the Earth-Europe, and the Transnational Institute. The American contingent included, in addition to the National Wildlife Federation, the US Climate Action Network, Defenders of Wildlife, and the Biodiversity Action Network.

To a considerable extent, this gap appears to be related to systematic differences in the links between NGOs and the state on either side of the Atlantic. While the American pattern is arms-length relationships, the European pattern follows the tradition of close consultative and financial relationships. The NGO community operated on the basis of the expectation of quasi-entitlement to funds at least to subsidize, if not fully finance, its participation. And European NGOs were not disappointed in this expectation. They received more or less open-ended funding on a continuing basis from the European Commission, a commitment corresponding to a line item introduced into the EU budget by the European Parliament. At least three influences were probably at work here. First there are the general inclusiveness and programmatic orientation that characterize the EU policy style. Second, to this general behavioral template must be added the specific mandates of the Fifth Environmental Action Program, "Towards Sustainability," which explicitly calls for a participatory approach. And third, there is the political weight assigned to the relationship with Washington. The European pattern encourages NGOs to find their place and cooperate with each other in a semiofficial consultative framework. Finally, the special opportunities afforded to political entrepreneurs in the EU multilevel system of governance may create additional incentives for NGOs in the Brussels system.

None of this applies to the American NGO scene, where organizations operate in a pluralistic, even slightly anarchic framework and networking is episodic and campaign based. This framework breeds a class of protest organizations that have no responsibility for designing and implementing policy. Funding is competitive and private, producing strong incentives to operate autonomously and few incentives to coordinate. This system creates an atmosphere of distrust between the US government and NGOs and among NGOs that becomes self-reinforcing as positions are sharply defined and polarized, partly as a matter of organizational marketing to gain potential supporters in a crowded field. On the other hand, as long as NGOs find favor in the funding pool, their voice will be heard.

Lack of Interest in International Issues

More pronounced in the US than in the EU, the lack of interest in international issues reinforced the disinclination to devote significant resources to the TAED. It also produced a reactive, defensive posture. In the US case, it probably derives from two interacting influences: the historically large role played by domestic conservation organizations and the nationally focused regulatory process designed to deal with pollution. In contrast, European integration revolves to a significant extent around reconciling different regulatory systems; in that sense, it is more outward looking. On the other hand, solutions reached through the ponderous Brussels policy-making process are scarcely negotiable beyond the EU. In fact, they create facts that are as intractable as anything produced in Washington.

NGO Preference for Global, not Transatlantic, Focus

The preference of NGOs for a global focus reflected significant ambivalence about the purpose of the TAED. The NGO community on both sides of the Atlantic would not contemplate substituting a transatlantic for a global approach because they doubted that solutions to international issues could be forged at that level. Moreover, they sensed that decisive negotiations on many issues would occur in global forums. That being the case, decisions about the commitment of organizational resources were made in favor of the latter. Therefore, despite the interest shown in the TAED process by the governments, NGO participants would not make TAED a central strand of their activities. Theirs was more a defensive than a transformative presence. This manifested itself inter alia in how much authority to give the NGO steering committee and who to send to represent the big organizations.

Demise

On November 21, 2000 the information arm of the TAED announced the suspension of its activities owing to lack of funding. Other e-mail traffic referred to the hold put on the Clinton administration's request for about $100,000 to support the US side of the TAED process by a Senate committee in deliberating the FY 2001 budget. Allegedly, funds were available in the twilight of the lame-duck government, after the presidential contest was decided in favor of the Bush campaign, but were turned down by the US environmental groups. The Bush administration made no effort to revive the TAED as such, but in the run-up to the US–EU summit in Stockholm it did invite the TAED to nominate individuals to be present at that event, scheduled for June 2001. Coming just weeks after the US withdrawal from the Kyoto Protocol and within days of the chaotic Genoa G8 meeting, this offer was spurned. In short, the unraveling of the TAED appears to have been an affair of American NGOs.

A major factor in the declining viability of the TAED was the mobilization of antiglobalization protesters. Trade liberalization has traditionally called forth protest from sectors that would encounter stiffer competition. What was new in the 1990s was that regulations and legislation not motivated by trade concerns at all were increasingly subject to review by international bodies. Many regulations concerned environmental goods. Their potential review caused alarm. Several cases involving international tribunals (e.g., the tuna-dolphin and sea turtle cases) highlighted the concern (see chapter 9). In the mid-1990s, NGOs campaigned against the Multilateral Agreement on Investment (MAI) sponsored by the Organization for Economic Cooperation and Development, which turned out to be a dry run for the WTO ministerial meeting planned for Seattle in December 1999.

Antiglobalization mobilizations merged with electoral politics in 2000 to narrow the NGOs' room to maneuver in consultations with government. Partisan politics highlighted issues that jeopardized their support on such issues as labeling of GMOs and finalizing the Kyoto Protocol to the climate change convention. That combination encouraged the NGO community in the United States to back away from the political arena until the election was settled. When it was, the positions of the new administration diverged too much from those of even many moderate NGOs to sustain a dialogue in the short term. By then, the TAED was already history and European counterparts were in no position to come to the rescue.

On the part of the NGOs, electoral politics may have been perceived as relevant to their strategy, but this would have been a misreading of the results of the 2000 election. Certainly, the campaign of Ralph Nader, nominated as the presidential candidate of the Green Party at its midsummer convention, took energy from and in turn energized antiglobalization mobilizations that followed the Seattle WTO ministerial meeting in the spring and summer of 2000. But it would be questionable to conclude that the Nader vote determined the election. The Nader electorate came to less than 3 percent nationwide and probably represented the net increase in turnout over 1996. And while it tightened races in several states, it was nowhere decisive. Rather, the reduction of the Reform Party from 6 percent in 1996 (and from nearly three times that amount in 1992) to a vanishing magnitude—owing to voters' abandonment of Patrick Buchanan—probably accounts for the Democrats' electoral quandary. A large portion of that swing vote, which accounts for candidate Clinton's victories in 1992 and 1996, presumably went to the Republican candidate instead of the Democrat in 2000.

However, the perception of mass antiglobalization radicalization probably affected the strategy of NGOs after Seattle. Street violence seemed to suggest that the strategy of NGO participation had not succeeded or had become too risky. At the very least, in 2000 the tension between more moderate and more militant NGOs increased in the United States.

Beyond the evolving political environment, other factors were probably at work in hastening the demise of the TAED. The question naturally arises as to the willingness of the NGOs to pick up the financing. They did not, and the project unwound altogether in 2001.

Conclusion

The TAED was an insiders' game, dominated from the start by part of the narrow segment of the NGO community concerned with managing international campaigns. It was far removed from the vision of a transatlantic learning community revolving around sharing local experiences drawn from different settings.

Reflecting its origins as an outreach mechanism of the governments' trade agenda and the splintering of consensus on "trade," the TAED was constituted as a kind of NGO consultative assembly. As such, it was polyphonic, although not inchoate. At one and the same time its participants were preoccupied with the transatlantic dimension of post-Uruguay international economic issues raised in the NTA and global issues from the Rio agenda, particularly global warming and biodiversity from the point of view of agricultural biotechnology.

The TAED delivered several goods to the participants. It educated a segment of environmental activists about some of the problems and priorities of their counterparts across the Atlantic. The Europeans were eager to report to American colleagues the Brussels lobbying activity of American business aimed at preventing various pieces of EU environmental legislation from being adopted or watering them down. The Americans were able to evaluate the workability of alternative solutions to environmental problems they share. Plenary discussions revealed the orientations of the participating organizations. Working groups afforded opportunities to discuss particular issues. Writing the texts of communiqués helped demonstrate shared interests and suggested common strategies. The exercise also included encounters between government officials and NGOs that revealed and subjected positions to debate.

Beyond positions on specific issues, the participants had an opportunity to observe how the policy process worked and how the NGO sector connected to it.

At both NGO and government levels the TAED revealed a certain asymmetry of approach that should be factored into future attempts to organize exchanges of views and common action. The Europeans came with an implicit development model of ecological modernization and through EU integration had developed an international environmental agenda. American environmentalists came as individual campaigners. And although America has an environmental vocation, its global responsibilities include a military dimension that is as important to it as the environment is to Europe. Far from being mutually exclusive positions, they are actually reinforcing.

Differences in the European and American systems of interest intermediation account for some of the asymmetry. The US–EU encounter in the TAED was approached by one side in terms of a set of individual interests and by the other with a prearranged equilibrium of interests. As is clear from the funding received by the EEB and programs that support individual environmental NGOs throughout the EU, the environmental community is part of the inclusive and programmatic decision-making apparatus in Brussels. Ongoing EU funding amounts to a condition of participation, and this encourages orientation to the long term. Guaranteed participation also allows the EU system to balance interests, e.g., business and NGOs, that are divergent in the short term.

Measured by American associational practices, the TAED was a novelty. American NGOs are self-reliant entities and active state orchestration does not exist. Each legislative act is a separate battle; little long-term strategy emerges. The grant enabling US NGOs to participate in the TAED was strictly limited. Funds would have to be diverted in NGO budgets or raised anew for further participation. Faced with the plurality of interests seeking representation in Congress, the executive had resorted to the deus ex machina of "fast track" authority in trade legislation. Certainly one of the reasons that such authority has been elusive since the ratification of the Uruguay Round in 1994 is that the winner take-all process undermines the kind of consensus required when trade liberalization means altering domestic legislation designed to protect the environment, workers, and consumers.

Finally, regarding NGO abandonment of the TAED during the US election campaign and its aftermath, it is hard to escape the impression that in the context of the new administration's initial policy positions this proved to be a major error. In destroying the forum, the basis for a transnational coalition was also undermined. As a result, the NGO world in environmental questions has been transported back to the pre-TAED era of national self-referentiality. Indeed, it can be argued that the frustration on the street during G8, WTO, and international financial institution meetings manifests the lack of effective mediation of interests transnationally. From this point of view, the unwinding of TAED in 2001 amounts to a missed opportunity stemming from a failure of vision, a failure of attention, and failure of nerve on the part of the NGOs.

Notes

1. The author expresses his own views and in no way speaks for the US Department of State or the US government in these pages.

2. See the pathbreaking work of Francesca Bignami and Steve Charnovitz, "Transatlantic Civil Society Dialogues," in Mark A. Pollack and Gregory C. Shaffer, eds., *Transatlantic Governance in the Global Economy* (Lanham, Md.: Rowman & Littlefield, 2001, pp. 255–284.) While Bignami and Charnovitz consider the entire genre of transatlantic dialogues, this effort describes insights from one of them.

V

Conclusions

14

Conclusion: The Necessary Dialogue

Michael G. Faure and Norman J. Vig

In our introduction we raised the question of whether the United States and the European Union are in fact following increasingly divergent paths on environmental policy, and what the sources of these tendencies might be. We also asked whether there are common trends or areas of policy convergence in which the two green giants are learning from each other. Finally, we asked what might be done to improve the state of transatlantic dialogue and cooperation in this field. In this chapter we review the contributions to the book and attempt to draw some tentative conclusions.

We first consider the general debate in part I over convergence and divergence. It is noteworthy that the European and American contributors take very different positions on whether divergence is actually occurring, and even on whether this is a meaningful question. While Theofanis Christoforou and Ludwig Krämer argue that the US and Europe have been diverging on most issues since the 1980s, Wiener challenges the entire framework of convergence and divergence and posits a more complex evolution toward international "hybridization." This section sets the terms of discussion for examining the more detailed evidence in parts II and III of the book.

The chapters in part II suggest that the US and EU are both searching for new, more effective regulatory approaches and instruments, and here we find evidence that mutual learning and hybridization seem to be taking place. The chapters in part III make it clear, however, that there are deep policy differences over global issues, such as climate change, international trade, and sustainable development. Americans and Europeans do appear to think differently about global environmental obligations; we try to summarize some of these differences at the end of the section on international policy divergence. We then briefly examine the two efforts to establish new transnational environmental dialogues within Europe and across the Atlantic covered in part IV of this book. This is followed by a summary of some of the principal institutional and political sources of the policy differences observed throughout the book. Finally, we attempt to suggest ways in which the US and the EU might begin to rebuild their partnership around common interests in stabilizing the global environment.

Convergence, Divergence, and Hybridization

It is not an easy task either to trace the political dynamics that have led to current transatlantic differences over environmental policy or to reach firm conclusions about the extent of these differences. Comparison of large and complex legal and policy systems is fraught with danger, as Jonathan Wiener argues in chapter 3. His warnings are well taken. Nevertheless, we think it is possible to reconstruct a relatively coherent picture of how the dynamics of European integration have led it to adopt policies and approaches to environmental protection that have taken on a character and direction of their own that differs from that of the United States. At the same time, domestic political developments within the US have exacerbated tensions at key junctures, particularly regarding issues that require multilateral solutions. In general, as European economic and political integration has progressed, environmental policy has come to occupy an increasingly central place as a core legitimizing purpose of the Union. By contrast, in the US environmental policy has been highly sensitive to changes in presidential administrations and partisan control of Congress, resulting in rather abrupt shifts in the priority attached to environmental protection as well as sharp changes of direction in foreign policy.[1]

The first two chapters in part I present European perspectives on the specific sources of these differences. In chapter 1, Theofanis Christoforou traces the emergence and application of the precautionary principle as a foundation for European environmental policy. He argues that in the 1970s and 1980s the US and EC followed quite similar "precautionary approaches" to protecting public health (including environmental threats to human health) when scientific evidence on causes and effects was uncertain. He cites numerous legal cases in which such approaches have been upheld in US as well as European courts. However, Christoforou then shows that the precautionary principle (which was first invoked in Germany as the *Vorsorgeprinzip* in the 1970s) came to be adopted as a justification for EC environmental policy per se during the 1980s. Moreover in 1989, when the EC first banned growth hormones in meat, the Community began to restrict import of products it considered potentially unsafe, leading to claims by the US that such restrictions were disguised trade barriers. Christoforou argues to the contrary that EC standards are based on the current state of scientific knowledge (which is often highly uncertain) and are clearly intended to protect health and the environment. He admits that Europeans may be more risk averse than Americans because of past regulatory failures, industrial accidents, and food scandals, but he also argues that it is their fundamental right to enact higher health and environmental standards than their competitors if, after open and democratic public debate, they find the potential risks unacceptable. He further points out that the precautionary principle was formally incorporated into the Treaty on European Union in 1992 and has been upheld by the European Court of Justice in various cases involving national and EC environmental legislation that impose some restraint on trade. This "constitutionalization" of the principle creates a positive obligation to exercise precaution in all environmental policy. Thus Christoforou sees the EC as enacting progressively more restrictive environmental legislation than the US since about 1990.

In any case, whereas the EU has institutionalized the precautionary principle and now considers it a part of customary international law, the US "insists that it has no legal status, but is only an 'approach' that can be used in certain narrow circumstances."[2] The US thus opposed its inclusion in the Rio Declaration on Environment and Development and has only reluctantly, and usually with modifications in wording, accepted precautionary language in international treaties.[3] Christoforou argues that the reason for this stance is that the US fears that other countries will exercise their sovereign right to enact protective legislation that might restrict American economic trade and investment opportunities.

Ludwig Krämer gives an even more critical analysis of US motivations in chapter 2. Like Christoforou, he sees a pattern of growing divergence since the 1980s. He argues, first, that European environmental policy can only be understood as part of the larger dynamic of European integration. Thus, while the US has a constitutional structure enabling it to enact and enforce centralized environmental regulations from 1970 on, it took considerable time for the EC to develop a basis for Community-wide legislation that limited national sovereignty. Thus it was not until the Treaty of Rome was modified by the Single European Act, which entered into force in 1987, that the EC was in a strong position to act.

However, Krämer argues that even before 1987 the roots of transatlantic divergence had been sown by US opposition to the right of regional organizations such as the EC to participate as a unit in international diplomatic negotiations (the US insisting instead that only individual countries be recognized). This disagreement, which came to a head in negotiations over the Montreal Protocol, was regarded as an attempt to limit European influence, and was one of the reasons the Single European Act not only included explicit language granting competence to enact environmental law within the EC, but also authorized the Community to take an active role in promoting international environmental agreements. The result was a flood of new environmental legislation between 1987 and 1992 and outright conflict with the US at the Rio Summit.

Krämer thus sees a systematic divergence between the US and the EC since the mid-1980s that has continued to the present. The fact that the Clinton–Gore administration did little to reverse this trend convinces Krämer that deeper political and cultural factors are involved. He points to the influence of free-market ideology, the predominance of cost-benefit analysis and preference for market solutions, and the power of economic interests in international negotiations as the primary factors shaping US environmental policies. He cites a litany of cases in which the US government has appeared to put economic and trade considerations ahead of environmental protection; indeed, he sees US positions as almost totally captured by business (even in bilateral negotiations during the Clinton administration). As a result, he argues that Europeans no longer regard the US as seriously committed to advancing environmental protection, while in Europe there is an increasingly strong consensus (even within business) that maintaining the environment is an integral part of the larger project of economic and social integration. Whether this view of US policy is justified or not, it appears to be a widely held perception in Europe.

In chapter 3 Jonathan Wiener, a leading American scholar, responds to both Christoforou and Krämer by rejecting their arguments that environmental policies in the US and EU have followed an increasingly divergent course and that Europe has generally adopted more stringent policies since the 1980s. Part of his argument is based on what he views as the dangers of overly broad and simplistic comparison. He argues that while it is relatively easy to find differences over particular policies, it is necessary to look at the entire legal structure on each side to compare the effective regulatory context and the actual consequences of policies. He argues, for example, that under traditional tort law procedures and recent environmental statutes that allow citizen suits, polluters may be held to stricter standards in the US than in Europe. But, more generally, he argues that the concept of convergence versus divergence fails to capture the dynamic nature of ongoing developments on both sides of the Atlantic and the interactions between them.

Wiener holds instead that divergence, convergence, and "hybridization" are all taking place at the same time, often at different levels of government and in different policy sectors. He argues that a model of hybridization, involving continual interaction, borrowing, and evolution of standards and practices on each side, corresponds far more closely to the actual world of "complex, multinodal webs or networks, with multiple actors pursuing multiple directions at once and interacting across system boundaries in many places at once" than do one-dimensional concepts such as convergence and divergence. From this perspective, Wiener rejects the notion that the US system has stagnated or that the EU has universally enacted more stringent environmental policies than the US in recent years. Instead he sees a mixed pattern in which, for whatever reasons, each side has chosen to regulate certain kinds of risks more strictly than others. Thus, he claims, neither system is categorically more "precautionary" than the other; they are simply more or less precautionary about different things (including, he points out, the threat of terrorism). Furthermore, both sides use many of the same tools for environmental standard-setting, including risk assessment and cost-benefit analysis, and both are beginning to use market-oriented approaches which, Wiener argues in response to Krämer, are not necessarily preferred by business or beneficial to established interests. Although there are still differences in regulatory styles and procedures, these variations are also diminishing as a result of mutual learning and hybridization. Finally, while conceding there are differences over certain issues such as climate change and genetically modified foods, Wiener suggests that the widely held perception that the US and Europe have "flip-flopped" or undergone a role reversal in environmental leadership is simply false. Rather, he sees a continuing process of balancing some risks against other risks and the gradual evolution of standards and procedures on both sides as hybridization occurs.

We find Wiener's arguments compelling in part. We agree that comparison of regulatory outcomes is extremely difficult and that it is probably impossible to say whether, in looking at the entire context of environmental law and administration, the EU or US is more protective or "precautionary" than the other. We also agree that transatlantic and global interactions are constantly producing regulatory innovations and adaptations; indeed that is one of themes of this book. Wiener's concept of hybridization provides a very useful tool for analyzing these dynamic trends.

With that said, however, we do not think that Wiener's analysis necessarily contradicts the larger picture of policy divergence drawn by Christoforou and Krämer. It does appear that environmental protection has become a more conscious and integral goal in European policymaking than in the US. Wiener himself states that Americans distrust centralized authority and general policy mandates such as the precautionary principle more than do Europeans, who since 1957 have pursued ever higher levels of cooperation through a supranational pooling of sovereignty in the European Community. In this process of planned integration (especially since 1987), environmental protection has arguably been given greater and more consistent priority than is the case in the United States. The incorporation of the precautionary principle and the goal of sustainable development into the EU Treaty in the 1990s implies a deeper philosophical commitment than any recent administration in the US has been willing to make (see chapter 11). We will return to this theme later.

Regulatory Trends: Mutual Learning and Hybridization

Part II contains a diverse set of chapters that analyze common issues and trends in the implementation of environmental policy within the US and EU. They indicate that both systems of regulation have struggled with many of the same problems of policy effectiveness, giving rise to similar administrative strategies for improving policy implementation and enforcement and to parallel experiments with new policy instruments. These trends suggest a potential for mutual policy learning and hybridization that might reduce transatlantic regulatory conflicts in the future.

One area in which considerable research has been done in recent years involves the question of "environmental federalism," or how environmental problems should be regulated within federal systems.[4] Although it is easy to argue that there are fundamental differences in the "constitutional" structures of the US and EU (as Krämer, for example, docs), there are also many similarities from a comparative perspective. There is obviously a major institutional difference between the EU and the US in that the central authorities of the EU can only enforce Treaty law and secondary legislation against member states, not against individual offenders; whereas in the US federal agencies such as the Environmental Protection Agency (EPA) can directly enforce national law against private parties.[5] Despite these important differences, however, it can be argued that the "federal" nature of the EU as well as the US creates incentives for a similar allocation of authority among levels of government and for adoption of common strategies for ensuring that decentralized agents charged with implementing federal law will in fact do so.[6] Thus even though there is no federal office comparable to the EPA in the European Union, central authorities may use similar means to ensure compliance with Community policy.

This is essentially the thesis of R. Daniel Kelemen in his careful study of trends in environmental federalism in chapter 4. The rationale for central regulation is similar in both systems. Individual states cannot deal with transboundary pollution, and uniform standards are necessary to prevent a "race to the bottom" among competing states. In the EU, harmonization of environmental law was also necessary to avoid trade distortions as economic integration proceeded.[7] Once "federal" laws and regulations were established, however, EU authorities faced an "implementation gap" owing to reliance on member states to effectively implement these policies. In the US, federal legislation also required the EPA to rely heavily on the states for implementation. Both systems thus developed decentralized implementation strategies. The European Commission and the EPA have different tools at their disposal for this purpose. While on the surface the EPA has much stronger enforcement tools (legal authority to intervene in states as well as fiscal resources with which to encourage state cooperation), it came to rely heavily on litigation by private parties (e.g., environmental groups) to get judicial backing for enforcement. (However, recent Supreme Court rulings have placed limits on litigation against states.) Kelemen argues that the European Commission has also come to rely more heavily on infringement suits against states before the European Court of Justice and the potential for cases to be brought by private parties in national courts against states for failing to implement EU law. Although starting from different points, therefore, it appears that the regulatory styles of the two systems are moving toward hybridization.

Christoph Demmke also suggests that regulatory enforcement is moving in parallel directions in the EU and US, but for different reasons. He argues that legal enforcement based on deterrent sanctions has not been very effective in either system owing to the lack of information on compliance and the unwillingness of authorities to impose large fines and penalties. Because of the ineffectiveness of these traditional command-and-control approaches, both the US and EU are increasingly relying on more flexible and noncoercive methods of enforcement.[8] These include positive incentives and assistance to encourage companies to improve their environmental performance beyond statutory requirements in return for greater flexibility in choice of means and reduced regulatory supervision. Voluntary self-auditing and environmental management systems are now used quite widely on both sides of the Atlantic. Demmke concludes, however, that "it is still not at all clear whether cooperative and other flexible approaches are cheaper and more effective than traditional approaches." There is still a surprising lack of information on the effectiveness of different policy instruments. Hence it appears that a mix of traditional deterrence-based and newer performance-based regulatory tools will continue to characterize both systems. Nevertheless, the globalization of business and trade, the emergence of new transnational compliance networks, and the involvement of a greater range of nongovernmental policy actors are likely to lead to mutual learning and hybridization of regulatory practices as more information becomes available.

There has been extensive discussion of the use of new "second-generation" instruments such as environmental taxes, ecolabels, ecoauditing, and voluntary agreements to supplement traditional command-and-control approaches in Europe as well as in the US.[9] The Fifth Environmental Action Program (1992–2000) of the EU recommended expanded use of such instruments, and recent US legislation has introduced new approaches such as emissions trading. However, most experimentation with new policy instruments to date has occurred at the national (EU) or state (US) level rather than at the "federal" level. The EU itself has only begun to implement such methods, largely because it lacks the authority to impose taxes and cannot directly regulate polluters. In the US the EPA has experimented with a variety of new programs under the rubric of "reinventing government," but these initiatives have been carried out by administrative means and lack clear legislative authority.[10] Still, it appears that there is considerable potential for transatlantic policy learning in this area.

Grimeaud's analysis of "negotiated environmental agreements" (NAs) in the EU and US reaches some interesting conclusions in this regard. Although many of the ideas for voluntary agreements originated in Europe, the US appears to have learned its lessons and moved beyond the original models (an excellent example of hybridization). In fact, with the partial exception of the Netherlands, which pioneered the use of industrial covenants, European NAs in other EU member states cover only very limited sectors and fail to meet the guidelines for voluntary agreements set out by the European Commission.[11] The EU itself has accepted only a limited number of nonbinding self-regulation agreements, the most important of which is a 1998 agreement with the European, Japanese, and Korean automobile manufacturers to reduce CO_2 emissions from new cars. In the US, by contrast, the EPA has negotiated more than fifty Project XL agreements with individual firms. Grimeaud finds that the Project XL model is superior to most European agreements in that it requires legally binding, transparent contracts; demands higher environmental performance; has stronger monitoring and reporting requirements; and requires much more extensive public and stakeholder participation. Grimeaud thus suggests that Project XL provides a desirable convergence model, though again, data are scarce on the actual results of these projects.[12] Conversely, if the US wishes to move toward more voluntary regulation at the state level (as President Bush has proposed), the effectiveness of the Dutch covenants and other national contracts in Europe should be studied.[13]

Another area in which the Europeans may benefit from US experience concerns the EU Environmental Liability Directive that was proposed in 2002 after more than a decade of debate and controversy. In contrast to the previous White Paper on the subject, which would have authorized a broad range of citizen and NGO lawsuits in European courts against those responsible for damage or loss of environmental amenities, the new directive proposes simply that "qualified entities" in member states be allowed to require restoration and cost recovery for certain kinds of environmental damage, as does the Superfund program in the US. In chapter 7 Swanson and Kontoleon explain the change of focus in the directive and ask whether the EU "retreat" from the extension of liability monitoring and control to the general public is justified, based on experience with general environmental damage assessment in US courts.

In order to answer this question, Swanson and Kontoleon trace the development of litigation in the US dealing with award of compensation for extensive damage to natural resources resulting from such failures as oil spills. The major new issues arise over compensation for "nonuse values" (NUVs) to individuals who do not suffer traditional types of personal injury or harm, but who nevertheless are aggrieved by the loss or damage of environmental resources. Swanson and Kontoleon point out the practical difficulties of deciding what kind of compensation is appropriate in such cases, and question whether current techniques such as contingent valuation can provide a rational basis for determining compensation to diffuse categories of people who may have little prior knowledge or interest in the resource affected. They conclude that while these kinds of abstract losses would be better addressed by legislation than by individual liability suits, damage assessment techniques such as those pioneered in US courts may be required in extreme cases as a final (if imperfect) form of redress in Europe as well. It is thus suggested that US experience with nonuse valuation techniques may still be relevant in some instances even though the currently proposed directive shies away from this approach. Perhaps here again we see interesting potentials for mutual learning and hybridization.

Divergence on Global Issues

If trends in domestic environmental regulations show many parallels and tendencies toward convergence or hybridization, transatlantic differences over international issues do not. Although the policies of President George W. Bush have clashed openly with those of the EU on numerous issues, differences over international environmental policies have deeper roots. The chapters in part III of the book examine these divergences in detail.

Miranda Schreurs shows in chapter 8 that EU–US differences over climate change date from the beginning of negotiations on a framework convention in 1990. From the outset, the US rejected any agreement involving mandatory reductions in greenhouse gas (GHG) emissions on the grounds that scientific evidence of global warming was too uncertain and that much more research was needed to develop a mitigation strategy. Hence the US would adopt only "no regrets" measures that could be justified for other reasons, and urged other nations, including developing countries, to take voluntary actions. The EU, on the other hand, proposed specific targets and timetables for stabilization of atmospheric GHGs, and argued that the industrial countries should take the lead since they were the principal sources of the problem. The Framework Convention on Climate Change (FCCC) that emerged from the Rio Summit in 1992 reflected these disagreements. No binding targets and timetables were established, and developing country parties were exempted from obligations other than certain reporting requirements. Although US–EU differences narrowed somewhat during 1995–1997, allowing formation of the Kyoto Protocol, they continued to plague subsequent negotiations to the point of breakdown in 2000 and US withdrawal in 2001.

The EU's stance on climate change is, according to Schreurs, based on the precautionary principle. The US, on the other hand, has largely based its position on economic grounds, claiming, on the one hand, that control of GHGs would result in higher energy prices and other costs that would cripple the US economy; and on the other hand, that the US economy would be unfairly burdened unless developing countries were also required to limit their emissions. The Byrd–Hagel Resolution that passed by a 95–0 vote in the US Senate in July 1997 expressed these reservations, and they were reiterated by President Bush as justification for withdrawal from the Kyoto Protocol.[14] Europeans reject these arguments on the grounds that energy prices are already far higher in the EU than the US, and that developed countries are morally obligated to take the lead in GHG reduction since they account for most past and current emissions. The US argues in turn that developing country emissions are rising rapidly and that no regime that excludes them can be effective. It also argues that global warming is a long-term problem that cannot be prevented; rather, it will require a combination of human adaptation and technological innovation over the coming century.[15]

However, even if the US and EU could agree on goals and emission targets, they would still disagree over the most effective measures to achieve these objectives. The climate change negotiations have been deadlocked more often than not over allowable policy instruments and accounting mechanisms.[16] The US has argued consistently (in the Clinton administration as well as in both Bush administrations) for "flexible mechanisms" such as emissions trading, joint implementation, and the clean development mechanism that were ultimately permitted under the Kyoto Protocol. The Europeans have been far more skeptical of these market-oriented approaches and have tried to cap their use in meeting targets for emissions reductions. The US appears to have won this debate, but important differences remain between US preferences for nonregulatory approaches and European adherence to mandatory limits, timetables, and rules.[17] Whether the Kyoto Protocol can be implemented without the US remains to be seen.

Chapter 9 by David Vogel makes it clear that many of the same issues arise in US–EU negotiations over global trade rules. Although the Uruguay Round of negotiations on the General Agreement on Tariffs and Trade (GATT) which culminated in establishment of the World Trade Organization (WTO) in 1994 did not result in any special provisions for reconciling potential conflicts between the rules of trade and environmental protection, both the US and EU have formally supported efforts to clarify the relationship within the WTO's Committee on Trade and Environment (CTE). They disagree, however, over specific policies because both sides wish to maintain their own protective legislation. They also disagree on whether changes in GATT/WTO rules are needed to accommodate multilateral environmental agreements (MEAs) that allow import restrictions and trade sanctions. The US position is that judgments should be made on a case-by-case basis under existing rules on whether specific environmental laws and agreements violate the trade agreement or fall within the exceptions allowed under various protocols (e.g., the Agreement on Sanitary and Phytosanitary Measures, or SPS). The Europeans, on the other hand, have tried to amend WTO rules to explicitly safeguard the legal status of MEAs because they strongly support multilateral approaches to environmental problems and because particular treaties (such as the Biosafety Protocol) allow them to restrict imports or require labeling of certain products they consider unsafe. In effect, Vogel states, "the EU wants the precautionary principle to be incorporated into international trade law." The US prefers that disputes be judged under the existing SPS Agreement, which requires that trade restrictions for protection of health and natural resources be justified by scientific risk assessment. Again, the EU and US have different concepts of how decisions should be made and where the burden of proof lies.

A potential source of solidarity between the US and EU in global environmental politics is their relationship to the 150 or so developing countries of the South. The green giants share similar levels of industrialization, income, and technology; they both support trade liberalization; they both recognize the need for special economic aid and assistance to poorer countries; and in theory they both support sustainable patterns of development. However, as their positions on climate change policy already indicate, there are also significant divisions between the US and EU over the role that developed and developing countries are expected to play in addressing environmental problems. This becomes clearer when we look specifically at patterns of foreign aid and acceptance of the concept of sustainable development.

In his chapter on official development assistance (ODA) and burden sharing, Paul Harris demonstrates that although neither the US nor the EU have come close to meeting the goals called for by developing countries at the 1992 Rio Summit, Europe has done considerably better than the US. In quantitative terms, the US ranks dead last among major industrial countries in percentage of gross national product (GNP) devoted to ODA (less than one-tenth of 1 percent), while EU countries average more than three times this level and some (such as Denmark) give ten times as much in relation to GNP. European ODA is also more likely to be channeled through multilateral agencies, directed to the poorest countries, and earmarked for environmental or sustainable development projects than is US assistance. Moreover, the Europeans exhibit a much stronger sense of moral obligation to aid developing countries as a matter of fairness and equity. They have more fully embraced the principle of "common but differentiated responsibility" that was adopted as part of the Rio Declaration on Environment and Development as well as the FCCC, which implies that the developed countries have a greater degree of responsibility and financial obligation in addressing environmental problems than do poorer countries.[18] Harris states that although the Clinton administration recognized this idea in principle, both the elder and the younger George Bush have rejected it. Instead, Harris argues, the current Bush administration has attempted to shift the responsibility for global warming and other problems to the developing countries. By contrast, the EU has argued that it is unrealistic to expect developing countries to reduce GHG emissions until developed countries have demonstrated their willingness to do so, and has accepted the need for increased financial and technical assistance to help developing countries pursue sustainable development strategies.

In light of the foregoing it is not surprising to find that the US and EU also diverge sharply on acceptance of the concept of sustainable development—although Baker and McCormick argue that neither side has actually done much to implement sustainable development policies. Part of the problem, they argue, is that although the idea of sustainable development is not new, it still lacks concrete definition and practical application. Still, since the concept was formally adopted and incorporated into most of the documents at the Earth Summit in 1992, the EU has done far more to enshrine the concept in its environmental policy than has the US. Indeed, "sustainable development" was adopted as a fundamental objective of the Community in the Amsterdam Treaty revisions of 1997, and the EU has increasingly employed the concept to justify and coordinate its environmental policies, both internally and externally. By contrast, the rhetoric of sustainable development has had very little resonance within the US, especially at the federal level.

What is most disturbing in Baker and McCormick's analysis is their account of how little difference the principles of sustainable development have actually made in terms of practical policy applications. Certainly the EU has done more "on the ground" than has the US.[19] Beginning with the Fifth Environmental Action Program (1992–2000), the EU has launched a series of processes to integrate environmental sustainability into policy-making in other sectors: transport, energy, agriculture, industry, tourism, and fisheries. Policies to promote energy efficiency and develop alternative energy supplies have been adopted. Yet the authors conclude that basic patterns of consumption, transport, pollution, and so on continue to worsen, with few concrete results to show to date. Indeed, Baker and McCormick make the interesting observation that "the EU integration process [itself] continues to result in the encouragement, stimulation, and funding of obstacles to sustainable development." In the US, on the other hand, there is no general policy for sustainable development. Although President Clinton appointed a Council on Sustainable Development that met from 1994 to 1999, it was not taken seriously by Congress or other agencies of the government. Given this lack of support in Washington, the emphasis of the council was on local initiatives to make communities (neighborhoods, cities, counties, etc.) more "livable" and "sustainable"; and while there are many examples of these "microlevel" projects, it is difficult to point to concrete achievements.[20]

Nevertheless, Baker and McCormick argue that commitment to the *idea* of sustainable development is playing an important role in defining the identity of the EU, both internally and externally, and in legitimizing the entire European integration project. They argue that sustainable development reflects "a deep-seated ethos of collective social responsibility for the welfare of the community as a whole" that "has allowed Europeans to see environmental protection as part of the protection of the common good," and that the legal commitment to sustainable development "provides a framework for the reconciliation of ecological, economic, and social goals." This is especially important now that the EU is expanding to central and eastern Europe, an area that has heretofore lagged considerably behind in environmental policy. It also "allows the EU to act as a normative power (as opposed to military power) in international politics . . . a major difference with the US." Thus, despite limited policy achievements to date, it can be argued that the declaratory values of the EU will put it on an increasingly divergent course with the US in the future.

Transnational Actors

A final trend we see is the development of new transnational networks for environmental policy. These networks bring new actors into the process of formulating official environmental policy or engage "civil society" organizations in dialogue with each other and with government over future policy directions. In Europe, especially, there has been a proliferation of formal and informal networks as environmental policy has risen to the top of the EU agenda in the past decade. In the US, environmental NGOs have continued their traditional activities of fundraising, lobbying, and mobilizing public opinion around particular issues, but have been less likely to join together in broad campaigns to shape environmental policy or promote sustainable development. In part, this is a function of how NGOs relate to government in the EU and in the US. In Europe, interest groups are often subsidized by national governments or by the European Commission and are recognized as having a permanent role in the policy-making process. In the case of environmental groups, the EU subsidizes an umbrella coalition of more than

160 environmental NGOs in Brussels—the European Environmental Bureau—and has established various environmental policy committees and forums as part of the normal policy-making process.[21] However, initiatives have also come from transnational networks such as the European Environmental Advisory Councils described in chapter 12. Although there is no comparable body in the United States, the US did take the lead in promoting transatlantic policy dialogues between selected NGOs and government officials during the Clinton administration. The findings in chapters 12 and 13 bring home some of the basic differences between the roles played by NGOs in the US and in the EU.

Chapter 12 by Macrory and Niestroy in many ways illustrates the differences. In this case, official but independent government environmental policy advisory councils have formed an unofficial network to promote transnational dialogue and coordination on environmental policy questions. What is interesting, first, is the broad range of such councils found throughout not only western Europe but in central and eastern Europe as well. They include traditional scientific advisory bodies composed of experts, but also broader councils representing NGOs and civic stakeholders that advise their governments on sustainable development as well as specific environmental issues. These councils are also increasingly active on the regional level; e.g., in attempting to influence EU legislation and policy. The most dramatic example detailed by Macrory and Niestroy involved the drafting of a joint policy statement on "greening" the EU's sustainable development strategy that was prepared for the European Summit meetings in Stockholm in 2001. Overall, this experience seems to indicate not only a much greater priority for sustainable development issues in Europe than in the US, but also greater openness to diverse sources of expertise than is typical in the US, where advisory councils and committees are normally composed of scientific experts in specific fields.[22]

Chapter 13 provides a final example of potentials for joint policy learning. In this case the US government initiated the Transatlantic Environmental Dialogue (TAED) as part of the New Transatlantic Agenda (NTA), which was launched in December 1995. Lankowski points out that although "people-to-people" dialogues were intended in several areas of policy, including labor, consumer, and environmental affairs, business groups quickly took the initiative in establishing the Transatlantic Business Dialogue (TABD) as a forum for negotiations on regulatory cooperation. By 1998, when the TAED was finally started, its motive was less to promote genuine dialogue and understanding among citizen actors than it was to reassure NGOs that were increasingly unhappy with the role of business in shaping the transatlantic trade agenda. The TAED provided a unique forum for NGO representatives to discuss and endorse environmental policy statements that were conveyed to EU and US government officials, but produced no tangible results.[23]

More significantly, it brought out the differences between the agendas and capabilities of NGOs on the two sides. US environmental organizations appeared to have a much more difficult time in coordinating their efforts and in focusing on transatlantic issues than did their European counterparts from the European Environmental Bureau. This reflects the "arms-length" relationship that environmental NGOs have with the different branches of government in the US compared with the established consultative role they have with the European Commission, as well as the absence of a consensual environmental agenda on the American side comparable to that of the EU. These asymmetries, as well as those between the success of the business dialogue and the TAED, further substantiate the bias toward representation of economic interests in US foreign policy. They also suggest the need for much more substantial dialogue on environmental issues between the US and EU if policy conflicts are to be reduced.

Sources of Divergence

As suggested in the introduction, there are multiple sources of the observed divergences. Some are short term and idiosyncratic, such as the attitudes of individual US presidents; some are more permanent, such as institutional structures; and still others seem deeply embedded in legal and philosophical traditions. We agree with Wiener that both the US and the EU are dynamic systems, although the EU clearly has been changing more rapidly as a result of ongoing revisions of the Treaty of Rome. We also agree with Wiener that some of the divergence is more rhetorical than actual; for example, as Baker and McCormick point out, the

Europeans are much better at talking about sustainable development than are Americans, yet they have arguably not yet done much more to implement sustainable policies than the US has. With this said, however, we have noted divergent trends in the substance and objectives of several major policies—especially those requiring international cooperation—and we think the reasons for these different trajectories are important since they point to increasing divisions unless greater efforts are made to bridge them.

Political trends in the US have not been favorable to environmental policy, and especially international environmental agreements, over the past two decades. Presidents Ronald Reagan, George H. W. Bush, and George W. Bush have opposed most new domestic and international environmental policy measures and have insisted on strict cost-benefit tests for regulatory actions.[24] The elder and younger Bush have had an especially marked personal impact on US climate change policy. President Clinton was more favorably disposed to environmental policy, including support of the Kyoto Protocol, the Convention on Biodiversity, the Treaty on Persistent Organic Pollutants, and other international agreements. However, Clinton was limited by hostility in the US Congress to the Kyoto Protocol and other treaties, especially after 1994 when the Republicans gained control of both houses. The Republican Party has become increasingly conservative on environmental issues.[25] In contrast, virtually all political parties and leaders in EU member states supported these agreements and the expansion of EU environmental policies during the period after 1987. The European Parliament also contained strong proenvironment majorities during 1989–1999, which generally pushed the European Commission toward adopting stronger environmental measures.[26]

Aside from partisan and ideological differences, the sheer timing of policy developments on each side of the Atlantic has contributed to the perception of divergence. In the US the most active period of environmental legislation came relatively early (1970–1980) and was followed by a period of reevaluation and in some cases retrenchment of environmental policy during the Reagan and first Bush administrations (1981–1993). By contrast, the greatest legislative expansion in Europe occurred during construction of the single market (1987–1992) and carried over into the "constitutionalization" processes in the 1990s, when environmental policy was relatively stagnant in the US. Thus asynchronous policy cycles are part of the explanation for divergence despite offsetting tendencies toward transatlantic cooperation and hybridization.

Asymmetric institutional processes also appear to play a significant role in explaining divergent policy trends. In the US, divisions between the executive branch and Congress have obviously made policy agreement difficult. The same party controlled the presidency and both houses of Congress for only one brief period between 1980 and 2000 (1993–1994). The fact that treaty ratification requires a two-thirds vote in the Senate has been an almost insuperable obstacle to US ratification of international environmental agreements; indeed President Clinton did not even submit the Convention on Biological Diversity and the Kyoto Protocol to the Senate because of certain defeat. As Schreurs points out, the 95–0 vote on the Byrd–Hagel Resolution in 1997 doomed US participation in the climate change regime. Vogel and Harris also point to the restrictions placed by Congress on international trade and development assistance policy. The gridlock between the White House and Congress (and within Congress) has largely blocked enactment of domestic environmental legislation since 1990 as well.[27]

In sharp contrast, the evolving institutional structure of the EU has facilitated environmental policy. The Single European Act of 1987 formally authorized the EC to ensure a "high level" of environmental protection and to participate in international environmental agreements. It provided qualified majority voting for measures adopted by the council of ministers and gave the European Parliament an expanded role in legislation.[28] Subsequent treaty revisions have strengthened these provisions; e.g., most environmental legislation is now adopted by qualified majority voting by the council and under procedures giving the parliament co-decision powers.[29] At the same time, as Krämer points out in chapter 2, the European Commission (which formally initiates EU legislation) strengthened its role in the environmental field and increasingly represented the EU in international environmental forums.

One of the most important differences between the EU and US is that EU legislation is considered and adopted by councils of ministers representing each policy sector; hence, environmental legislation is enacted (with the assent of the parliament) by a council composed of environment ministers from the fifteen member states. As Krämer argues, this allows environmental policy to be considered separately from other legislation (e.g., economic and trade policy). And, once agreement is reached by the council and parliament, the commission and member states are bound by these decisions in external relations. In the US, by contrast, not only is domestic environmental legislation subject to all of the counterpressures from other interests in Congress, but foreign policy is considered separately and often deviates from domestic policy. The president has a great deal of control over which interests are represented since he is charged with conducting foreign policy. Krämer is certainly correct in arguing that American diplomacy is heavily influenced by concerns for business interests, as reflected in the fact that environmental negotiations are often conducted by the economic departments of the US government rather than by the Environmental Protection Agency or other environmental officials.[30]

An interesting variable to emerge from several chapters is the comparative role of interest groups or NGOs. Schreurs, for example, points to the strong opposition to the Kyoto Protocol from industry lobbies in the US, whereas in Europe business and industry groups have supported or at least not actively opposed climate change agreements. European economic interests appear to be more open to voluntary agreements, ecoauditing, and environmental management schemes than American businesses.[31] Environmental NGOs, on the other hand, appear to have more strongly institutionalized consultative roles in EU policy-making than do environmental groups in the US. Although business lobbies of all kinds are active in Brussels—and far outnumber those from any other sector[32]—environmental, consumer, human rights, and other "public interest" organizations appear to have more balanced representation there than in Washington.

Finally, environmental policy divergence undoubtedly reflects deeper cultural and philosophical traditions across the Atlantic. Europeans (and different nations within Europe) and Americans have different sensibilities regarding particular societal risks, as Wiener, Christoforou, and Vogel point out. Europeans and Americans also have different attitudes toward political economy and state intervention in private markets, as several authors emphasize. Harris argues, for example, that Europeans have a much stronger sense of obligation to aid developing countries than Americans, which in turn mirrors the more generous welfare programs within European countries compared with those in the US. American and European (except British) legal systems stem from different roots, which are reflected in European support for comprehensive rules and treaties, in contrast to the US preference for settling individual disputes under adversarial tort law procedures. Attitudes toward international law are especially different. The EU itself is the result of multilateral bargaining and agreement and a gradual pooling of sovereignty, a model that Europe is said to wish to extend to the world.[33] The US, by contrast, has long been suspicious of any international agreements or institutions that limit its freedom and sovereignty, a tendency that has been reinforced since the end of the cold war.[34] The EU's strong support for normative principles such as the precautionary principle, common but differentiated responsibility, and sustainable development to ensure intergenerational equity reflects a worldview quite different from the great power assumptions that appear to govern Washington, especially under the present neoconservative administration.[35]

Potentials for Cooperation

It should be emphasized that complete "convergence" of environmental policy and regulatory practices is neither possible nor desirable. Differences of legal and administrative traditions alone preclude transatlantic harmonization, but it is also the case that environmental problems differ in each region and that public concerns about particular risks vary considerably. As Wiener suggests, one of the most interesting questions is why different societies choose to regulate different risks more strictly than others.[36] Another reason for regulatory pluralism is simply that we do not know how effective different policy instruments are, and continuing experimentation is probably wise at this juncture.

Moreover, as Wiener points out, there is constant interaction between US and EU over the details of environmental regulation, and hence a good deal of mutual learning and hybridization is already occurring. Much of this occurs at the working level in normal business and government transactions, and it is likely that continuing trade liberalization will encourage further standardization and cooperation.[37] As first-generation command-and-control regulation gives way to second-generation "smart regulation" in the form of more flexible, market-oriented instruments, there is undoubtedly a great deal more that can be learned from a mutual exchange of information. It will be interesting to compare, for example, the success of the new European emissions trading system for carbon dioxide (p. 224) with emissions trading systems in the US.[38] As Demmke and Grimeaud indicate, the relative efficiency and effectiveness of new instruments such as these or the use of voluntary industry agreements is not at all certain. Issues such as quantifying measures of environmental damage for assessing liability and the development of concrete indicators of sustainable development require joint consultation and research. Further exchanges on methods of risk assessment, cost-benefit analysis, life-cycle analysis, and other techniques for improving the scientific basis of decision-making can only help. More carefully planned "civil society" dialogues such as the TAED could also prove useful in identifying points of common interest.

We are also concerned, however, with the deeper policy divergences in the transatlantic alliance. It has recently been argued that in the post-cold war, post-September 11 world, the strategic priorities of the US and Europe are starkly different; the two sides have different concepts of world order and consequently different goals and agendas.[39] The US is said to have achieved an unprecedented degree of military superiority and hegemony in the world, allowing it to do what it will without fear of countervailing force.[40] Its priorities are military and economic security, in which close relations with Europe seem less critical than protecting and enhancing US interests in other regions such as the Middle East, Asia, and Latin America. The Bush administration thus seems to have downgraded transatlantic relations and to view Europe as exercising only "soft" power that does not contribute much to the "hard" military mission of the US.[41] In this context, environmental policy ranks near the bottom of US priorities, whereas for the EU it has become a major component of foreign policy.

If this scenario bears some resemblance to reality, the chances for enhanced cooperation to address "third-generation" environmental issues—essentially global problems such as climate change, loss of biodiversity, deforestation, depleting fish stocks, shortages of fresh water, diffusion of toxic chemicals, the spread of human and animal diseases, and sustainable development generally—do not appear to be good. Without US and EU collaboration, it is unlikely that much international progress will be made.[42] On the other hand, cooperation is clearly necessary in the longer term. As even the prophets of "American primacy" put it,

Washington also needs to be concerned about the level of resentment that an aggressive unilateral course would engender among its major allies. After all, it is influence, not power, that is ultimately most valuable. The further one looks beyond the immediate short term, the clearer become the many issues—the environment, disease, migration, and the stability of the world economy, to name a few—that the United States cannot solve on its own. Such issues entail repeated dealings with many partners over many years. Straining relationships now will lead only to a more challenging policy environment later on.[43]

It is arguable, therefore, that the present tensions between the US and Europe may create new opportunities for reconciliation.

The Atlantic Alliance, which has been crucial for maintaining international peace and security over the past half-century, is at present in tatters over the US-led war in Iraq. However, the United States and Europe still share most basic values and will need to rebuild their historic partnership in the years to come.[44] We believe one of the primary avenues for rebuilding confidence in the transatlantic relationship could be in the field of environmental policy. Despite recent patterns of divergence, the US and EU have a unique opportunity to seize the leadership in addressing emerging global environmental problems. Indeed, these problems are increasingly coming to be seen as part of the international security context. In recently announcing a remarkable commitment to reduce carbon dioxide emissions in the United Kingdom by 60 percent over the next 5 decades, Prime Minister Tony Blair has stated that "there will be no genuine security if the planet is ravaged by climate change."[45]

The differing approaches of the US and the EU to global warming could provide the starting point for a new transatlantic dialogue. If the US is serious about lowering greenhouse gas intensity through scientific research and technological advances, it should make a credible case to the EU and the rest of the world. It has not done so to date.[46] The Europeans, in turn, need to spell out how the Kyoto targets and timetables are in fact to be implemented, what the costs are likely to be, and how they will actually affect climate change if no restraints are placed on developing countries. Both sides need to reexamine their financial obligations to the poorest areas of the world if population growth and demands for rising consumption are not to overwhelm local and regional environmental resources.

Rather than competing paradigms, US and EU approaches to economic growth and environmental sustainability could be considered complementary pillars. Research and technological advances by the US could provide part of the foundation for a new strategic partnership. European concepts of precaution, prevention, shared responsibility, equitable allocation of costs, cross-media and cross-sectoral policy integration, ecological modernization, and multilateral collaboration deserve careful and respectful consideration by the United States. Only if a new strategic transatlantic environmental partnership of this kind is forged at the highest levels of governance is it likely that genuine progress can be made in averting global environmental and social disasters in the coming decades.

Notes

1. See Norman J. Vig and Michael E. Kraft, eds., *Environmental Policy: New Directions for the Twenty-First Century*, 5th ed. (Washington, D.C.: CQ Press, 2003), especially chaps. 5 and 6.

2. David Hunter, James Salzman, and Durwood Zaelke, *International Environmental Law and Policy*, 2nd ed. (New York: Foundation Press, 2002), p. 408.

3. Principle 15 of the Rio Declaration is the most widely recognized statement of this principle: "In order to protect the environment, the precautionary approach shall be widely applied by States according to their capabilities. Where there are threats of serious or irreversible damage, lack of scientific certainty shall not be used as a reason for postponing cost-effective measures to prevent environmental degradation" (ibid., p. 406). Ironically, this wording is a compromise reflecting US influence. The principle is found in different forms in several recent international treaties, including the Framework Convention on Climate Change, the Treaty on Biological Diversity, the Cartagena Protocol on Biosafety, and the Stockholm Convention on Persistent Organic Pollutants.

4. For excellent overviews of this literature, see Richard L. Revesz, "Federalism and Interstate Environmental Externalities," *University of Pennsylvania Law*

Review 144 (1996): 2341–1416; and Revesz, "Federalism and Environmental Regulation: An Overview," in Richard Revesz, Philippe Sands, and Richard Stewart, eds., *Environmental Law, the Economy, and Sustainable Development: The United States, the European Union and the International Community* (Cambridge: Cambridge University Press, 2000), pp. 37–79. On the federalism debate in Europe, see Roger Van den Bergh, "Economic Criteria for Applying the Subsidiarity Principle in European Environmental Law," in Revesz et al., *Environmental Law, the Economy, and Sustainable Development*, pp. 80–95.

5. This has wide-ranging implications; for example, the European Commission cannot enforce European-wide air quality standards, but can try to get member states to enforce common emissions limits.

6. On principal-agent problems in federal regulation, see Giandomenico Majone, *Regulating Europe* (London and New York: Routledge, 1996), chap. 2.

7. Van den Bergh, "Economic Criteria for Applying the Subsidiarity Principle."

8. For a good survey, see Neil Gunningham and Peter Grabosky, *Smart Regulation: Designing Environmental Policy* (Oxford: Clarendon Press, 1998).

9. See, e.g., Richard B. Stewart, "Economic Incentives for Environmental Protection: Opportunities and Obstacles," in Revesz et al., *Environmental Law, the Economy, and Sustainable Development*, pp. 171–245.

10. See Walter A. Rosenbaum, "Still Reforming after All These Years: George W. Bush's 'New Era' at the EPA," in Vig and Kraft, eds., *Environmental Policy*, pp. 175–199.

11. Communication on Environmental Agreements [COM (96) 561, November 27, 1996].

12. See Alfred A. Marcus, Donald A. Geffen, and Ken Sexton, *Reinventing Environmental Regulation: Lessons from Project XL* (Washington, D.C.: Resources for the Future, 2002).

13. See E. W. Orts and K. Deketelaere, eds., *Environmental Contracts: Comparative Approaches to Regulatory Innovation in the United States and Europe* (The Hague: Kluwer Law International, 2000).

14. See, e.g., David E. Sanger, "Bush Will Continue to Oppose Kyoto Pact on Global Warming," *New York Times*, June 12, 2001.

15. Bush's program was announced on February 14, 2002. See White House, "Fact Sheet: President Bush Announces Clear Skies & Global Climate Change Initiatives," February 14, 2002; Eric Pianan, "Bush Touts Greenhouse Gas Plan," *Washington Post*, February 14, 2002; Andrew C. Revkin, "Climate Plan is Criticized as a Risky Bet," *New York Times*, February 26, 2002. In his 2003 State of the Union address Bush proposed a research program to develop hydrogen-powered cars in 15–20 years; see Danny Hakim, "Hydrogen Cars Remain Decades in the Future Under New Budget," *New York Times*, February 5, 2003.

16. See, e.g., David G. Victor, *The Collapse of the Kyoto Protocol and the Struggle to Slow Global Warming* (Princeton, N.J.: Princeton University Press, 2001).

17. This is not to imply that the EU doesn't favor the use of market mechanisms in other areas as well; see e.g., Pamela M. Barnes and Ian G. Barnes, *Environmental Policy in the European Union* (Cheltenham, UK: Edward Elgar, 1999); Jonathan Golub, ed., *New Instruments for Environmental Policy in the EU* (London: Routledge, 1998); and Duncan Liefferink and Mikael Skou Andersen, eds., *The Innovation of European Environmental Policy* (Oslo: Scandinavian University Press, 1997).

18. Principle 7 of the Rio Declaration states in part: "In view of the different contributions to global environmental degradation, States have common but differentiated responsibilities. The developed countries acknowledge the responsibility that they bear in the international pursuit of sustainable development in view of the pressures their societies place on the global environment and the technologies and financial resources they command." Quoted in Hunter et al., *International Environmental Law and Policy*, p. 402. For a full discussion, see Anita Halvorssen, *Equality Among Unequals in International Environmental Law: Differential Treatment for Developing Countries* (Boulder, Col.: Westview Press, 1999).

19. Some individual EU countries have done more than others; see William M. Lafferty and James Meadowcroft, eds., *Implementing Sustainable Development: Strategies and Initiatives in High Consumption Societies* (Oxford: Oxford University Press, 2000).

20. See Daniel A. Mazmanian and Michael E. Kraft, eds., *Toward Sustainable Communities: Transitions and Transformations in Environmental Policy* (Cambridge, Mass.: MIT Press, 1999); and Gary C. Bryner, "The United States: 'Sorry—Not Our Problem,'" in Lafferty and Meadowcroft, *Implementing Sustainable Development*, pp. 273–302.

21. This is not to say that environmental groups generally prevail over business organizations in EU policy formation; see Wyn Grant, Duncan Matthews, and Peter Newell, *The Effectiveness of European Union Environmental Policy* (London: Macmillan, 2000), chap. 2. On the role of EU interest groups in general, see Jeremy J. Richardson, *European Union: Power and Policy Making*, 2nd ed. (London: Routledge, 2001), chap. 11; and J. Greenwood, *Representing Interests in the European Union* (London: Macmillan, 1997).

22. See, e.g., Sheila Jasanoff, *The Fifth Branch: Science Advisers as Policymakers* (Cambridge, Mass.: Harvard University Press, 1994). Advice on high-level policy issues is usually dominated by political interests rather than scientific experts; e.g., the Bush administration primarily consulted business and industry in designing its energy and environmental policies.

23. See also Fancesca Bignami and Steve Charnovitz, "Transatlantic Civil Society Dialogues," in Mark A. Pollack and Gregory C. Shaffer, eds., *Transatlantic Governance in the Global Economy* (Lanham, Md.: Rowman & Littlefield, 2001), pp. 255–284.

24. Norman Vig, "Presidential Leadership and the Environment," in Vig and Kraft, eds., *Environmental Policy*, pp. 103–125.

25. Charles R. Shipan and William R. Lowry, "Environmental Policy and Party Divergence in Congress," *Political Research Quarterly* 54 (2001):245–263.

26. David Judge, "Predestined to Save the Earth? The Environment Committee of the European Parliament," in David Judge, ed., *A Green Dimension for the European Community: Political Issues and Processes* (London: Frank Cass, 1993), pp. 186–212. This may no longer be the case since the 1999 elections, which produced a center-right majority; see Grant, Mathews, and Newell, *Effectiveness of European Union Environmental Policy*, p. 36.

27. See Michael E. Kraft, "Environmental Policy in Congress: From Consensus to Gridlock," in Vig and Kraft, eds., *Environmental Policy*, pp. 127–150.

28. Philipp M. Hildebrand, "The European Community's Environmental Policy, 1957 to '1992': From Incidental Measures to an International Regime?" in Judge, *Green Dimension for the European Community*, pp. 13–44; and John McCormick, *Environmental Policy in the European Union* (Basingstoke, UK.: Palgrave, 2001), pp. 55–61.

29. McCormick, *Environmental Policy in the European Union*, pp. 61–68; Barnes and Barnes, *Environmental Policy in the European Union*, pp. 46–56.

30. However, sometimes it is in the interests of both environmental and industry groups to promote domestic rules and regulations overseas ("Baptist-bootlegger" coalitions). See Elizabeth R. DeSombre, *Domestic Sources of International Environmental Policy: Industry, Environmentalists, and U.S. Power* (Cambridge, Mass.: MIT Press, 2000). On the other hand, some US critics, including Wiener, have argued that the Kyoto Protocol was designed to favor European business interests; see Jonathan B. Wiener, "On the Political Economy of Global Environmental Regulation," *Georgetown Law Journal* 87 (1999):749–794.

31. See Marc Allen Eisner, "The Market Is Not Enough: The State, Corporate Environmentalism, and Self-Regulation in Comparative Perspective," paper presented at the annual meeting of the American Political Science Association, San Francisco, September 1, 2001.

32. On the overrepresentation of business in Brussels, see James A. Caporaso, *The European Union: Dilemmas of Regional Integration* (Boulder, Col.: Westview Press, 2000), pp. 68–69.

33. Robert Kagan, "Power and Weakness," *Policy Review* 113 (2002):17–18.

34. Recent Republican administrations have been especially contemptuous of international law; see, e.g., Burns H. Weston, "The Reagan Administration Versus International Law," *Case Western Reserve Journal of International Law* 19 (1987):295–302. See also Robert Kagan, *Of Power and Paradise* (New York: Knopf, 2003), pp. 76–84, on the post-cold war pursuit of American interests.

35. See Ivo H. Daalder and James M. Lindsay, *America Unbound: The Bush Revolution and Foreign Policy* (Washington, D.C.: Brookings, 2003).

36. On this question generally, see Mary Douglas and Aaron Wildavsky, *Risk and Culture: An Essay on the Selection of Technical and Environmental Dangers* (Berkeley: University of California Press, 1982); and Norman J. Vig and Herbert Paschen, eds., *Parliaments and Technology: The Development of Technology Assessment in Europe* (Albany, New York: SUNY Press, 2000).

37. See George A. Berman, Matthias Herdegen, and Peter L. Lindseth, eds., *Transatlantic Regulatory Cooperation: Legal Problems and Political Prospects* (Oxford: Oxford University Press, 2000) for details on regulatory cooperation in other fields.

38. See, e.g., Wolfgang Strobele, Bernhard Hillebrand, and Eric Christian Meyer, eds., *CO_2 Emissions Trading Put to the Test: Design Problems of the EU Proposal for an Emissions Trading System in Europe* (Münster: Lit Verlag, 2003).

39. See Robert Kagan, *Of Paradise and Power*, for the full argument.

40. Stephen G. Brooks and William C. Wohlforth, "American Primacy in Perspective," *Foreign Affairs* 81(4) (2002):20–33.

41. Kagan, "Power and Weakness."

42. Ivo H. Daalder, "Are the United States and Europe Headed for Divorce?" *International Affairs* 77 (2001):565.

43. Brooks and Wohlforth, "American Primacy in Perspective," pp. 32–33. Reprinted by permission of *Foreign Affairs*. Copyright (2002) by the Council on Foreign Relations.

44. See Philip S. Gordon, "Bridging the Atlantic Divide," *Foreign Affairs* 82(1) (2003):70–83; and William Wallace, "Europe, the Necessary Partner," *Foreign Affairs* 80(3) (2001):16–34, on shared values and common strategic interests.

45. Lizette Alvarez, "Blair Outlines Plans to Slash Emissions Over 50 Years," *New York Times*, February 25, 2003.

46. An expert panel of the National Academy of Sciences recently criticized Bush's research program on climate change, saying that it "lacks most of the elements of a strategic plan." Andrew C. Revkin, "Experts Fault Bush's Proposal to Examine Climate Change," *New York Times*, February 26, 2003.

About the Editors and Contributors

Susan Baker is Reader in the Cardiff School of Social Sciences, Cardiff University, Wales. For the academic year 2003–04 she will hold the King Carl Gustaf's Professorship in Environmental Science in Sweden. Her recent publications include *The Politics of Sustainable Development: Theory, Policy and Practice within the European Union*, co-edited with M. Kousis, D. Richardson, and S. Young (1997).

Theofanis Christoforou has been a legal adviser with the Legal Service of the European Commission since 1984. He holds a J.D. from the University of Thessaloniki, Greece (1976), an LL.M. from University College, London (1979), and an LL.M. from Harvard Law School (1989). He participated in the General Agreement in Tariffs and Trade Uruguay Round negotiations and has handled cases before World Trade Organization dispute settlement panels and the appellate body.

Christoph Demmke is associate professor for comparative public administration at the European Institute of Public Administration in Maastricht, the Netherlands. He is author of *Approaches in Implementing and Enforcing European Environmental Law and Policy* (2001), co-author of *European Environmental Policy: The Administrative Challenge for Member States* (2001), and co-editor of *Managing European Environmental Policy: The Role of Member States in the Policy Process* (1997), among other publications.

Michael G. Faure is a professor of comparative and international environmental law at Maastricht University, the Netherlands. He is also academic director of the Maastricht Institute for Transnational Legal Research (METRO) and attorney at law at the Antwerp (Belgium) bar. He is co-editor with J. Vervaele and A. Weale of *Environmental Standards in the European Union in an Interdisciplinary Framework* (1994); with R. Van den Bergh of *Essays in Law and Economics* (1989); and with K. Deketelaere of *Environmental Law in the United Kingdom and Belgium from a Comparative Perspective* (1999).

David J. E. Grimeaud (LL.M.) is currently working as a research associate at the Institute for Transnational Legal Research at the University of Maastricht, The Netherlands. He is writing a Ph.D. thesis on "Instruments for Climate Change Policy."

Paul G. Harris   is an associate professor of politics at Lingnan University, Hong Kong, and a senior lecturer in international relations at London Metropolitan University. Dr. Harris's books include *International Environmental Cooperation: Politics and Diplomacy in Pacific Asia* (2002); *International Equity and Global Environmental Politics* (2001); *The Environment, International Relations, and U.S. Foreign Policy* (2000); and *Climate Change and American Foreign Policy* (2000).

R. Daniel Kelemen   is University Lecturer and Fellow of Lincoln College, University of Oxford. He is author of *The Rules of Federalism: Institutions and Regulatory Politics in the EU and Beyond* (Harvard University Press, forthcoming) as well as recent articles in journals including *International Organization*, *Comparative Political Studies*, and *West European Politics*.

Andreas Kontoleon   is a lecturer in environmental economics in the Department of Land Economy at the University of Cambridge.

Ludwig Krämer   is head of the unit on Environmental Governance of the Directorate-General Environment in the Commission of the European Communities. He has worked with the European Commission on environmental law for 30 years. He is also honorary professor at Bremen University, visiting professor at University College, London, and lecturer at the College of Europe, Bruges. He has published several books, including *EC Environmental Law* (1999), *Focus on European Environmental Law*, 2nd ed. (1997), and *European Environmental Law Casebook* (2002), as well as more than 100 articles on EU environmental law.

Carl Lankowski   is deputy director of area studies and coordinator for European area studies in the Foreign Service Institute of the US State Department. He has taught at the School of International Service at American University in Washington, D.C. and has held visiting professorships in Germany, Denmark, and Belgium. Among other publications, Dr. Lankowski is co-editor with Alan Cafruny of *Europe's Ambiguous Unity: Conflict and Consensus in the Post-Maastricht Era* (1997).

John McCormick   is a professor and chairman of the Department of Political Science at the Indianapolis campus of Indiana University. He is author of *Environmental Policy in the European Union* (2001), *The European Union: Politics and Policies*, 3rd ed. (forthcoming), and other books on environmental politics.

Richard Macrory   is a barrister and professor of environmental law at University College, London, where he directs a new Centre for Law and the Environment. He is a member of the Royal Commission for Environmental Pollution and was chair of the steering committee of the European Environmental Advisory Councils (EEAC) in 2001–2002. Professor Macrory is also editor-in-chief of the *Journal of Environmental Law*.

Ingeborg Niestroy   has been the executive secretary of the Focal Point of the European Environmental Advisory Councils (EEAC) since 1999, currently located in The Hague. She has done research on strategic environmental assessment in California and Germany, and received a Doctor of Engineering degree from Berlin Technical University, Faculty of Environment and Society.

Miranda A. Schreurs   is an associate professor in the Department of Government and Politics, University of Maryland, College Park. She is author of *Environmental Politics in Japan, Germany, and the United States: Competing Paradigms* (2002) and co-editor with Elizabeth Economy of *The Internationalization of Environmental Protection* (1997).

Timothy Swanson   is professor of law and economics in the Department of Economics and the Centre for Social and Economic Research on the Global Environment (CSERGE) at University College, London. He is co-author of *Global Economic Problems and International Environmental Agreements* (1999) and editor of *Law and Economics of Environmental Policy* (2002).

Norman J. Vig   is the Winifred and Atherton Bean Professor of Science, Technology and Society, Emeritus, at Carleton College in Minnesota. He is author, co-editor, or co-author of twelve books, including *The Global Environment: Institutions, Law and Policy* (1999), *Environmental Policy: New Directions for the Twenty-first Century*, 5th ed. (2003), and *Parliaments and Technology: The Development of Technology Assessment in Europe* (2000).

David Vogel   is professor in the Department of Political Science and the Haas School of Business at the University of California, Berkeley. His books include *Benefits or Barriers? Regulation in Transatlantic Trade* (1999), *Kindred Strangers: The Uneasy Relationship Between Business and Politics in America* (1996), and *Trading Up: Consumer and Environmental Regulation in a Global Economy* (1995).

Jonathan B. Wiener   is professor of law and of environmental policy at Duke University, Durham, North Carolina; faculty director of the Duke Center for Environmental Solutions, and a University Fellow of Resources for the Future. He is a co-organizer of the Transatlantic Dialogues on "The Reality of Precaution: Comparing Approaches to Risk and Regulation," and is co-author, with John D. Graham, of the book *Risk vs. Risk: Tradeoffs in Protecting Health and the Environment* (1995). Professor Wiener has served on the President's Council of Economic Advisers and Office of Science and Technology Policy and worked at the Justice Department, in both the first Bush and the Clinton administrations.

Index

Note: Pages where chapters in the book begin are listed in boldface type by author(s). Commonly used abbreviations may be used in the index. For example:

CO_2 Carbon dioxide
EC   European Commission
EPA   Environmental Protection Agency
EU   European Union
US   United States
WTO   World Trade Organization

Aarhus Convention, UN, 126, 134n52
Advisory bodies. *See also* European Environmental Advisory Councils (EEACs)
Agency for International Development (USAID), US, 257
Agenda 21, UNCED, 256
Aggregation issues, 195–199
Agreement on Sanitary and Phytosanitary Measures (SPS), 23, 48n63, 234, 359
Agreement on Subsidies and Countervailing Measures, 232
Agricultural Directorate-General, EU, 284
Agriculture
agricultural subsidies, 28–29
sustainable agriculture, 285–286, 328n31
Air pollution *See also* Emissions
strict US domestic regulation of, 74, 87, 90
transboundary, 64
Alpharma Inc v. Council of the EU, 83, 95
American Trader case, 188
Amsterdam Treaty, 119, 282, 322
Angelopharm case, 31
Animal health and welfare regulations, 32, 65
Antibiotics use, 48n62
Anticommandeering principle, 123, 125
Antiglobalization movement, 340–342
Atlantic Alliance, 370
need to rebuild, 370–371
Audit policy, EPA, 149
Austria
Clean Air Commission, 310t, 312
environmental advisory councils in, 310t

Austria (cont.)
  greenhouse gas emissions trends in,
  226t, 227

Baker, Susan and John McCormick,
  277, 361–363, 364–365, 377, 378
Basel Convention on the Export of
  Hazardous Wastes, 26, 64, 240,
  249
Beef hormone ban, EU, 246
Belgium. *See also* Flanders
  environmental advisory councils in,
  308, 310t
  greenhouse gas emissions trends in,
  226t, 227
Benzene decision, 82–83
Bilateral environmental meetings, 60,
  66
Biodiversity
  EU efforts to sustain, 240, 250, 286
  Montreal Protocol on, 61–62, 240,
  350
  threats to, 295
  valuation of harm to, 190, 199–200
Biodiversity Action Network, 338
Biosafety, Cartagena Protocol on, 22,
  26, 65, 240, 359
Blair, Tony, 370
Bovine spongiform encephalitis
  (BSE), 24, 74, 83, 90, 91–92,
  327n17
Britain. *See* United Kingdom (UK)
Brundtland Commission, 277, 278,
  280. *See also* Sustainable
  development
Brundtland, Gro Harlem, 280
Btu tax, 213
Buchanan, Patrick, 341
Burden sharing, 253, 261–262,
  266–268, 360–362. *See also*
  Developing countries; Development
  assistance
  Bush on, 263–264
  Clinton on, 258, 261–263
Bush, George H. W., 81, 211, 262,
  365

Bush, George W., 2, 81, 92, 124,
  258–259, 268, 340
  and risk of terrorism, 92–93
  applying precautionary principle to
  terrorism, 92–93
  on burden sharing by developing
  countries, 254–255, 262
  energy plan focusing on supply,
  219–220
  position on climate change issues,
  211, 220–221
  withdrawal from the Kyoto
  Protocol, 2, 65, 208, 218, 219,
  249, 263–264, 271, 292, 358
Business sector, 159, 164, 350. *See
  also* Market-based approaches
  environmental proactivism in, 163
  industry-based coalitions and
  lobbying, 214, 367
Byrd-Hagel Resolution, 214, 358

California energy crisis, 219
Canada, 32
Carbon dioxide (CO_2) emissions,
  263–264
  reduction goals, 208, 210–211
Carbon sequestration, 219
Cardiff Process, 284
Carson, Rachel, *The Silent Spring*,
  53
Cartagena Protocol on Biosafety, 22,
  26, 65, 240, 359
Cheney, Dick, energy plan, 219–220
China, CO_2 emissions of, 263–264
Chirac, Jacques, 218
Chlorofluorocarbons (CFCs), 19,
  222, 234
Christoforou, Theofanis, 17, 85–86,
  89–90, 347, 349, 377. *See also*
  Divergence in US/EU environmental
  policies; Precautionary principle
Citizen involvement. *See also*
  Stakeholder involvement
  EU backtracking from White Paper
  position, 191–192
  in the EU, 143–144, 191–192

in international agreements, 178n11
in the US, 14, 143, 144, 152
Citizens' right-to-sue, 120, 180n38, 184, 191, 199, 351
Civil society, 268–269, 332
Claussen, Eileen, 225
Clean Air Act (CAA), 1990 amendments to, 75, 76, 211
Clean Water Act, US, 19
Climate Action Report 2002, US, 221
Climate Change Convention. *See* Framework Convention on Climate Change (FCCC)
Climate change negotiations, 99, 210–217. *See also* Greenhouse gas (GHG) emissions; Kyoto Protocol
EU/US divergence on, 65–66, 207, 209, 227–228, 254, 266–268, 370–371
leading to the Kyoto Protocol, 212–217
role of developing countries, 211–215, 222, 253–254
Clinton, William J. (Bill)
on developing countries and burden sharing, 258, 261–263
negotiated agreements under, 162–163, 171–172, 213
positions regarding climate change treaty, 213–216, 258, 365
promoting transatlantic dialogue, 1–2, 330–333
signing the Kyoto Protocol, 208
on sustainable development, 291–292
on trade-environmental linkages, 232–235
Cohesion Fund, EU, 128
Collaborative approaches. *See* Convergence in US/EU environmental policies
Command-and-control (CAC) regulation, 5, 159, 175n2, 268, 369
lack of data to enforce, 354

Commerce, US Dept. of, representing the US in trade matters, 57, 60, 62–63
Committee on the Trade and the Environment (CTE), 231, 233. *See also* World Trade Organization (WTO)
Common but differentiated responsibility, 261, 360
Bush rejection of, 264
Principle 7 of Rio Declaration, 373n18
Community funding for development, 260–261. *See also* Microlevel policy initiatives
Company environmental plans, 164, 170–173
Comparison, hazards of, 75–78
Comparative risk. *See* Risk assessment
Compliance incentives, 145–147
compliance assistance approach, 147–149, 151
compliance incentive approach, 149–151
economic incentives, 55, 95
Comprehensive Environmental Response, Compensation, and Liability Act (CERCLA), US, 183, 185
Congress, US, 7, 121, 236, 270, 343, 366
House of Representatives, 236–237
party shifts in, 6–7, 61, 341, 348, 365
Senate, 214, 235, 358
Conservation, 279
early US conservation movement, 289–290
multiple use and sustained yield policies, 290
Consumers, European, preferences, 36, 38
Consumption levels, 223, 293, 295
Convention on Biological Diversity, 26

Convention on International Trade in Endangered Species (CITES), 23, 59–60

Convergence in US/EU environmental policies. *See also* Hybridization model

early convergence of US and EU policies, 18–19, 348–349

future potential for, 368–369, 371

in risk management and standard setting, 86–90

Convergence/divergence models. *See also* Convergence in US/EU environmental policies; Divergence in US/EU environmental policies

critique of, 78–80

Coregulation, 144

Cost-benefit analysis, 37–38, 56, 86, 87, 224, 296

greater reliance on by the US, 41

scientific uncertainty and, 49n68, 68, 69

Cost-effective measures, 164

Costs of pollution abatement, 164

of precautionary and regulatory measures, 33, 38, 88

Council for Environmental Quality (CEQ), US, 305–306

Council of Ministers, EU

directives, 19, 183

role of, 12n1

Country Studies Program, US, 258

Court cases, EU, 31, 83–85

Alpharma case, 83, 95

Danish Bottles case, 24, 31

Hormone Beef case, 31, 47n51

Nestucca case, 196

Pfizer case, 83–85

Court cases, US, 20, 23, 98

American Trader case, 188

Exxon Valdez case, 187

Montrose Chemical Corp v. Superior Court, 196

New York v. United States, 123

Court of Justice, EU, 20, 24–25, 48n62, 91, 349. *See also* Legal implementation in the EU

Court of First Instance, 43–44n19, 83

infringement rulings, 222–223

principle of state liability, 127

Court system, US, 22–23, 187, 196, 368. *See also* Legal implementation in the US

Supreme Court, 82, 125, 354

Croatia, environmental advisory council in, 311t

Crosscutting requirements, 127, 128

Czech Republic, environmental advisory council in, 309, 310t

Dangerous substances, EC Council directive 67/548/EEC, 19

Danish Bottles case, EU, 24, 31

Davis, Gray, 225

Delaney clause (US), 19, 20

Demmke, Christoph, **135**, 354–355, 369, 377. *See also* Policy instruments

Democratic legitimacy issue, EU, 29–30

Democratic Party, 116–117, 121

Denmark

agreement on greenhouse gas emissions, 180n32

Compensation for Environmental Damage Act, 189

environmental advisory council in, 310t

greenhouse gas emissions trends in, 226t, 227

official development assistance, 257, 360

Deregulation, 56

Deterrent approaches, 150

Developing countries

environmental burden sharing issues regarding, 253, 259, 261–262, 266–268

role in climate change agreements debated, 211–215, 222, 254

Developing Country Climate Change
Initiative, US, 258
Development assistance, 212, 253,
360–362
community funding for, 260–261
levels of by developed countries,
255–257, 259–261
Diesel emissions, 88, 91
Disclosure requirements, 90–91, 95
Divergence in US/EU environmental
policies, 2–3, 17, 103n32,
347–348, 350–351
differing perspectives on global
environmental issues, 26–27,
66–68, 70, 205, 347
future considerations, 369–371
re: status of precautionary principle
in international law, 25–27, 40–42,
241–242, 268
sources of, 67–70, 364–368
Doha Round, WTO, 250
Dolphin-safe tuna dispute, 231–232,
233, 243
Domestic politics. *See* Political culture
Dutch covenants, 164, 170–171, 174,
180n35, 355–356. *See also*
Netherlands
later adopted in US court case, 98
as a model for European NAs,
171

EAPs. *See* Environmental Action
Programs (EAPs)
Earth Council, 292
Earth Summit. *See* UNCED
Eastern Europe, 3, 286–287
Ecoefficiency, 286. *See also* Natural
resources
Ecolabeling, 5, 25, 244–245
Eco-Management and Audit Scheme,
EU, 147
Economic assessment requirements,
56–57
Economic development, 322–323
Economic incentives, 95, 119–120,
127. *See also* Ecotaxation

Economic policy. *See also* Trade
agreements
of the EU viewed as linked with
environmental policy, 57–58, 66,
67–68, 71n2
of the US viewed as dominant over
environmental, 3, 57, 60, 62–63,
64, 66, 70
Ecotaxation, 69, 213, 223. *See also*
Economic incentives
EEAC network, 316–317
emergence of, 316–317, 327n23
"Greening Sustainable Development
Strategies" paper, 320, 321–322,
324–325
policy influence of, 321–322,
324–325
regional expansion of, 318–319
structure of, 320–321
Emergency Planning and Community
Right-to-Know Act (EPCRA), US,
167
Emissions. *See also* Greenhouse gas
(GHG) emissions
CO_2 emissions, 263–264
vehicle emissions, 19
Emissions reduction targets, 357–358
CO_2 emissions reduction goals, 208,
210–211
emissions caps issue, 217–219
greenhouse gas emissions national
inventories and targets, 217t, 226t,
263–264, 266
voluntary vs. binding, 63, 262,
263–264
Emissions trading, 5, 69, 94, 216,
224, 358
proposed EU system, 224
Endangered Species Act, US, 19
Energy Department (DOE), US, 293
Energy Framework Program, EU,
285
Energy taxes, 213, 223
Enforcement issues. *See*
Implementation of environmental
policy

"Environmental 2010" (EAP report), 161

Environmental Action Programs (EAPs), 162, 282–284

Environmental Council of the States, US, 117

Environmental federalism. See also Institutional factors
analysis of federal structures in terms of environmental policy, 113, 129–130, 353
constitutional structures of the EU, 117–119, 353, 366
constitutional structures of the US, 113, 116–117, 353
decentralized enforcement, 124–127
fragmentation of power and, 120–122
role of the states in US environmental policy, 113, 115, 225–226

Environmental groups, 163, 170, 373n21. See also Nongovernmental Organizations (NGOs)
status in litigation, 125, 127

Environmental impact statements (EIS), 95

Environmental liability. See Liability systems

Environmental Liability Directive, EU, 183

Environmental management systems, 147, 150

Environmental policy. See History of environmental policy development; Implementation of environmental policy; Institutional factors; Policy-making climate; Precautionary principle; Risk assessment; Trade-environmental linkages

Environmental Protection Agency (EPA), US, 19, 116
economic assessment requirements, 56–57
evaluation components of, 142
incentive policies of, 149
inspection and compliance monitoring of, 140
powers of enforcement against private parties, 67, 121, 353, 354
relationship with the states, 146
Success Management Report, 150

Environmental Valuation Resource Inventory (EVRI), 191

Epistemic communities, 7, 306

Estonia, environmental advisory council in, 308, 311t

Ethyl Corp. decision, 98

EU. See European Union (EU)

EU integration
environmental goals intrinsic to, 348, 351, 352, 365–367
incorporating the precautionary principle, 21, 28–30, 38–41, 81, 241–243, 349, 352–353
sustainable development themes, 281–289, 343

EU member nations
environmental policy implementation of, 138t, 139–140, 171
greenhouse gas emissions trends in, 226t
political culture of, 222–223, 270–271
violations of environmental laws, 142–143

European advisory councils. See European Environmental Advisory Councils (EEACs)

European Climate Change Program, multistakeholder approach of, 225

European Commission (EC), 12n1, 55
commitment to environmental policy development, 1, 282
Communication on the Precautionary Principle, 44n19, 99
dialogue with EEAC network, 317–318
Directorate-General for environmental affairs, 139
Forward Studies Unit (FSU), 322

global equity strategies suggested by, 264–265

international representation by, 58, 60

reforming the regulatory framework, 176–177n6

White Paper on Governance, 144–145

European Economic Community, 11–12n1, 71n2

European Environment Agency (EEA), 62, 81, 118

European Environmental Advisory Councils (EEACs). *See also* EEAC network

institutional diversity among, 213, 313f, 314, 317–318, 363

national councils for sustainable development (NCSDs), 308–309, 310–311t, 314, 319–321

European Environmental Bureau (EEB), 335, 337, 363

European Free Trade Association (EFTA), 232

European Network for the Implementation and Enforcement of Environmental Law (IMPEL), 140–143, 152–153

European Parliament, 7, 12n1, 121, 365

1999 elections, 374n26

ratification of Kyoto agreement, 219

European Union (EU), 12n1, 138t, 284. *See also* Environmental Action Programs (EAPs); EU integration; EU member nations; European Commission (EC); Single European Act (SEA)

balancing economic and environmental policy, 66–68

Council presidencies, 322, 332, 335

Eastern European participation planned in, 286–287

enforcement networks in, 140–143

Environmental Economics Unit, 322

greenhouse gas emissions reduction

goals of, 265–266

history of environmental policy development, 1, 54–55, 62, 117–119

implementation processes and powers, 139–140, 155n21

institutional constraints, 58–59, 139

international aid efforts, 259–261

Evaluation of negotiated environmental agreements (NAs), 165–168

Evaluation of policy implementation, 154n19

in the US, 141–142, 156n26

Executive Order 12291, US, 56

Executive Order 12866 on Regulatory Planning and Review, US, 98–99

Exxon Valdez case, 187

Fast-track negotiation authority, 235, 334, 343

Faure, Michael G. and Günther Heine, 142, 150

Faure, Michael G. and Norman J. Vig, 347

Federal Insecticide, Fungicide. and Rodenticide Act, US, 19

Federalism. *See* Environmental federalism

Fedesa case, 31

Feed additives, EC Council directive 70/524/EEC, 19

Feed/food regulations, 19, 26, 50–51n78

Finland, 138t

environmental advisory councils in, 308, 310t, 326–327n15

Fiscal incentives. *See* Economic incentives; Ecotaxation

Fischer, Joschka, 93

Flanders

decree on environmental covenants, 178n19, 179n25

Environment and Nature Council, 315

Flexible policy instruments, 145–146, 164, 173, 216, 354. *See also* Negotiated environmental agreements (NAs); Regulatory relief; Voluntary compliance
issue of emissions caps, 217–219
self-auditing, 354, 355–356
Food, Drug, and Cosmetic Act, US, 19
Food Quality Protection Act, US, 75
Food scandals, 28, 33
Foodstuffs labeling, EC Council directive 79/112/EEC, 19
Foreign aid. *See* Development assistance
Forest Service, US, 290
Framework Convention on Climate Change (FCCC), 81, 207, 211, 212–216, 217–219, 222, 262. *See also* Kyoto Protocol
France, 7, 138t
environmental advisory council in, 310t, 311
water protection policy of, 137–138
French water law of 1992, 180n30
Friends of the Earth, Europe, 338
Fuel economy standards, 247–248

G-8 meetings, 153, 340, 344
General Agreement on Tariffs and Trade (GATT), 63, 231–232, 233, 247. *See also* Uruguay Round
Generations of environmental problems, 5–6
Genetically modified organisms (GMOs)
EU restrictions on, 240, 247, 327n17
precautionary principle applied to, 25
traceability and labeling of, 35–36
Geneva Convention on Long-Range Transboundary Air Pollution (LRTAP), 59, 64
German Environment Agency study, 142–143

German Environmental Liability Act, 189
German League for Nature and the Environment, 338
German Marshall Fund, 334
Germany, 87, 138t, 213
compliance incentive policy of, 149
Council for Land Stewardship, 310t, 312
environmental advisory councils in, 309, 310t
international aid efforts, 258
reduction of greenhouse gas emissions, 179n29, 226–227, 226t
GHGs. See Greenhouse gases (GHGs)
Global Climate Change Coalition (GCC), 214
Global environmental conventions, issue of EU accession to, 57–60
Global environmental issues, 6, 357–358, 369–370
US/EU policy divergence on, 26–27, 66–68, 70, 205, 347
view that they are overstated, 297
Global warming. *See* Climate change agreements
GMOs. *See* Genetically modified organisms (GMOs)
Gore, Albert, 64, 213, 216, 262–263
Gothenburg summit, EU, 322
Greece
environmental advisory council in, 310t
greenhouse gas emissions trends in, 226t, 227
Greenhouse gas (GHG) emissions. *See also* Climate change negotiations; Kyoto Protocol
CO_2 emissions reduction goals, 208, 210, 211, 265–266
Danish agreement on, 180n32
Germany's reduction of, 179n29, 226–227, 226t
individual states in the US active in reducing, 225, 264
national inventories and targets,

217t, 226t, 263–264, 266
reduction targets, 211, 216, 217t,
221, 225
"Greening Sustainable Development
Strategies" (EEAC paper), 320,
321–322, 324–325
Green parties
European, 7, 29, 69, 223
representation in government, 7
US Green Party, 341
Grimeaud, David J. E., **159**, 355,
369, 377. *See also* Negotiated
environmental agreements (NAs);
Policy instruments

Habitat equivalency analysis,
201–202n16
Habitats directive, EU, 129
Harris, Paul G., **253**, 368, 378. *See
also* Development assistance;
Political culture
Hazardous and Solid Waste
Amendments, US, 75
Hazardous waste disposal, 19, 90,
179n28, 240, 247, 249
Helsinki Summit, EU, 322
History of environmental policy
development
in the EU, 54–55, 117–119
in the US, 1, 53–54, 56–57, 60–61,
75–76, 82, 115–117
Hormone Beef case, 31, 47n51
Hormones
meat hormones, 28, 31–34
milk-enhancing hormones, 31–32
Hungary, environmental advisory
council in, 311t
Hybridization model, 73, 100, 347,
348, 351–352, 368–369, 379
concept defined, 78–79
differing from convergence, 80, 99
examples of hybridization, 98–100,
353–357
transatlantic process of, 74–75,
78–80
Hydrogen fuel technologies, 220

Implementation of environmental
policy, 135, 151, 154n19. *See also*
Legal implementation in the EU;
Legal implementation in the US;
Regulatory systems; Risk regulation
compliance incentives, 145–150
EU enforcement limitations, 119,
122, 135–138, 354
by EU member nations, 138t,
139–140, 171
evaluation of in the US, 141–142,
156n26
fragmentation in, 95–96
indicators for enforcement, 143
internationalization of, 153
parallel developments in, 18–21
US and EU compared on, 18, 67,
151–153, 353–354
Incentives. *See* Compliance incentives
India, CO_2 emissions of, 263–264
Information/data
availability to citizens, 152, 167,
333
disclosure requirements, 90–91, 95
on compliance, 141, 145
knowledge networks, 306–307
lack of, 354–355
role in valuation of natural
resources, 189, 191, 197, 204
transparency, 144, 237
Inland Waterways Commission, US,
290
Inspections, 140
Institutional factors, 97, 270–277,
296–297, 366–367. *See also*
Environmental federalism;
Sovereignty
of the US and EU compared, 7, 55,
113–114, 120–122, 366–367
Integrated pollution control (IPC), 96
Integrated Pollution Prevention and
Control (IPPC) Directive, EU, 146
Interest groups, 223, 269, 363–364.
See also Nongovernmental
Organizations (NGOs)
Interior Department (DOI), US, 185

International Civil Aviation
Organization, 64
International Network for
Environmental Compliance and
Enforcement (INECE), 153
International Standard on
Environmental Management
Systems, 147
International Undertaking on Plant
Genetic Resources, 26
INTERPOL, 152
IPCC. See Intergovernmental Panel
on Climate Change (IPCC)
Ireland
environmental advisory councils in,
309, 310t
greenhouse gas emissions trends in,
226t, 227
Italy, greenhouse gas emissions trends
in, 226t, 227

Japan
international aid efforts, 255
negotiation and ratification of
climate change agreement, 209,
218
Joint implementation, 358
Judicial independence, 120–121

Kagan, Robert, 104n42, 374n33–35,
375n39
Kagan, Robert A., 87
Kantor, Mickey, 232–233
Kelemen, R. Daniel, 113, 353–354,
378. See also Environmental
federalism
Knowledge networks, 306–307. See
also information/data
Kohl, Helmut, 213
Krämer, Ludwig on US/EU policy
divergence, 53, 347, 350–351, 367,
378. See also Divergence in US/EU
environmental policy; EU
integration
Kyoto Protocol, 217t See also
Climate change negotiations;

Framework Convention on Climate
Change (FCCC); Greenhouse gas
(GHG) emissions
Bush administration withdrawal
from, 2, 65, 208, 218–219, 221,
249, 263–264, 271, 292, 358
Clinton administration support for,
208
precautionary principle and, 210
ratification of by EU member
nations, 207–209, 219, 240, 249
US-EU differences, 65–66, 212–216,
222

Lankowski, Carl, **329**, 378
Larsson, Kjell, 218
Lead Industry v. EPA, 20
Legal implementation in the EU,
76–77, 135, 177n7. *See also* Court
cases, EU; Court of Justice, EU
Legal implementation in the US, 56,
76–78, 98–99. *See also* Court
cases, US; Court system, US
citizens' right-to-sue, 120, 180n38,
184, 191, 199, 351
tort law procedures, 119, 120,
126–127, 351, 368
Legal standing and aggregation
issues, 195–199
Legitimization values, 288–289
Liability systems, 183, 198–200
compensatory restoration, 202n17
EU retreat from White Paper,
183–184, 189, 191–192, 355,
356–357
evolving approaches, 188–189
fault-based liability, 190
least cost options, 191
Natural resource damage assessment
(NRDA) guidelines, 188–189
nonuse values and, 186–187, 198
US approaches (*see also* Tort law),
187–189
White Paper on Environmental
Liability, EC, 183, 189–191,
197–198, 200

LIFE (*L'Instrument Financier pour l'Environnement*), 128
Limits to Growth report, 297
Lisbon process, EU, 323–324
Lithuania, environmental advisory council in, 309, 311t
Litigation. *See* Court cases; Tort law, US
Local/regional level. *See* Microlevel policy initiatives
Lomé Convention, 260
Long-range Transboundary Air Pollution (LRTAP), 59, 64
Lujan decisions, 125
Luxembourg, greenhouse gas emissions trends in, 226t, 227

Maastricht Treaty, 1, 12n1, 29, 122, 282
Macrory, Richard and Ingeborg Niestroy, 305, 378–379
Mad cow disease, 24, 74, 83, 90–92, 327n17
Maine v. Taylor, 23
Market-based approaches
emissions trading, 5, 69, 94, 216, 224, 358
environmental taxes, 69, 213, 223
EU use of, 373n17
US advocacy of, 67–69, 211, 268, 350
Mazmanian, Daniel A. and Michael E. Kraft, 293–294
MEAs. *See* Multilateral environmental agreements (MEAs)
Meat hormones, 28, 31–34
Meat Hormones case, WTO, 23, 33–34
Microlevel policy initiatives, 296, 324, 361
community funding for development, 260–261
some US states active in reducing greenhouse gas emissions, 225, 264
Milk-enhancing hormones, 31–32
Monitoring, reporting and

verification, 167
Montreal Protocol on Biodiversity, 61–62, 240, 350
Montreal Protocol on Substances That Deplete the Ozone Layer, 26, 61, 210
beginning of conflict with US, 61–62
Montrose Chemical Corp v. Superior Court, 196
Moral considerations, 282, 368. *See also* Valuation of natural resources
legitimization values, 288–289
nonmarket values, 189
perceived benefits of environmental protection, 68–70
value, 191–192
Multilateral Agreement on Investment (MAI), 341
Multilateral environmental agreements (MEAs), 239–240, 250
Multinational corporations, 223
Multistakeholder environmental advisory councils, 313f, 314–316, 326n14
Mutual recognition agreement (MRA), 331, 333

Nader, Ralph, 341
NAFTA. *See* North America Free Trade Agreement (NAFTA)
NAs. *See* Negotiated environmental agreements (NAs)
National Climate Change Technology Initiative, 220
National councils for sustainable development (NCSDs), 308–309, 310–311t, 314, 319–321. *See also* European Environmental Advisory Council (EEACs)
National Environmental Policy Act (NEPA), Netherlands, 170–171
National Marine Sanctuaries Act, US, 185
National Oceanic and Atmospheric Administration (NOAA), US, 185–186, 188, 193

National Wildlife Federation (NWF), 334–335, 338

Natural resource damage assessment (NRDA) guidelines, 188–189

Natural resources. *See also* Agriculture; Conservation; Energy sector; Sustainable development
allocation and management of, 279, 286
as public trust, 185
utilization of, 238, 281–282

Natural Resources Defense Council (NRDC), 220

Negotiated environmental agreements (NAs). *See also* Dutch covenants; Project XL
benefits of, 161–165
defined, 159–160
design and evaluation of, 165–168
in European environmental policy, 161–162, 168–171, 177n7
evaluation of, 165–170
legal aspects of, 167, 172
in US environmental policy, 159, 162–163, 177n9
of the US and Europe compared, 174–175, 355–356

Nestucca case, 196

Netherlands, 45n29, 138t. *See also* Dutch covenants
environmental advisory councils in, 310t, 315
greenhouse gas emissions trends in, 226t, 227
voluntary compliance principles in, 147
water protection policy of, 137–138

Networks. *See also* EEAC network
IMPEL network, 140–143, 152–153
International Network for Environmental Compliance and Enforcement, 153
regional networking, 318–319
transnational policy networks, 7, 277, 303, 355, 362–364

New Delhi Conference, 266

New York v. United States, 123

NGOs. *See* Nongovernmental Organizations (NGOs)

Nixon, Richard, 116

Noise pollution, 64

Nongovernmental Organization (NGOs), 339–340, 342
environmental groups, 163, 170, 373n21
interest groups, 269
power to sue considered for, 183
stakeholder roles accorded to, 332–333
of the US and EU compared, 337–339, 343–344, 362–364, 367

Nonmarket values, 189

Nonuse values (NUVs), 186–187
legal standing and aggregation issues, 195–199
US experience with, 357

North America Free Trade Agreement (NAFTA), 63, 232, 235, 334

Occupational Safety and Health Administration (OSHA), 82–83

ODA. *See* Development assistance

Ohio vs. US Department of Interior, 187

Oil disposal, EC Council directive (75/439/EEC), 24

Oil Pollution Act, US, 185

Oil spills, 187, 196

Organization for Economic Cooperation and Development (OECD), 341

Ozone depletion, 222

Pacific Northwest, 292

Performance objectives, 146, 172–173, 180n37, 181n40

Persistent organic pollutants (POPs), 65. *See also* Stockholm Convention

Pesticide restrictions, 19, 24–25
EC Council directive 76/117/EEC, 19

Pew Center on Global Climate Change, 221, 225

Pfizer Animal Health SA v. Council of the EU, 83–85
Pinchot, Gifford, 289–290
Plant-level agreements, 164, 170–173
Poland, environmental advisory council in, 311t
Policy frames, 6
Policy instruments, 369. *See also* Command-and-control (CAC) regulation; Flexible policy instruments
Policy-making climate, 337, 339
nongovernmental organizations of the US and EU compared, 337–339, 343–344, 362–367
political culture of the US and EU compared, 222, 268–270, 293–298, 343–344
Political culture, 114. *See also* EU integration; US political culture
domestic and international perspectives linked, 368
proportionality/proportional representation, 7, 29, 50n74
and risk selection, 94
US efforts to export domestic standards, 234–235, 243–245
of the US and EU compared, 222, 268–270, 293–298, 343–344
Political pluralism, 268–270
Pollution. *See* Emissions; Water pollution
Pollution abatement costs, 33, 38, 88, 164
Pollution control technologies, 238. *See also* Technology development focus
Portugal, 138t
environmental advisory council in, 311t, 327n23
greenhouse gas emissions trends in, 226t, 227
Precautionary principle, 17, 81, 348–349. *See also* Risk assessment
application to environmental protection, 21–25

application to public health risks, 19–25, 349
applied by US to terrorism, 92–94
defined, 17, 18, 22–24, 36, 210–211
EU constitutionalization of, 21, 28–30, 38–41, 81, 241–243, 349, 352–353
perceived societal benefits of protection, 68–70
US/EU divergence over status of in international law, 25–27, 40–42, 241–242, 268
US opposing inclusion in Rio declaration, 63, 349
Preference-based valuation, 187, 194–196
President's Council on Sustainable Development (PCSD), 291–292, 305
Principle 15 of the Rio declaration, 371n3
Prodi group, 322–323
Project XL, EPA
model aspects of, 356
negotiated agreement initiative, 164–165, 167, 171–173, 175
performance objectives in, 146, 172–173, 180, 180n37, 181n40
Proportionality/proportional representation, 7, 29, 50n74
Public Citizen v. Young, 20
Public health risks, 19–25, 349
Public opinion
European consumer preferences, 36, 38
in Europe supporting environmental protection, 69
perceptions of risk, 85–86
in the US, 224–225, 267
Public trusteeship, 185, 190, 200n4, 290–291. *See also* Citizen involvement

rBST ban, 33
Reagan, Ronald, 56, 116, 117, 365

Reform Party, US, 341
Regional environmental integration, 292–293, 296
Regional networking, 318–319
Regulatory failures, 30–33
Regulatory relief, 146, 164, 172, 181n42. *See also* Flexible policy instruments
Regulatory systems, 183, 354. *See also* Legal implementation
in the European Union, 96
in the US, 96, 114–117
Regulatory trends, 15, 111, 144–145, 162–163, 368–370. *See also* Negotiated environmental agreements (NAs)
"Reinventing Environmental Regulation" (Clinton administration report), 162, 171
Remediation, 192
Renewable energy, 285
Representatives, US House of, 236–237. *See also* Congress, US
Republican Party, 235–236, 334, 365
Reserve Mining Co. v. EPA, 20
Resources. *See* Natural resources
Restoration, 192, 202n17
Restrictions on Hazardous Substances in Electronics and Electronic Equipment, EU, 247
Rio Conference. *See* UNCED
Rio Declaration on Environment and Development, 63, 81, 349, 360, 371n3, 373n18
Risk, public perceptions of, 85–86
Risk assessment, 17, 18–21, 23, 34
countervailing risks and tradeoffs, 88–90, 93
normative aspects of, 36–37, 48n63, 85
quantitative approach, 82–83
risk selection and, 82–85, 90–94
role of science in, 83–85
uncertainty in, 39–40, 49n68
zero-risk policies, 37

Risk management, standard setting, 86–90
Risk reduction, EC Council directive 75/442/EEC, 19
Risk regulation, 73
EU as more risk averse than the US, 27–28, 34, 35–36, 39, 349
US and EU as differing on risk selection, 74
Roosevelt, Franklin D., 290
Roosevelt, Theodore, 290

Safe Drinking Water Act, US, 75
Sandoz BV case (EU), 20
SAVE, 285
Sbragia, A. M. and C. Damro, 270
Scaling analysis, 188–189
Schreurs, Miranda A., **207**, 357–358, 379
Schröder, Gerhard, 2
Science
diverging US/EU perspectives on, 17, 30–31, 34–35, 68–69, 220, 247, 268
role of in risk assessment, 83–85
scientific environmental advisory councils, 312, 313f, 314–316, 326n13
scientific uncertainty, 39–40, 49n68
Senate, US
Byrd-Hagel Resolution, 214, 358
positions on environmental pacts, 214, 235
Shabecoff, Philip, 262
Shrimp-turtle cases, 243–244
Sierra Club, 220
Simon, Julian and Herman Kahn, 297
Single European Act (SEA), 1987. *See also* European Union (EU)
environmental mandate a part of, 1, 28, 30, 40, 54–55, 57, 62, 118, 281–282, 337, 366
Site specificity, 164, 170–173
Slovak Republic, environmental advisory council in, 311t

Slovenia, environmental advisory
council in, 311t
Smitch case, 23
Smoking risks, 90
Social-democratic parties, 223
Sovereignty
doctrine of sovereign immunity, 123,
125–126 (*see also* States, US)
of the EU a mixture of nation and
supranational, 58–59, 368
of the US and international
agreements, 67, 368
Spain, 138t
environmental advisory councils in,
311t, 327n23
greenhouse gas emissions trends in,
226t, 227
SPS agreement, WTO, 242
Stakeholder involvement *See also*
Business sector; Citizen
involvement; Information/data
in establishing liability, 190
multistakeholder approach, 225
multistakeholder environmental
advisory councils, 313f, 314–316,
326n14
in negotiated agreements, 159, 166,
170, 172, 174
of NGOs, 332–333
Standards Code, 232, 234
States, US
doctrine of sovereign immunity, 123,
125–126
environmental policy role(s) of, 113,
115, 225–226
some active in reducing greenhouse
gas emissions, 225, 264
states rights vs. centralizing policy,
68, 96, 117, 123–124
Steel Co. decision, 125
Stockholm Conference on the Human
Environment, 279
Stockholm Convention on Persistent
Organic Pollutants (POPs) US
failure to ratify, 65, 81
Subsidies, 55

agricultural subsidies, 28–29
Superfund, 183, 185. *See also*
Comprehensive Environmental
Response, Compensation and
Liability Act (CERCLA)
Supplementary Agreement on the
Environment (SAE), 235
Supreme Court, US, 82, 125, 354
Sustainable agriculture, 285–286,
328n31
Sustainable development *See also*
Brundtland Commission;
Conservation; Economic
development; Natural resources;
Treaty of Amsterdam
consumption levels and, 223, 293,
295
early US conservation movement
initiating idea of, 289–290
EU integration incorporating themes
of, 281–289
evolution of the concept of,
279–281, 323
and foreign aid, 260, 360–361
lacking clear definition, 277–278,
361
language vs. reality regarding,
361–362, 364
measurable results lacking,
293–295, 297–298
national councils for sustainable
development (NCSDs), 308–309,
310–311t, 314, 319–321
political culture affecting policies of,
293, 294–298
"Towards Sustainability" (EAP
report), 162
World Summit on Sustainable
Development, 328n38
Swanson, Timothy and Andreas
Kontoleon, 183, 356–357,
378–379
Sweden, 322
Environment Council, 311t, 315
greenhouse gas emissions trends in,
226t

TAED. *See* Transatlantic
  Environmental Dialogue (TAED)
Tariffs on environmental goods, 238
Technology development focus, 220,
  222–223, 238
  Best Available Technology (BAT)
  approaches, 94
  "The Reality of Precaution" project,
  80
Tort law, US, 119–120, 126–127,
  351, 368
  citizens' right-to-sue, 180n38, 184,
  191, 199
Toxics Release Inventory, 167
Trade agreements. *See also* Economic
  policy; North American Treaty
  Organization (NAFTA); World
  Trade Organization (WTO)
  Agreement on Sanitary and
  Phytosanitary Measures (SPS), 23,
  48n63, 234
  Agreement on Subsidies and
  Countervailing Measures, 232
  Commerce Dept. representing the
  US in negotiating, 57, 60, 62–63
  SPS agreement, 242
  Standards Code, 232, 234
  Supplementary Agreement on the
  Environment (SAE), 235
  Technical Barriers to Trade
  Agreement (TBT), 245
Trade associations, 178n18
Trade disputes
  dolphin-safe tuna dispute, 231–232,
  233, 243
  shrimp-turtle cases, 243–244
Trade-environmental linkages
  disputes over process and
  production issues, 243–245
  EU proposals on MEAs. *See*
  Multilateral environmental
  agreements (MEAs)
  EU support for strengthening, 240,
  246–247
  perceived as protectionism, 236,
  245–248, 349

US/EU divergence regarding, 231,
  233, 246–250
US initiatives, 234–235, 237–238
Transatlantic Environmental Dialogue
  (TAED), 342–344, 364
  demise of, 340–344
  formation of, 329–335
  NGO roles differing in US and EU,
  337–339, 343–344
  organizational structure of,
  336–338, 342
  participation gap in, 338–339
  US/EU divergence within, 73,
  333–334, 343, 369
Transboundary air pollution, 64
Transnational environmental
  dialogue, 277, 329, 347
Transnational policy networks, 7,
  277, 303, 355, 362–364. *See also*
  EEAC network
Transparency, 144, 237. *See also*
  Information/data
Transportation policy, 284–285
  fuel economy standards, 247–248
Treaty of Amsterdam, 119, 282, 322
Treaty on European Union, 12n1,
  349. *See also* Maastricht Treaty
Treaty of Rome, 11–12n1, 350
Tripartate contracts, 144
Trusteeship, 185, 190, 200n4

UNCED (UN Conference on
  Environment and Development), 1,
  210–211, 253, 256, 262, 265,
  280–281, 308, 332
Unfunded mandates, 117
United Kingdom (UK), 138t
  English Nature (EN), 309, 311t
  environmental advisory councils in,
  309, 311t
  greenhouse gas emissions trends in,
  226t
  plan to cut greenhouse gases by 60
  percent, 370
  Royal Commission on
  Environmental Pollution, 314–315

voluntary compliance approach of, 148–149
water protection policy of, 137–138
United Nations (UN), 210, 279–280. *See also* Brundtland Commission; Framework Convention on Climate Change
Commission on Sustainable Development (CSD), 282
Conference on Environment and Development (UNCED), 253, 256, 308, 332
Conference on the Human Environment, 279
Economic and Social Council (ECOSOC), 279
Environment Programme (UNEP), 61, 264
United States (US). *See also* Congress, US; Presidential administrations; States, US; US political culture; *and by department or agency*
addressing ozone depletion, 222
citizen's right-to-sue in, 180n38
economic focus of foreign policy, 60, 62–63, 349, 367
efforts to export domestic environmental standards, 234–235, 243–245
environmental advisory body in, 311t
evaluation of policy implementation in, 141–142, 156n
global hegemony of, 63, 369–371
greenhouse gas emissions trends in, 226t, 227
having no cabinet level environmental department, 57, 60, 62–63
history of environmental policy development, 1, 53–54, 56–57, 60–61, 75–76, 82, 115–117
international aid efforts, 255, 257–259
negotiations on climate change agreement, 213–214

political culture, 266–270
political trends, 235–237, 365
reluctant to support new international initiatives, 1, 67, 81, 270
role of citizen involvement in, 14, 143, 144, 152
Uruguay Round, 63, 231, 232, 234, 330, 332. *See also* World Trade Organization (WTO)
US Information Agency (USIA), 335
US political culture, 222–224, 266–270. *See also* Congress, US; Presidential administrations, US
Democratic Party, 116–117, 121
gridlock, 366
party politics and ideological shifts, 6–7, 61, 341, 348, 365
Republican Party, 235–236, 334, 365

Valuation of natural resources, 192–194, 198. *See also* Nonuse values (NUVs)
contingent valuation method (CV), 186–188, 191, 193
cost of studies, 194, 198
economic valuation, 190, 194–195
of harm to biodiversity, 190, 199–200
knowledge and data component of, 189, 191, 197
legal standing and aggregation issues, 195–199
nonmarket values, 189
preference-based valuation, 187, 194–196
Veterinary medicinal products, EC Council directive 81/851/EEC, 19
Vienna Convention for Protection of the Ozone Layer, 23, 61
Vig, Norman J. and Michael G. Faure, **1**, 379
Vogel, David, 82, 86, **231**, 359, 379. *See also* Trade-environmental linkages

Voluntary compliance, 5, 146, 166,
    175n1, 354–355, 358–359. See
    also Flexible policy instruments;
    Negotiated environmental
    agreements (NAs)
    US focus on, 147–148

Wallström, Margot, 2, 92, 218, 222
Waste. See Hazardous waste disposal
Water Framework Directive,
    137–138, 138t
Watson, Harlan, 222
White Paper on Environmental
    Liability, EC, 183, 189–191,
    197–198, 200
    EU retreat from, 183–184, 189,
    191–192, 355, 356–357
Wiener, Jonathan B., 5, 73, 347,
    351–352, 368–369, 374n30, 379.
    See also Hybridization model; Risk
    regulation
Wirth, Timothy, 213–214
Working Party on Environmental
    Crime, INTERPOL, 152
World Environment Organization,
    proposed, 264
World Summit on Sustainable
    Development, 265–266, 281,
    328n38
World Trade Organization (WTO),
    231, 330, 359. See also Committee
    on the Trade and the Environment;
    Trade agreements; Uruguay Round
    dispute settlement process, 238, 246
    Doha Round, 250
    environmental issues related to, 23,
    237
World Wide Fund for Nature
    (WWF), 338

Zero-risk policies, 37
Zoeleck, Robert, 236, 243–244